HVAC
Design Data
Sourcebook

Other HVAC Books of Interest from McGraw-Hill

HVAC
Design Data
Sourcebook

Robert O. Parmley, P.E.
Editor-in-Chief

McGraw-Hill, Inc.

New York   San Francisco   Washington, D.C.   Auckland   Bogotá
Caracas   Lisbon   London   Madrid   Mexico City   Milan
Montreal   New Delhi   San Juan   Singapore
Sydney   Tokyo   Toronto

Library of Congress Cataloging-in-Publication Data

HVAC design data sourcebook / Robert O. Parmley, editor-in-chief.
      p.   cm.
   Includes index.
   ISBN 0-07-048572-0 (alk. paper)
   1. Heating—Equipment and supplies—Design and construction.
2. Ventilation—Equipment and supplies—Design and construction.
3. Air conditioning—Equipment and supplies—Design and
construction. I. Parmley, Robert O.
TH7345.H85   1994
697—dc20   94-4200
                                                        CIP

  2 3 4 5 6 7 8 9 0 DOH/DOH 9 0 9 8 7 6 5 4

ISBN 0-07-048572-0

*The sponsoring editor for this book was Robert W. Hauserman,
the editing supervisor was Stephen M. Smith, and the production
supervisor was Pamela A. Pelton. It was set in Century
Schoolbook by North Market Street Graphics.*

Printed and bound by R. R. Donnelley & Sons Company.

This book is printed on acid-free paper.

Dedicated to my family

Contents

Section 2. Heat Loss Estimating   57

Section 3. Building Material Properties   69

Section 4. Heating Fuel Data   101

Section 5. Heating Systems   143

Section 8. Piping Design   233

Section 9. Air Distribution Design   257

Section 10. Attic Ventilation   271

Section 11. Sheet Metal Duct Design   311

Section 12. Energy Conservation   337

Section 13. Climatic Data   347

Preface

The atmosphere in which we live, many times, requires artificial modification to provide healthy, safe, and livable conditions. Man has struggled with a variety of adverse air conditions since time immemorial. Originally, the basic need to heat a small area of air for protection from the inclement cold weather was achieved by simply burning forest products. As time passed, the need for more refined heating facilities evolved and led to the necessity of additional fuel sources, such as coal, gas, oil, hydroelectric, solar, and nuclear.

When the Industrial Revolution came, the need for adequate ventilation systems for the workplace became necessary. In this current century, regulatory agencies developed codes for heating and ventilating systems for most buildings. Air conditioning came of age by the mid-twentieth century and now is common in most modern buildings.

Knowledge comes slowly, especially technical knowledge. The hidden secrets of nature continuously resist discovery. Each scrap of information must be painstakingly extracted, analyzed, and preserved in order to build a library of technical data. The HVAC field was no exception to this rule. Credit is due to those many unsung individuals who have made significant contributions to this field throughout the centuries. Their practical insight and innovative talent have generally gone unheralded. From the first crude attempts to warm air in a confined space by burning wood, until the eighteenth century, little "scientific" or significant research into the nature of thermodynamics was conducted. Most accumulated knowledge in the field was acquired with the "trial and error" method by artisans. It was not until the end of the 1700s that two scientists took a closer look into the nature of heat. They were American-born Benjamin Thompson and the British chemist Sir Humphry Davy, who raised doubts about the age-old caloric theory. Their observations triggered research resulting in the understanding that heat is a form of energy. During the mid-1800s, this fact was proven by three gentlemen—James Joule, a British physicist, and J. R. von Mayer and Hermann von Helmholtz, both German physicists. In 1748, Dr. William Cullen at the University of Glasgow was the first to demonstrate artificial refrigeration by evaporating ether in a partial vacuum. His work led to further study and research into chemical and mechanical systems that developed low temperatures. Early in the 1800s, closed-cycle compression refrigeration was proposed. A well-documented trail of events followed, resulting in practical air-conditioning by the 1920s. However, residential air conditioning did not become cost-effective until the 1950s.

As knowledge began to accumulate at an ever-increasing rate, heating and ventilating technology evolved and many innovative designs were perfected. In the mid-1970s, fuel efficiency became a major factor to combat the worldwide energy crunch. This condition produced many new heating innovations and required designers to consider heat loss of building envelopes more seriously in order to conserve energy. During this time frame, regulatory agencies and code requirements became more restrictive on industrial ventilation and effluent from smokestacks. Today, almost everyone is well aware of how important HVAC technology is to the citizens of the modern world.

I believe that any engineer or designer worthy of his or her salt will continuously collect essential technical material relative to the field or fields in which he or she practices. Recognizing this fact, this sourcebook has attempted to extract the basics from the voluminous HVAC publications and arrange the material into a compact, practical, and user-friendly format.

It is impossible to present the full range of HVAC technology in a single volume and certainly no claim is remotely suggested that this is an all-inclusive book. However, there are basic data that I personally have collected over three decades that are essential for HVAC understanding, and this material has been assembled herein. One or two blank Notes pages are provided at the end of each section to allow the user the opportunity to tailor this sourcebook to his or her individual preference.

It is imperative that the reader consult all applicable local, state, and federal codes prior to the final design stage to avoid conflict with mandated regulations.

Minimal material on electrical and solar heating is provided because they represent specialized skills not normally provided by general HVAC contractors and the majority of A/E firms. Nuclear energy data have been completely omitted for obvious reasons. I suggest that the users of this book consult other sources for these special areas of technology.

In the final analysis, it was my mission to provide a data sourcebook that everyday, shirt-sleeved engineers and designers could turn to for basic, authoritative answers on general HVAC questions without searching through scores of books and manuals: a one-stop location where quick, reliable answers can easily be obtained. Hopefully, I have been reasonably successful. This concept is an extension of my first effort in this area, the *HVAC Field Manual*, published in 1988. The Contents and Index have been specially edited to provide quick location of specific information.

My respect and appreciation go out to the many sources that participated in this project by amply providing data, material, and illustrations. I salute these technical societies, research institutions, manufacturers, firms, and individuals who contributed to this endeavor. Their spirit of cooperation made the editorial process enjoyable. Each source has been given proper credit at the appropriate place in the book so the reader will know the specific contribution origin.

A special thanks to my wife, Lana, and my secretary, Ethne, for their typing assistance.

Robert O. Parmley, P.E.

ABOUT THE EDITOR-IN-CHIEF

Robert O. Parmley, P.E., CMfgE, CSI, is President and Principal Consulting Engineer of Morgan & Parmley, Ltd., Professional Consulting Engineers, Ladysmith, Wisconsin. He is also a member of the National Society of Professional Engineers, the American Society of Mechanical Engineers, the American Society of Heating, Refrigerating, and Air-Conditioning Engineers, the Construction Specifications Institute, the American Design Drafting Association, and the Society of Manufacturing Engineers, and is listed in the AAES *Who's Who in Engineering.* Mr. Parmley holds a BSME and a MSCE from Columbia Pacific University and is a registered professional engineer in Wisconsin, California, and Canada. He is also a certified manufacturing engineer under SME's national certification program and a certified wastewater plant operator. In a career covering more than three decades, Mr. Parmley has worked on the design and construction supervision of a wide variety of structures, systems, and machines—from dams and bridges to municipal sewage facilities, HVAC systems, and water projects. The author of over 40 technical articles, he is also the Editor-in-Chief of the *Standard Handbook of Fastening and Joining,* Second Edition, the *HVAC Field Manual,* the *Field Engineer's Manual,* the *Hydraulics Field Manual,* and the *Mechanical Components Handbook,* all published by McGraw-Hill.

Basic Data and Fundamentals

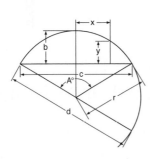

Circumference = 6.28318 r = 3.14159 d
Diameter   = 0.31831 circumference
Area   = 3.14159 r²

Arc $a = \dfrac{\pi r A°}{180°} = 0.017453\; r\, A°$

Angle $A° = \dfrac{180°\, a}{\pi r} = 57.29578\; \dfrac{a}{r}$

Radius $r = \dfrac{4\,b^2 + c^2}{8\,b}$

Chord $c = 2\sqrt{2\,br - b^2} = 2\,r\sin\dfrac{A}{2}$

Rise $b = r - \frac{1}{2}\sqrt{4\,r^2 - c^2} = \dfrac{c}{2}\tan\dfrac{A}{4}$

           $= 2\,r\sin^2\dfrac{A}{4} = r + y - \sqrt{r^2 - x^2}$

      $y = b - r + \sqrt{r^2 - x^2}$

      $x = \sqrt{r^2 - (r + y - b)^2}$

Diameter of circle of equal periphery as square = 1.27324 side of square
Side of square of equal periphery as circle   = 0.78540 diameter of circle
Diameter of circle circumscribed about square = 1.41421 side of square
Side of square inscribed in circle   = 0.70711 diameter of circle

---

CIRCULAR SECTOR

r = radius of circle   y = angle ncp in degrees

Area of Sector ncpo = ½ (length of arc nop × r)

              $= \text{Area of Circle} \times \dfrac{y}{360}$

              $= 0.0087266 \times r^2 \times y$

CIRCULAR SEGMENT

r = radius of circle   x = chord   b = rise

Area of Segment nop = Area of Sector ncpo − Area of triangle ncp

             $= \dfrac{(\text{Length of arc nop} \times r) - x\,(r - b)}{2}$

Area of Segment nsp = Area of Circle − Area of Segment nop

VALUES FOR FUNCTIONS OF π
$\pi = 3.14159265359, \quad \log = 0.4971499$

$\pi^2 = 9.8696044, \log = 0.9942997$ $\dfrac{1}{\pi} = 0.3183099, \log = \overline{1}.5028501$ $\sqrt{\dfrac{1}{\pi}} = 0.5641896, \log = \overline{1}.7514251$

$\pi^3 = 31.0062767, \log = 1.4914496$ $\dfrac{1}{\pi^2} = 0.1013212, \log = \overline{1}.0057003$ $\dfrac{\pi}{180} = 0.0174533, \log = \overline{2}.2418774$

$\sqrt{\pi} = 1.7724539, \log = 0.2485749$ $\dfrac{1}{\pi^3} = 0.0322515, \log = \overline{2}.5085504$ $\dfrac{180}{\pi} = 57.2957795, \log = 1.7581226$

Note: Logs of fractions such as $\overline{1}.5028501$ and $\overline{2}.5085500$ may also be written 9.5028501 − 10 and 8.5085500 − 10 respectively.

FIGURE 1-1 Properties of the circle. (*Courtesy: American Institute of Steel Construction.*)

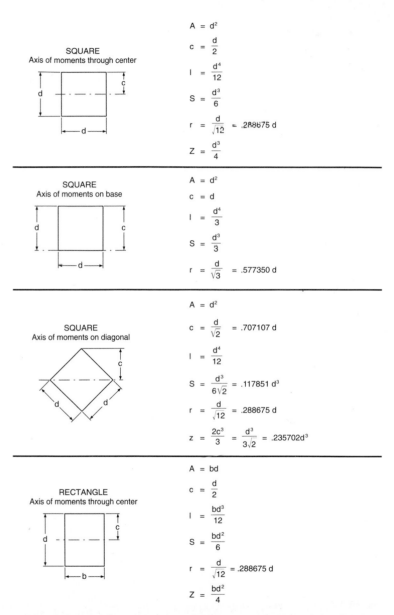

SQUARE
Axis of moments through center

$$A = d^2$$

$$c = \frac{d}{2}$$

$$I = \frac{d^4}{12}$$

$$S = \frac{d^3}{6}$$

$$r = \frac{d}{\sqrt{12}} = .288675\ d$$

$$Z = \frac{d^3}{4}$$

SQUARE
Axis of moments on base

$$A = d^2$$

$$c = d$$

$$I = \frac{d^4}{3}$$

$$S = \frac{d^3}{3}$$

$$r = \frac{d}{\sqrt{3}} = .577350\ d$$

SQUARE
Axis of moments on diagonal

$$A = d^2$$

$$c = \frac{d}{\sqrt{2}} = .707107\ d$$

$$I = \frac{d^4}{12}$$

$$S = \frac{d^3}{6\sqrt{2}} = .117851\ d^3$$

$$r = \frac{d}{\sqrt{12}} = .288675\ d$$

$$z = \frac{2c^3}{3} = \frac{d^3}{3\sqrt{2}} = .235702 d^3$$

RECTANGLE
Axis of moments through center

$$A = bd$$

$$c = \frac{d}{2}$$

$$I = \frac{bd^3}{12}$$

$$S = \frac{bd^2}{6}$$

$$r = \frac{d}{\sqrt{12}} = .288675\ d$$

$$Z = \frac{bd^2}{4}$$

FIGURE 1-2   Properties of geometric sections. (*Courtesy: American Institute of Steel Construction.*)

RECTANGLE
Axis of moments on base

$A = bd$

$c = d$

$I = \dfrac{bd^3}{3}$

$S = \dfrac{bd^2}{3}$

$r = \dfrac{d}{\sqrt{3}} = .577350\,d$

RECTANGLE
Axis of moments on diagonal

$A = bd$

$c = \dfrac{bd}{\sqrt{b^2 + d^2}}$

$I = \dfrac{b^3 d^3}{6\,(b^2 + d^2)}$

$S = \dfrac{b^2 d^2}{6\,\sqrt{b^2 + d^2}}$

$\quad = \dfrac{bd}{\sqrt{6\,(b^2 + d^2)}}$

RECTANGLE
Axis of moments any line
through center of gravity

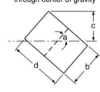

$A = bd$

$c = \dfrac{b \sin a + d \cos a}{2}$

$I = \dfrac{bd\,(b^2 \sin^2 a + d^2 \cos^2 a)}{12}$

$S = \dfrac{bd\,(b^2 \sin^2 a + d^2 \cos^2 a)}{6\,(b \sin a + d \cos a)}$

$r = \sqrt{\dfrac{b^2 \sin^2 a + d^2 \cos^2 a}{12}}$

HOLLOW RECTANGLE
Axis of moments through center

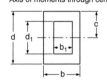

$A = bd - b_1 d_1$

$c = \dfrac{d}{2}$

$I = \dfrac{bd^3 - b_1 d_1^3}{12}$

$S = \dfrac{bd^3 - b_1 d_1^3}{6d}$

$r = \sqrt{\dfrac{bd^3 - b_1 d_1^3}{12\,A}}$

$Z = \dfrac{bd^2}{4} - \dfrac{b_1 d_1^2}{4}$

FIGURE 1-2   (*Continued*)

$$A = b (d - d_1)$$

$$c = \frac{d}{2}$$

EQUAL RECTANGLES
Axis of moments through
center of gravity

$$I = \frac{b (d^3 - d_1^3)}{12}$$

$$S = \frac{b (d^3 - d_1^3)}{6d}$$

$$r = \sqrt{\frac{d^3 - d_1^3}{12(d - d_1)}}$$

$$Z = \frac{b}{4} (d^2 - d_1^2)$$

$$A = bt + b_1 t_1$$

UNEQUAL RECTANGLES
Axis of moments through
center of gravity

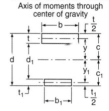

$$c = \frac{\frac{1}{2} bt^2 + b_1 t_1 (d - \frac{1}{2} t_1)}{A}$$

$$I = \frac{bt^3}{12} + bty^2 + \frac{b_1 t_1^3}{12} + b_1 t_1 y_1^2$$

$$S = \frac{I}{c} \qquad S_1 = \frac{I}{c_1}$$

$$r = \sqrt{\frac{I}{A}}$$

$$Z = \frac{A}{2} \left[d - \left(\frac{t + t_1}{2} \right) \right]$$

TRIANGLE
Axis of moments through
center of gravity

$$A = \frac{bd}{2}$$

$$c = \frac{2d}{3}$$

$$I = \frac{bd^3}{36}$$

$$S = \frac{bd^2}{24}$$

$$r = \frac{d}{\sqrt{18}} = .235702 \, d$$

TRIANGLE
Axis of moments on base

$$A = \frac{bd}{2}$$

$$c = d$$

$$I = \frac{bd^3}{12}$$

$$S = \frac{bd^2}{12}$$

$$r = \frac{d}{\sqrt{6}} = .408248 \, d$$

FIGURE 1-2 *(Continued)*

TRAPEZOID
Axis of moments through
center of gravity

$$A = \frac{d(b + b_1)}{2}$$

$$c = \frac{d(2b + b_1)}{3(b + b_1)}$$

$$I = \frac{d^3(b^2 + 4bb_1 + b_1^2)}{36(b + b_1)}$$

$$S = \frac{d^2(b^2 + 4bb_1 + b_1^2)}{12(2b + b_1)}$$

$$r = \frac{d}{6(b + b_1)}\sqrt{2(b^2 + 4bb_1 + b_1^2)}$$

CIRCLE
Axis of moments
through center

$$A = \frac{\pi d^2}{4} = \pi R^2 = .785398\,d^2 = 3.141593\,R^2$$

$$c = \frac{d}{2} = R$$

$$I = \frac{\pi d^4}{64} = \frac{\pi R^4}{4} = .049087\,d^4 = .785398\,R^4$$

$$S = \frac{\pi d^3}{32} = \frac{\pi R^3}{4} = .096175\,d^3 = .785398\,R^3$$

$$r = \frac{d}{4} = \frac{R}{2}$$

$$Z = \frac{d^3}{6}$$

HOLLOW CIRCLE
Axis of moments
through center

$$A = \frac{\pi(d^2 - d_1^2)}{4} = .785398\,(d^2 - d_1^2)$$

$$c = \frac{d}{2}$$

$$I = \frac{\pi(d^4 - d_1^4)}{64} = .049087\,(d^4 - d_1^4)$$

$$S = \frac{\pi(d^4 - d_1^4)}{32d} = .086175\,\frac{d^4 - d_1^4}{d}$$

$$r = \frac{\sqrt{d^2 + d_1^2}}{4}$$

$$Z = \frac{d^3}{6} - \frac{d_1^3}{6}$$

HALF CIRCLE
Axis of moments through
center of gravity

$$A = \frac{\pi R^2}{2} = 1.570796\,R^2$$

$$c = R\left(1 - \frac{4}{3\pi}\right) = .575587\,R$$

$$I = R^4\left(\frac{\pi}{8} - \frac{8}{9\pi}\right) = .109757\,R^4$$

$$S = \frac{R^3}{24}\frac{(9\pi^3 - 64)}{(3\pi - 4)} = .190687\,R^3$$

$$r = R\frac{\sqrt{9\pi^3 - 64}}{6\pi} = .264336\,R$$

FIGURE 1-2 *(Continued)*

PARABOLA

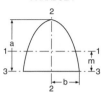

$$A = \frac{4}{3} ab$$

$$m = \frac{2}{5} a$$

$$I_1 = \frac{16}{175} a^3 b$$

$$I_2 = \frac{4}{15} ab^3$$

$$I_3 = \frac{32}{105} a^3 b$$

HALF PARABOLA

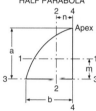

$$A = \frac{2}{3} ab$$

$$m = \frac{2}{5} a$$

$$n = \frac{3}{8} b$$

$$I_1 = \frac{8}{175} a^3 b$$

$$I_2 = \frac{19}{480} ab^3$$

$$I_3 = \frac{16}{105} a^3 b$$

$$I_4 = \frac{2}{15} ab^3$$

COMPLEMENT OF HALF
PARABOLA

$$A = \frac{1}{3} ab$$

$$m = \frac{7}{10} a$$

$$n = \frac{3}{4} b$$

$$I_1 = \frac{37}{2100} a^3 b$$

$$I_2 = \frac{1}{80} ab^3$$

PARABOLIC FILLET IN
RIGHT ANGLE

$$a = \frac{t}{2\sqrt{2}}$$

$$b = \frac{t}{\sqrt{2}}$$

$$A = \frac{1}{6} t^2$$

$$m = n = \frac{4}{5} t$$

$$I_1 = I_2 = \frac{11}{2100} t^4$$

FIGURE 1-2　(*Continued*)

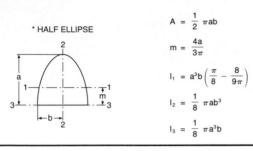

* HALF ELLIPSE

$$A = \frac{1}{2}\pi ab$$

$$m = \frac{4a}{3\pi}$$

$$I_1 = a^3b\left(\frac{\pi}{8} - \frac{8}{9\pi}\right)$$

$$I_2 = \frac{1}{8}\pi ab^3$$

$$I_3 = \frac{1}{8}\pi a^3b$$

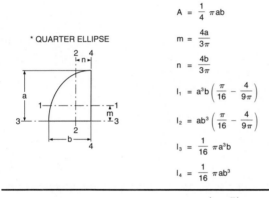

* QUARTER ELLIPSE

$$A = \frac{1}{4}\pi ab$$

$$m = \frac{4a}{3\pi}$$

$$n = \frac{4b}{3\pi}$$

$$I_1 = a^3b\left(\frac{\pi}{16} - \frac{4}{9\pi}\right)$$

$$I_2 = ab^3\left(\frac{\pi}{16} - \frac{4}{9\pi}\right)$$

$$I_3 = \frac{1}{16}\pi a^3b$$

$$I_4 = \frac{1}{16}\pi ab^3$$

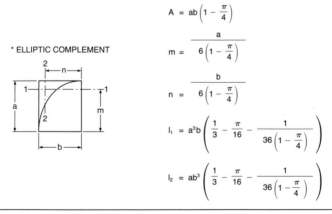

* ELLIPTIC COMPLEMENT

$$A = ab\left(1 - \frac{\pi}{4}\right)$$

$$m = \frac{a}{6\left(1 - \frac{\pi}{4}\right)}$$

$$n = \frac{b}{6\left(1 - \frac{\pi}{4}\right)}$$

$$I_1 = a^3b\left(\frac{1}{3} - \frac{\pi}{16} - \frac{1}{36\left(1 - \frac{\pi}{4}\right)}\right)$$

$$I_2 = ab^3\left(\frac{1}{3} - \frac{\pi}{16} - \frac{1}{36\left(1 - \frac{\pi}{4}\right)}\right)$$

* To obtain properties of half circle, quarter circle and circular complement substitute a = b = R.

FIGURE 1-2 *(Continued)*

n = Number of sides

$$\phi = \frac{180°}{n}$$

REGULAR POLYGON
Axis of moments
through center

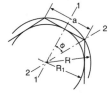

$$a = 2\sqrt{R^2 - R_1^2}$$

$$R = \frac{a}{2 \sin \phi}$$

$$R_1 = \frac{a}{2 \tan \phi}$$

$$A = \frac{1}{4} na^2 \cot \phi = \frac{1}{2} nR^2 \sin 2\phi = nR_1^2 \tan \phi$$

$$I_1 = I_2 = \frac{A(6R^2 - a^2)}{24} = \frac{A(12R_1^2 + a^2)}{48}$$

$$r_1 = r_2 = \sqrt{\frac{6R^2 - a^2}{24}} = \sqrt{\frac{12R_1^2 + a^2}{48}}$$

ANGLE
Axis of moments through
center of gravity

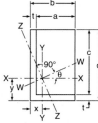

Z-Z is axis of minimum 1

$$\tan 2\phi = \frac{2K}{I_Y - I_X}$$

$$A = t(b + c) \quad x = \frac{b^2 + ct}{2(b + c)} \quad y = \frac{d^2 + at}{2(b + c)}$$

K = Product of Inertia about X-X & Y-Y

$$= \mp \frac{abcdt}{4(b + c)}$$

$$I_X = \frac{1}{3}\left(t(d - y)^3 + by^3 - a(y - t)^3\right)$$

$$I_Y = \frac{1}{3}\left(t(b - x)^3 + dx^3 - c(x - t)^3\right)$$

$$I_Z = I_X \sin^2\theta + I_Y \cos^2\theta + K \sin2\theta$$

$$I_W = I_X \cos^2\theta + I_Y \sin^2\theta - K \sin2\theta$$

K is negative when heel of angle, with respect to c. g., is in 1st or 3rd quadrant, positive when in 2nd or 4th quadrant.

BEAMS AND CHANNELS
Transverse force oblique
through center of gravity

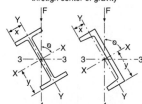

$$I_3 = I_X \sin^2\phi + I_Y \cos^2\phi$$

$$I_4 = I_X \cos^2\phi + I_Y \sin^2\phi$$

$$f_b = M\left(\frac{y}{I_X} \sin\phi + \frac{x}{I_Y} \cos\phi\right)$$

where M is bending moment due to force F.

FIGURE 1-3   Properties of geometric sections and structural shapes. (*Courtesy: American Institute of Steel Construction.*)

TRIGONOMETRIC FUNCTIONS

Radius AF $= 1$
$= \sin^2 A + \cos^2 A = \sin A \ \mathrm{cosec}\ A$
$= \cos A \sec A = \tan A \cot A$

Sine A $= \dfrac{\cos A}{\cot A} = \dfrac{1}{\mathrm{cosec}\ A} = \cos A \tan A = \sqrt{1 - \cos^2 A} = BC$

Cosine A $= \dfrac{\sin A}{\tan A} = \dfrac{1}{\sec A} = \sin A \cot A = \sqrt{1 - \sin^2 A} = AC$

Tangent A $= \dfrac{\sin A}{\cos A} = \dfrac{1}{\cot A} = \sin A \sec A = FD$

Cotangent A $= \dfrac{\cos A}{\sin A} = \dfrac{1}{\tan A} = \cos A \ \mathrm{cosec}\ A = HG$

Secant A $= \dfrac{\tan A}{\sin A} = \dfrac{1}{\cos A} = AD$

Cosecant A $= \dfrac{\cot A}{\cos A} = \dfrac{1}{\sin A} = AG$

RIGHT ANGLED TRIANGLES

$a^2 = c^2 - b^2$
$b^2 = c^2 - a^2$
$c^2 = a^2 + b^2$

Known	Required					
	A	B	a	b	c	Area
a, b	$\tan A = \dfrac{a}{b}$	$\tan B = \dfrac{b}{a}$			$\sqrt{a^2 + b^2}$	$\dfrac{ab}{2}$
a, c	$\sin A = \dfrac{a}{c}$	$\cos B = \dfrac{a}{c}$		$\sqrt{c^2 - a^2}$		$\dfrac{a\sqrt{c^2 - a^2}}{2}$
A, a		$90° - A$		$a \cot A$	$\dfrac{a}{\sin A}$	$\dfrac{a^2 \cot A}{2}$
A, b		$90° - A$	$b \tan A$		$\dfrac{b}{\cos A}$	$\dfrac{b^2 \tan A}{2}$
A, c		$90° - A$	$c \sin A$	$c \cos A$		$\dfrac{c^2 \sin 2A}{4}$

OBLIQUE ANGLED TRIANGLES

$s = \dfrac{a + b + c}{2}$

$K = \sqrt{\dfrac{(s - a)(s - b)(s - c)}{s}}$

$a^2 = b^2 + c^2 - 2bc \cos A$
$b^2 = a^2 + c^2 - 2ac \cos B$
$c^2 = a^2 + b^2 - 2ab \cos C$

Known	Required					
	A	B	a	b	c	Area
a, b, c	$\tan \frac{1}{2} A = \dfrac{K}{s - a}$	$\tan \frac{1}{2} B = \dfrac{K}{s - b}$	$\tan \frac{1}{2} C = \dfrac{K}{s - c}$			$\sqrt{s(s - a)(s - b)(s - c)}$
a, A, B			$180° - (A + B)$	$\dfrac{a \sin B}{\sin A}$	$\dfrac{a \sin C}{\sin A}$	
a, b, A		$\sin B = \dfrac{b \sin A}{a}$			$\dfrac{b \sin C}{\sin B}$	
a, b, C	$\tan A = \dfrac{a \sin C}{b - a \cos C}$				$\sqrt{a^2 + b^2 - 2ab \cos C}$	$\dfrac{ab \sin C}{2}$

FIGURE 1-4 Trigonometric formulas. (*Courtesy: American Institute of Steel Construction.*)

TABLE 1-1   Decimals of a Foot for Each 32nd of an Inch

Inch	0	1	2	3	4	5
0	0	.0833	.1667	.2500	.3333	.4167
1/32	.0026	.0859	.1693	.2526	.3359	.4193
1/16	.0052	.0885	.1719	.2552	.3385	.4219
3/32	.0078	.0911	.1745	.2578	.3411	.4245
1/8	.0104	.0938	.1771	.2604	.3438	.4271
5/32	.0130	.0964	.1797	.2630	.3464	.4297
3/16	.0156	.0990	.1823	.2656	.3490	.4323
7/32	.0182	.1016	.1849	.2682	.3516	.4349
1/4	.0208	.1042	.1875	.2708	.3542	.4375
9/32	.0234	.1068	.1901	.2734	.3568	.4401
5/16	.0260	.1094	.1927	.2760	.3594	.4427
11/32	.0286	.1120	.1953	.2786	.3620	.4453
3/8	.0313	.1146	.1979	.2812	.3646	.4479
13/32	.0339	.1172	.2005	.2839	.3672	.4505
7/16	.0365	.1198	.2031	.2865	.3698	.4531
15/32	.0391	.1224	.2057	.2891	.3724	.4557
1/2	.0417	.1250	.2083	.2917	.3750	.4583
17/32	.0443	.1276	.2109	.2943	.3776	.4609
9/16	.0469	.1302	.2135	.2969	.3802	.4635
19/32	.0495	.1328	.2161	.2995	.3828	.4661
5/8	.0521	.1354	.2188	.3021	.3854	.4688
21/32	.0547	.1380	.2214	.3047	.3880	.4714
11/16	.0573	.1406	.2240	.3073	.3906	.4740
23/32	.0599	.1432	.2266	.3099	.3932	.4766
3/4	.0625	.1458	.2292	.3125	.3958	.4792
25/32	.0651	.1484	.2318	.3151	.3984	.4818
13/16	.0677	.1510	.2344	.3177	.4010	.4844
27/32	.0703	.1536	.2370	.3203	.4036	.4870
7/8	.0729	.1563	.2396	.3229	.4063	.4896
29/32	.0755	.1589	.2422	.3255	.4089	.4922
15/16	.0781	.1615	.2448	.3281	.4115	.4948
31/32	.0807	.1641	.2474	.3307	.4141	.4974

SOURCE: American Institute of Steel Construction.

TABLE 1-1   Decimals of a Foot for Each 32nd of an Inch (*Continued*)

Inch	6	7	8	9	10	11
0	.5000	.5833	.6667	.7500	.8333	.9167
$\frac{1}{32}$	.5026	.5859	.6693	.7526	.8359	.9193
$\frac{1}{16}$	.5052	.5885	.6719	.7552	.8385	.9219
$\frac{3}{32}$	.5078	.5911	.6745	.7578	.8411	.9245
$\frac{1}{8}$	.5104	.5938	.6771	.7604	.8438	.9271
$\frac{5}{32}$	.5130	.5964	.6797	.7630	.8464	.9297
$\frac{3}{16}$	.5156	.5990	.6823	.7656	.8490	.9323
$\frac{7}{32}$	.5182	.6016	.6849	.7682	.8516	.9349
$\frac{1}{4}$	.5208	.6042	.6875	.7708	.8542	.9375
$\frac{9}{32}$	.5234	.6068	.6901	.7734	.8568	.9401
$\frac{5}{16}$	.5260	.6094	.6927	.7760	.8594	.9427
$\frac{11}{32}$	.5286	.6120	.6953	.7786	.8620	.9453
$\frac{3}{8}$	.5313	.6146	.6979	.7813	.8646	.9479
$\frac{13}{32}$	.5339	.6172	.7005	.7839	.8672	.9505
$\frac{7}{16}$	.5365	.6198	.7031	.7865	.8698	.9531
$\frac{15}{32}$	.5391	.6224	.7057	.7891	.8724	.9557
$\frac{1}{2}$	.5417	.6250	.7083	.7917	.8750	.9583
$\frac{17}{32}$	.5443	.6276	.7109	.7943	.8776	.9609
$\frac{9}{16}$	.5469	.6302	.7135	.7969	.8802	.9635
$\frac{19}{32}$	.5495	.6328	.7161	.7995	.8828	.9661
$\frac{5}{8}$	.5521	.6354	.7188	.8021	.8854	.9688
$\frac{21}{32}$	.5547	.6380	.7214	.8047	.8880	.9714
$\frac{11}{16}$	.5573	.6406	.7240	.8073	.8906	.9740
$\frac{23}{32}$	.5599	.6432	.7266	.8099	.8932	.9766
$\frac{3}{4}$	.5625	.6458	.7292	.8125	.8958	.9792
$\frac{25}{32}$	.5651	.6484	.7318	.8151	.8984	.9818
$\frac{13}{16}$	.5677	.6510	.7344	.8177	.9010	.9844
$\frac{27}{32}$	.5703	.6536	.7370	.8203	.9036	.9870
$\frac{7}{8}$	.5729	.6563	.7396	.8229	.9063	.9896
$\frac{29}{32}$	.5755	.6589	.7422	.8255	.9089	.9922
$\frac{15}{16}$	.5781	.6615	.7448	.8281	.9115	.9948
$\frac{31}{32}$	.5807	.6641	.7474	.8307	.9141	.9974

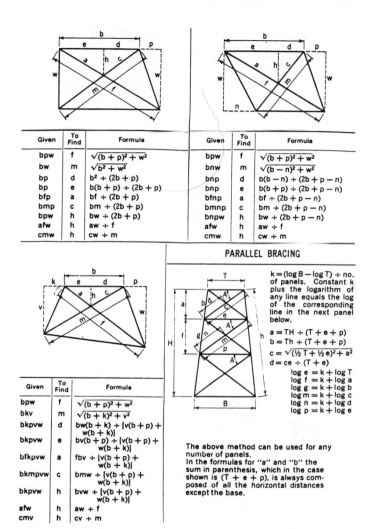

FIGURE 1-5 Bracing formulas. (*Courtesy: American Institute of Steel Construction.*)

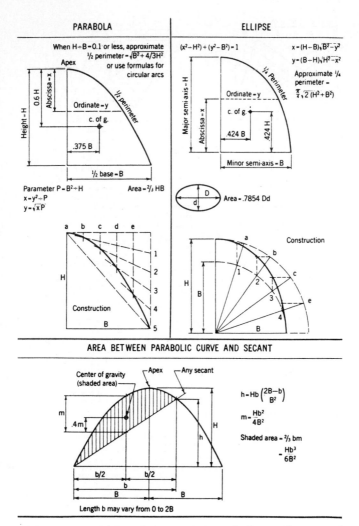

FIGURE 1-6 Properties of the parabola and ellipse. (*Courtesy: American Institute of Steel Construction.*)

TABLE 1-2   Weights and Measures—United States Customary System (USCS)

LINEAR MEASURE

Inches	Feet		Yards		Rods		Furlongs		Miles
1.0 =	.08333	=	.02778	=	.0050505	=	.00012626	=	.00001578
12.0 =	1.0	=	.33333	=	.0606061	=	.00151515	=	.00018939
36.0 =	3.0	=	1.0	=	.1818182	=	.00454545	=	.00056818
198.0 =	16.5	=	5.5	=	1.0	=	.025	=	.003125
7920.0 =	660.0	=	220.0	=	40.0	=	1.0	=	.125
63360.0 =	5280.0	=	1760.0	=	320.0	=	8.0	=	1.0

SQUARE AND LAND MEASURE

Sq. Inches	Square Feet		Square Yards		Sq. Rods		Acres		Sq. Miles
1.0 =	.006944	=	.000772						
144.0 =	1.0	=	.111111						
1296.0 =	9.0	=	1.0	=	.03306	=	.000207		
39204.0 =	272.25	=	30.25	=	1.0	=	.00625	=	.0000098
	43560.0	=	4840.0	=	160.0	=	1.0	=	.0015625
			3097600.0	=	102400.0	=	640.0	=	1.0

AVOIRDUPOIS WEIGHTS

Grains		Drams		Ounces		Pounds		Tons
1.0	=	.03657	=	.002286	=	.000143	=	.0000000714
27.34375	=	1.0	=	.0625	=	.003906	=	.00000195
437.5	=	16.0	=	1.0	=	.0625	=	.00003125
7000.0	=	256.0	=	16.0	=	1.0	=	.0005
14000000.0	=	512000.0	=	32000.0	=	2000.0	=	1.0

DRY MEASURE

Pints		Quarts		Pecks		Cubic Feet		Bushels
1.0	=	.5	=	.0625	=	.01945	=	.01563
2.0	=	1.0	=	.125	=	.03891	=	.03125
16.0	=	8.0	=	1.0	=	.31112	=	.25
51.42627	=	25.71314	=	3.21414	=	1.0	=	.80354
64.0	=	32.0	=	4.0	=	1.2445	=	1.0

LIQUID MEASURE

Gills		Pints		Quarts		U.S. Gallons		Cubic Feet
1.0 =		.25 =		.125 =		.03125	=	.00418
4.0 =		1.0 =		.5 =		.125	=	.01671
8.0 =		2.0 =		1.0 =		.250	=	.03342
32.0 =		8.0 =		4.0 =		1.0	=	.1337
						7.48052	=	1.0

SOURCE: American Institute of Steel Construction.

TABLE 1-3   Weights and Measures—International System of Units (SI)[a] (Metric Practice)

BASE UNITS			SUPPLEMENTARY UNITS		
Quantity	Unit	Symbol	Quantity	Unit	Symbol
length	metre	m	plane angle	radian	rad
mass	kilogram	kg	solid angle	steradian	sr
time	second	s			
electric current	ampere	A			
thermodynamic temperature	kelvin	K			
amount of substance	mole	mol			
luminous intensity	candela	cd			

DERIVED UNITS (WITH SPECIAL NAMES)

Quantity	Unit	Symbol	Formula
force	newton	N	$kg \cdot m/s^2$
pressure, stress	pascal	Pa	N/m^2
energy, work, quantity of heat	joule	J	$N \cdot m$
power	watt	W	J/s

DERIVED UNITS (WITHOUT SPECIAL NAMES)

Quantity	Unit	Formula
area	square metre	m^2
volume	cubic metre	m^3
velocity	metre per second	m/s
acceleration	metre per second squared	m/s^2
specific volume	cubic metre per kilogram	m^3/kg
density	kilogram per cubic metre	kg/m^3

SI PREFIXES

Multiplication Factor	Prefix	Symbol
$1\ 000\ 000\ 000\ 000\ 000\ 000 = 10^{18}$	exa	E
$1\ 000\ 000\ 000\ 000\ 000 = 10^{15}$	peta	P
$1\ 000\ 000\ 000\ 000 = 10^{12}$	tera	T
$1\ 000\ 000\ 000 = 10^{9}$	giga	G
$1\ 000\ 000 = 10^{6}$	mega	M
$1\ 000 = 10^{3}$	kilo	k
$100 = 10^{2}$	hecto[b]	h
$10 = 10^{1}$	deka[b]	da
$0.1 = 10^{-1}$	deci[b]	d
$0.01 = 10^{-2}$	centi[b]	c
$0.001 = 10^{-3}$	milli	m
$0.000\ 001 = 10^{-6}$	micro	μ
$0.000\ 000\ 001 = 10^{-9}$	nano	n
$0.000\ 000\ 000\ 001 = 10^{-12}$	pico	p
$0.000\ 000\ 000\ 000\ 001 = 10^{-15}$	femto	f
$0.000\ 000\ 000\ 000\ 000\ 001 = 10^{-18}$	atto	a

[a] Refer to ASTM E380-79 for more complete information on SI.
[b] Use is not recommended.

SOURCE: American Institute of Steel Construction.

TABLE 1-4   SI Conversion Factors[a]

Quantity	Multiply	by	to obtain	
Length	inch	[b]25.400	millimetre	mm
	foot	[b] 0.304 800	metre	m
	yard	[b] 0.914 400	metre	m
	mile (U.S. Statute)	1.609 347	kilometre	km
	millimetre	$39.370\ 079 \times 10^{-3}$	inch	in
	metre	3.280 840	foot	ft
	metre	1 093 613	yard	yd
	kilometre	0.621 370	mile	mi
Area	square inch	[b] $0.645\ 160 \times 10^3$	square millimetre	mm^2
	square foot	[b] $0.092\ 903$	square metre	m^2
	square yard	0.836 127	square metre	m^2
	square mile (U.S. Statute)	2.589 998	square kilometre	km^2
	acre	$4.046\ 873 \times 10^3$	square metre	m^2
	acre	0.404 687	hectare	
	square millimetre	$1.550\ 003 \times 10^{-3}$	square inch	in^2
	square metre	10.763 910	square foot	ft^2
	square metre	1.195 990	square yard	yd^2
	square kilometre	0.386 101	square mile	mi^2
	square metre	$0.247\ 104 \times 10^{-3}$	acre	
	hectare	2.471 044	acre	
Volume	cubic inch	[b]$16.387\ 06 \times 10^3$	cubic millimetre	mm^3
	cubic foot	$28.316\ 85 \times 10^{-3}$	cubic metre	m^3
	cubic yard	0.764 555	cubic metre	m^3
	gallon (U.S. liquid)	3.785 412	litre	l
	quart (U.S. liquid)	0.946 353	litre	l
	cubic millimetre	$61.023\ 759 \times 10^{-6}$	cubic inch	in^3
	cubic metre	35.314 662	cubic foot	ft^3
	cubic metre	1.307 951	cubic yard	yd^3
	litre	0.264 172	gallon (U.S. liquid)	gal
	litre	1.056 688	quart (U.S. liquid)	qt
Mass	ounce (avoirdupois)	28.349 52	gram	g
	pound (avoirdupois)	0.453 592	kilogram	kg
	short ton	$0.907\ 185 \times 10^3$	kilogram	kg
	gram	$35.273\ 966 \times 10^{-3}$	ounce (avoirdupois)	oz av
	kilogram	2.204 622	pound (avoirdupois)	lb av
	kilogram	$1.102\ 311 \times 10^{-3}$	short ton	

[a] Refer to ASTM E380-79 for more complete information on SI.
[b] Indicates exact value.

SOURCE: American Institute of Steel Construction.

TABLE 1-4 SI Conversion Factors (*Continued*)

Quantity	Multiply	by	to obtain	
Force	ounce-force	0.278 014	newton	N
	pound-force	4.448 222	newton	N
	newton	3.596 942	ounce-force	
	newton	0.224 809	pound-force	lbf
Bending Moment	pound-force-inch	0.112 985	newton-metre	N·m
	pound-force-foot	1.355 818	newton-metre	N·m
	newton-metre	8.850 748	pound-force-inch	lbf·in
	newton-metre	0.737 562	pound-force-foot	lbf·ft
Pressure, Stress	pound-force per square inch	6.894 757	kilopascal	kPa
	foot of water (39.2 F)	2.988 98	kilopascal	kPa
	inch of mercury (32 F)	3.386 38	kilopascal	kPa
	kilopascal	0.145 038	pound-force per square inch	lbf/in^2
	kilopascal	0.334 562	foot of water (39.2 F)	
	kilopascal	0.295 301	inch of mercury (32 F)	
Energy, Work, Heat	foot-pound-force	1.355 818	joule	J
	[c]British thermal unit	$1.055\,056 \times 10^3$	joule	J
	[c]calorie	[b] 4.186 800	joule	J
	kilowatt hour	[b] $3.600\,000 \times 10^6$	joule	J
	joule	0.737 562	foot-pound-force	ft·lbf
	joule	$0.947\,817 \times 10^{-3}$	[c]British thermal unit	Btu
	joule	0.238 846	[c]calorie	
	joule	$0.277\,778 \times 10^{-6}$	kilowatt hour	kW·h
Power	foot-pound-force/second	1.355 818	watt	W
	[c]British thermal unit per hour	0.293 071	watt	W
	horsepower (550 ft. lbf/s)	0.745 700	kilowatt	kW
	watt	0.737 562	foot-pound-force/second	ft·lbf/s
	watt	3.412 141	[c]British thermal unit per hour	Btu/h
	kilowatt	1.341 022	horsepower (550 ft.·lbf/s)	hp
Angle	degree	$17.453\,29 \times 10^{-3}$	radian	rad
	radian	57.295 788	degree	
Temperature	degree Fahrenheit	t°C = (t°F − 32)/1.8	degree Celsius	
	degree Celsius	t°F = 1.8 × t°C + 32	degree Fahrenheit	

[a] Refer to ASTM E380-79 for more complete information on SI.
[b] Indicates exact value.
[c] International Table.

TABLE 1-5 Decimals of an Inch for Each 64th of an Inch, with Millimeter Equivalents

Fraction	¹⁄₆₄ths	Decimal	Millimeters (Approx.)	Fraction	¹⁄₆₄ths	Decimal	Millimeters (Approx.)
...	1	.015625	0.397	...	33	.515625	13.097
¹⁄₃₂	2	.03125	0.794	¹⁷⁄₃₂	34	.53125	13.494
...	3	.046875	1.191	...	35	.546875	13.891
¹⁄₁₆	4	.0625	1.588	⁹⁄₁₆	36	.5625	14.288
...	5	.078125	1.984	...	37	.578125	14.684
³⁄₃₂	6	.09375	2.381	¹⁹⁄₃₂	38	.59375	15.081
...	7	.109375	2.778	...	39	.609375	15.478
¹⁄₈	8	.125	3.175	⁵⁄₈	40	.625	15.875
...	9	.140625	3.572	...	41	.640625	16.272
⁵⁄₃₂	10	.15625	3.969	²¹⁄₃₂	42	.65625	16.669
...	11	.171875	4.366	...	43	.671875	17.066
³⁄₁₆	12	.1875	4.763	¹¹⁄₁₆	44	.6875	17.463
...	13	.203125	5.159	...	45	.703125	17.859
⁷⁄₃₂	14	.21875	5.556	²³⁄₃₂	46	.71875	18.256
...	15	.234375	5.953	...	47	.734375	18.653
¹⁄₄	16	.250	6.350	³⁄₄	48	.750	19.050
...	17	.265625	6.747	...	49	.765625	19.447
⁹⁄₃₂	18	.28125	7.144	²⁵⁄₃₂	50	.78125	19.844
...	19	.296875	7.541	...	51	.796875	20.241
⁵⁄₁₆	20	.3125	7.938	¹³⁄₁₆	52	.8125	20.638
...	21	.328125	8.334	...	53	.828125	21.034
¹¹⁄₃₂	22	.34375	8.731	²⁷⁄₃₂	54	.84375	21.431
...	23	.359375	9.128	...	55	.859375	21.828
³⁄₈	24	.375	9.525	⁷⁄₈	56	.875	22.225
...	25	.390625	9.922	...	57	.890625	22.622
¹³⁄₃₂	26	.40625	10.319	²⁹⁄₃₂	58	.90625	23.019
...	27	.421875	10.716	...	59	.921875	23.416
⁷⁄₁₆	28	.4375	11.113	¹⁵⁄₁₆	60	.9375	23.813
...	29	.453125	11.509	...	61	.953125	24.209
¹⁵⁄₃₂	30	.46875	11.906	³¹⁄₃₂	62	.96875	24.606
...	31	.484375	12.303	...	63	.984375	25.003
½	32	.500	12.700	1	64	1.000	25.400

SOURCE: American Institute of Steel Construction.

TABLE 1-6   Equivalent Listing

Fraction–Decimal–Millimeter

Fract. inch.	Inch decimal equiv.	Millimeter equiv.	Fract. inch.	Inch decimal equiv.	Millimeter equiv.
1/64	.0156	.397	33/64	.5156	13.097
1/32	.0312	.794	17/32	.5312	13.494
3/64	.0469	1.191	35/64	.5469	13.891
1/16	.0625	1.588	9/16	.5625	14.288
5/64	.0781	1.984	37/64	.5781	14.684
3/32	.0937	2.381	19/32	.5937	15.081
7/64	.1094	2.778	39/64	.6094	15.478
1/8	.1250	3.175	5/8	.6250	15.875
9/64	.1406	3.572	41/64	.6406	16.272
5/32	.1562	3.969	21/32	.6562	16.669
11/64	.1719	4.366	43/64	.6719	17.066
3/16	.1875	4.763	11/16	.6875	17.463
13/64	.2031	5.159	45/64	.7031	17.859
7/32	.2187	5.556	23/32	.7187	18.256
15/64	.2344	5.953	47/64	.7344	18.653
1/4	.2500	6.350	3/4	.7500	19.050
17/64	.2656	6.747	49/64	.7656	19.447
9/32	.2812	7.144	25/32	.7812	19.844
19/64	.2969	7.541	51/64	.7969	20.241
5/16	.3125	7.938	13/16	.8125	20.638
21/64	.3281	8.334	53/64	.8281	21.034
11/32	.3437	8.731	27/32	.8437	21.431
23/64	.3594	9.128	55/64	.8594	21.828
3/8	.3750	9.525	7/8	.8750	22.225
25/64	.3906	9.922	57/64	.8906	22.622
13/32	.4062	10.319	29/32	.9062	23.019
27/64	.4219	10.716	59/64	.9219	23.416
7/16	.4375	11.113	15/16	.9375	23.813
29/64	.4531	11.509	61/64	.9531	24.209
15/32	.4687	11.906	31/32	.9687	24.606
31/64	.4844	12.303	63/64	.9844	25.003
1/2	.5000	12.700	1	1.0000	25.400

Millimeters–Inches

Millimeters	Inches	Millimeters	Inches
1	.0394	16	.6299
2	.0787	17	.6693
3	.1181	18	.7087
4	.1575	19	.7480
5	.1968	20	.7874
6	.2362	21	.8268
7	.2756	22	.8661
8	.3150	23	.9055
9	.3543	24	.9449
10	.3937	25	.9842
11	.4331	26	1.0236
12	.4724	27	1.0630
13	.5118	28	1.1024
14	.5512	29	1.1417
15	.5905	30	1.1811

SOURCE: Boston Gear, Quincy, MA.

TABLE 1-7 General Conversions, USCS and Metric

Multiply	*by*	*to obtain*
acres	43,560	square feet
acres	4047	square meters
acres	1.562×10^{-3}	square miles
acres	4840	square yards
amperes	1/10	abamperes
atmospheres	76.0	cms. of mercury
atmospheres	29.92	inches of mercury
atmospheres	33.90	feet of water
atmospheres	14.70	lbs. per sq. inch
British thermal units	0.2520	kilogram-calories
British thermal units	777.5	foot-pounds
British thermal units	3.927×10^{-4}	horse-power-hours
British thermal units	1054	joules
British thermal units	107.5	kilogram-meters
British thermal units	2.928×10^{-4}	kilowatt-hours
B.t.u. per min.	12.96	foot-pounds per sec.
B.t.u. per min.	0.02356	horse-power
B.t.u. per min.	0.01757	kilowatts
B.t.u. per min.	17.57	watts
B.t.u. per sq. ft. per min.	0.1220	watts per sq. inch
bushels	1.244	cubic feet
bushels	2150	cubic inches
centimeters	0.3937	inches
centimeters	0.01	meters
centimeter-grams	980.7	centimeter-dynes
centimeters of mercury	0.01316	atmospheres
centimeters of mercury	0.4461	feet of water
centimeters per second	1.969	feet per minute
centimeters per second	0.03281	feet per second
cubic centimeters	3.531×10^{-5}	cubic feet
cubic centimeters	6.102×10^{-2}	cubic inches
cubic feet	2.832×10^{4}	cubic cms.
cubic feet	1728	cubic inches
cubic feet	0.02832	cubic meters
cubic feet	0.03704	cubic yards
cubic feet	7.481	gallons
cubic feet	62.43	pounds of water

SOURCE: Dietzgen Company.

TABLE 1-7   General Conversions, USCS and Metric (*Continued*)

Multiply	by	to obtain
cubic feet per minute...	472.0	cubic cms. per sec.
cubic feet per minute...	0.1247	gallons per sec.
cubic feet per minute.	0.4720	liters per second
cubic feet per minute...	62.4	lbs. of water per min.
cubic inches	16.39	cubic centimeters
cubic inches	5.787×10^{-4}	cubic feet
cubic yards	27	cubic feet
cubic yards	0.7646	cubic meters
cubic yards per minute..	0.45	cubic feet per sec.
degrees (angle)	60	minutes
degrees (angle)	0.01745	radians
degrees (angle)	3600	seconds
dynes	7.233×10^{-5}	poundals
ergs	2.390×10^{-11}	kilogram-calories
feet	30.48	centimeters
feet	12	inches
feet	0.3048	meters
feet	.36	varas
feet	1/3	yards
feet of water	0.02950	atmospheres
feet of water	0.8826	inches of mercury
feet of water	304.8	kgs. per sq. meter
feet of water	62.43	pounds per sq. ft.
feet of water	0.4335	pounds per sq. inch.
foot-pounds	1.286×10^{-3}	British thermal units
foot-pounds	1.356×10^{7}	ergs
foot-pounds	5.050×10^{-7}	horse-power-hours
foot-pounds	1.356	joules
foot-pounds per min.	1.286×10^{-3}	B.t. units per minute
foot-pounds per min.	0.01667	foot-pounds per sec.
foot-pounds per min.	3.030×10^{-5}	horse-power
foot-pounds per min.	2.260×10^{-5}	kilowatts
foot-pounds per sec.	7.717×10^{-2}	B.t.units per minute
gallons	8.345	pounds of water
gallons	231	cubic inches
gallons	3.785	liters
gallons per minute	2.228×10^{-3}	cubic feet per sec.
grains (troy)	0.06480	grams

TABLE 1-7   General Conversions, USCS and Metric (*Continued*)

Multiply	by	to obtain
grams	980.7	dynes
grams	0.03527	ounces
horse-power	42.44	B.t.units per minute
horse-power	33,000	foot-pounds per min.
horse-power	550	foot-pounds per sec.
horse-power	1.014	horse-power (metric)
horse-power	10.70	kg.-calories per min.
horse-power	0.7457	kilowatts
horse-power	745.7	watts
horse-power (boiler)	33,520	B.t.u. per hour
horse-power (boiler)	9.804	kilowatts
horse-power-hours	2547	British thermal units
horse-power-hours	1.98×10^6	foot-pounds
inches	2.540	centimeters
inches of mercury	0.03342	atmospheres
inches of mercury	1.133	feet of water
inches of water	0.002458	atmospheres
inches of water	0.03613	pounds per sq. inch
kilograms	980,665	dynes
kilograms	2.2046	pounds
kilogram-calories	3.968	British thermal units
kilogram-calories	3086	foot-pounds
k.g.-calories per min.	51.43	foot-pounds per sec.
k.g.-calories per min.	0.06972	kilowatts
kilometers	10^5	centimeters
kilometers	3281	feet
kilometers	10^3	meters
kilometers	0.6214	miles
kilometers	1093.6	yards
kilowatts	56.92	B.t.units per min.
kilowatts	4.425×10^4	foot-pounds per min.
kilowatts	737.6	foot-pounds per sec.
kilowatts	1.341	horse-power
kilowatts	14.34	kg.-calories per min.
kilowatts	10^3	watts
kilowatt-hours	3415	British thermal units
kilowatt-hours	2.655×10^6	foot-pounds

TABLE 1-7   General Conversions, USCS and Metric (*Continued*)

Multiply	by	to obtain
kilowatt-hours	1.341	horse-power-hours
kilowatt-hours	3.6×10^6	joules
kilowatt-hours	860.5	kilogram-calories
kilowatt-hours	3.671×10^5	kilogram-meters
$\log^{10} N$	2.303	$\log_e N$ or $\ln N$
$\log_e N$ or $\ln N$	0.4343	$\log_{10} N$
meters	100	centimeters
meters	3.2808	feet
meters	39.37	inches
miles	5280	feet
miles	1.6093	kilometers
miles per hour	88	feet per minute
ounces	28.35	grams
ounces per sq. inch	0.0625	pounds per sq. inch
pints (dry)	33.60	cubic inches
pints (liq.)	28.87	cubic inches
pounds	453.6	grams
pounds	16	ounces
pounds of water	0.01602	cubic feet
pounds of water	27.68	cubic inches
pounds of water	0.1198	gallons
pounds of water per min.	2.669×10^{-4}	cubic feet per sec.
pounds per cubic foot	5.787×10^{-4}	pounds per cubic in.
pounds per sq. foot	0.01602	feet of water
pounds per sq. inch	0.06804	atmospheres
pounds per sq. inch	2.307	feet of water
quarts	32	fluid ounces
quarts (dry)	67.20	cubic inches
quarts (liq.)	57.75	cubic inches
rods	16.5	feet
square centimeters	0.1550	square inches
square inches	6.452	square centimeters
square miles	640	acres
square miles	27.88×10^6	square feet
square yards	0.8361	square meters
temp. (degs. C.)+17.8	1.8	temp. (degs. Fahr.)
temp. (degs. F.)−32	5/9	temp. (degs. Cent.)
yards	.9144	meters

TABLE 1-8 Temperature Conversion Table—Celsius (Centigrade) to Fahrenheit

C	F	C	F	C	F	C	F	C	F
-30	-22.0	20	68.0	70	158.0	120	248.0	170	338.0
-29	-20.2	21	69.8	71	159.8	121	249.8	171	339.8
-28	-18.4	22	71.6	72	161.6	122	251.6	172	341.6
-27	-16.6	23	73.4	73	163.4	123	253.4	173	343.4
-26	-14.8	24	75.2	74	165.2	124	255.2	174	345.2
-25	-13.0	25	77.0	75	167.0	125	257.0	175	347.0
-24	-11.2	26	78.8	76	168.8	126	258.8	176	348.8
-23	-9.4	27	80.6	77	170.6	127	260.6	177	350.6
-22	-7.6	28	82.4	78	172.4	128	262.4	178	352.4
-21	-5.8	29	84.2	79	174.2	129	264.2	179	354.2
-20	-4.0	30	86.0	80	176.0	130	266.0	180	356.0
-19	-2.2	31	87.8	81	177.8	131	267.8	181	357.8
-18	-0.4	32	89.6	82	179.6	132	269.6	182	359.6
-17	+1.4	33	91.4	83	181.4	133	271.4	183	361.4
-16	3.2	34	93.2	84	183.2	134	273.2	184	363.2
-15	5.0	35	95.0	85	185.0	135	275.0	185	365.0
-14	6.8	36	96.8	86	186.8	136	276.8	186	366.8
-13	8.6	37	98.6	87	188.6	137	278.6	187	368.6
-12	10.4	38	100.4	88	190.4	138	280.4	188	370.4
-11	12.2	39	102.2	89	192.2	139	282.2	189	372.2
-10	14.0	40	104.0	90	194.0	140	284.0	190	374.0
-9	15.8	41	105.8	91	195.8	141	285.8	191	375.8
-8	17.6	42	107.6	92	197.6	142	287.6	192	377.6
-7	19.4	43	109.4	93	199.4	143	289.4	193	379.4
-6	21.2	44	111.2	94	201.2	144	291.2	194	381.2
-5	23.0	45	113.0	95	203.0	145	293.0	195	383.0
-4	24.8	46	114.8	96	204.8	146	294.8	196	384.8
-3	26.6	47	116.6	97	206.6	147	296.6	197	386.6
-2	28.4	48	118.4	98	208.4	148	298.4	198	388.4
-1	30.2	49	120.2	99	210.2	149	300.2	199	390.2

SOURCE: The Trane Company, LaCrosse, WI. Reproduced by permission.

TABLE 1-8 Temperature Conversion Table—Celsius (Centigrade) to Fahrenheit (*Continued*)

C	F	C	F	C	F	C	F	C	F
0	32.0	50	122.0	100	212.0	150	302.0	200	392.0
1	33.8	51	123.8	101	213.8	151	303.8	201	393.8
2	35.6	52	125.6	102	215.6	152	305.6	202	395.6
3	37.4	53	127.4	103	217.4	153	307.4	203	397.4
4	39.2	54	129.2	104	219.2	154	309.2	204	399.2
5	41.0	55	131.0	105	221.0	155	311.0	205	401.0
6	42.8	56	132.8	106	222.8	156	312.8	206	402.8
7	44.6	57	134.6	107	224.6	157	314.6	207	404.6
8	46.4	58	136.4	108	226.4	158	316.4	208	406.4
9	48.2	59	138.2	109	228.2	159	318.2	209	408.2
10	50.0	60	140.0	110	230.0	160	320.0	210	410.0
11	51.8	61	141.8	111	231.8	161	321.8	211	411.8
12	53.6	62	143.6	112	233.6	162	323.6	212	413.6
13	55.4	63	145.4	113	235.4	163	325.4	213	415.4
14	57.2	64	147.2	114	237.2	164	327.2	214	417.2
15	59.0	65	149.0	115	239.0	165	329.0	215	419.0
16	60.8	66	150.8	116	240.8	166	330.8	216	420.8
17	62.6	67	152.6	117	242.6	167	332.6	217	422.6
18	64.4	68	154.4	118	244.4	168	334.4	218	424.4
19	66.2	69	156.2	119	246.2	169	336.2	219	426.2

TABLE 1-9   Basic Heating and Ventilating Formulas

Formula	Legend
Btu $= TC/55$	Btu $=$ British thermal unit
Btu/h $= 60TC/55$	$T =$ temperature rise, °F
	$C =$ cubic feet of air delivery
DS $= 144C/60v$	DS $=$ cross section of duct, in
	$v =$ flow velocity, ft/min
	$\Delta T =$ final temperature of air—temperature of entering air
Btu/$C = 1.085 \times C \times \Delta T$	Btu/$C =$ heating capacity of coil, Btu
SHL $= 24hd\ (t_i - t_a)/t_i - t_e$	SHL $=$ seasonal heat loss, Btu
	$h =$ hourly building heat loss, Btu
	$d =$ total days of heating season
	$t_i =$ interior design temperature, °F
	$t_e =$ exterior design temperature, °F
	$t_a =$ average outside temperature for heating season, °F
	$24 =$ h/day
$U = 1/R$	$U =$ overall conductance (equal to number of Btu that will flow through 1 ft^2 of structure from air to air owing to a temperature difference of 1°F, in 1 h)
$R = 1/U$	$R =$ thermal resistance (reciprocal of conductance)
$R = 1/c$	$c =$ thermal conductance (heat flow through a given thickness of 1 ft^2 of material with 1°F temperature differential)
$Q = VN/60$	$Q =$ air changes, ft^3/min
	$V =$ volume of room, ft^3
	$N =$ number of hourly air changes
$q = 1.08QwT$	$q =$ heat required to warm up cold air, Btu/h
	$Qw =$ air to be warmed, expressed in cubic feet per minute

SOURCE: R. O. Parmley, *HVAC Field Manual,* McGraw-Hill, New York, 1988. Used with permission of publisher.

TABLE 1-10   Minimum Ventilation Air for Various Activities

Type of occupancy	Ventilation air, ft^3/min per person
Inactive, theaters	5
Light activity, offices	10
Light activity with some odor generations, restaurant	15
Light activity with moderate odor generations, bars	20
Active work, shipping rooms	30
Very active work, gymnasiums	50

SOURCE: R. O. Parmley, *HVAC Field Manual,* McGraw-Hill, New York, 1988. Used with permission of publisher.

TABLE 1-11   Application Formulas

TO OBTAIN	HAVING	FORMULA
Velocity (V) Feet Per Minute	Pitch Diameter (D) of Gear or Sprocket – Inches & Rev. Per Min. (RPM)	$V = .2618 \times D \times RPM$
Rev. Per Min. (RPM)	Velocity (V) Ft. Per Min. & Pitch Diameter (D) of Gear or Sprocket – Inches	$RPM = \dfrac{V}{.2618 \times D}$
Pitch Diameter (D) of Gear or Sprocket – Inches	Velocity (V) Ft. Per Min. & Rev. Per Min. (RPM)	$D = \dfrac{V}{.2618 \times RPM}$
Torque (T) In. Lbs.	Force (W) Lbs. & Radius (R) Inches	$T = W \times R$
Horsepower (HP)	Force (W) Lbs. & Velocity (V) Ft. Per Min.	$HP = \dfrac{W \times V}{33000}$
Horsepower (HP)	Torque (T) In. Lbs. & Rev. Per Min. (RPM)	$HP = \dfrac{T \times RPM}{63025}$
Torque (T)	Horsepower (HP) & Rev. Per Min. (RPM)	$T = \dfrac{63025 \times HP}{RPM}$
Force (W) Lbs.	Horsepower (HP) & Velocity (V) Ft. Per Min.	$W = \dfrac{33000 \times HP}{V}$
Rev. Per Min. (RPM)	Horsepower (HP) & Torque (T) In. Lbs.	$RPM = \dfrac{63025 \times HP}{T}$

SOURCE: Boston Gear, Quincy, MA.

TABLE 1-12   Basic Heating Data

1 Btu (British thermal unit) = heat required to raise 1 lb of water 1°F
Btu ÷ 3.413 = W
Specific heat of water = 1 Btu
Specific heat of air = 0.24 Btu
1 ft³ of air @ 32°F = 0.0807 lb
1 Btu will heat 1 ft³ of air 55°F
1 gal (U.S.) of water = 8.33 lb
1 gal (U.S.) of water = 231 in³ @ 39.2°F
1 ft³ of water = 7.48 gal
1 ft³ of water = 62.418 lb @ 39.2°F
1 ft² of steam radiation = 240 Btu
1 ft² of hot water radiation = 150 Btu
No. 2 fuel oil = 20,571 Btu/lb or 144,000 Btu/gal
Natural gas = 18,000 Btu/lb or 1,030 Btu/ft³
Propane = 21,564 Btu/lb or 2,572 Btu/ft³
Butane = 21,440 Btu/lb or 3,200 Btu/ft³
1 degree day = 1°F, in mean outdoor temperature, below 65°F (24-h avg.)
  per day
1 watt (W) = 1 joule (J) per second (s)
1 kWh = 1.341 hph
1 Btu = 778 ft·lb = 1055 J = 252 cal
1 hph = 2545 Btu = 0.7457 kWh
1 gal of LP gas = 36.6 ft³

SOURCE: R. O. Parmley, *HVAC Field Manual,* McGraw-Hill, New York, 1988. Used with permission of publisher.

TABLE 1-13 Pressure Facts

| | | | | | | | Vapor pressure, psig | | | | | | | | |
| | | | | | | | Outside temperature, °F | | | | | | | | |
	-30	-20	-10	0	10	20	30	40	50	60	70	80	90	100	110
100% Propane	6.8	11.5	17.5	24.5	34	42	53	65	78	93	110	128	150	172	204
70% Propane 30% Butane	...	4.7	9	15	20.5	28	36.5	46	56	68	82	96	114	134	158
50% Propane 50% Butane	...	...	3.5	7.6	12.3	17.8	24.5	32.4	41	50	61	74	88	104	122
70% Butane 30% Propane	...	...	...	2.3	5.9	10.2	15.4	21.5	28.5	36.5	45	54	66	79	93
100% Butane	...	...	...	...	...	...	...	3.1	6.9	11.5	17	23	30	38	47

SOURCE: R. O. Parmley, *HVAC Field Manual*, McGraw-Hill, New York, 1988. Used with permission of publisher.

TABLE 1-14   Power, Mass, Volume, and Weight of Water Formulas

Power	
1 Btu per h	= 0.293 W
	= 12.96 ft·lb/min
	= 0.00039 hp
1 ton refrigeration (U.S.)	= 288,000 Btu per 24 h
	= 12,000 Btu/h
	= 200 Btu/min
	= 83.33 lb ice melted per h from and at 32°F
	= 2000 lb ice melted per 24 h from and at 32°F
1 hp	= 550 ft·lb/s
	= 746 W
	= 2545 Btu/h
1 boiler hp	= 33,480 Btu/h
	= 34.5 lb water evap. per h from and at 212°F
	= 9.8 kW
1 kW	= 3413 Btu/h

Mass	
1 lb (avoir.)	= 16 oz (avoir.)
	= 7000 grain
1 ton (short)	= 2000 lb
1 ton (long)	= 2240 lb

Volume	
1 gal (U.S.)	= 128 fl oz (U.S.)
	= 231 in^3
	= 0.833 gal (Brit.)
1 ft^3	= 7.48 gal (U.S.)

Weight of Water
1 ft^3 at 50°F weighs 62.41 lb
1 gal at 50°F weighs 8.34 lb
1 ft^3 of ice weighs 57.2 lb
Water is at its greatest density at 39.2°F
1 ft^3 at 39.2°F weighs 62.43 lb

SOURCE: R. O. Parmley, *HVAC Field Manual,* McGraw-Hill, New York, 1988. Used with permission of publisher.

TABLE 1-15   Water Pressure Conversion: Pounds per Square Inch to Feet Head and Feet Head to Pounds per Square Inch

Water Pressure to Feet Head

Pounds per square inch	Feet head	Pounds per square inch	Feet head
1	2.31	100	230.90
2	4.62	110	253.98
3	6.93	120	277.07
4	9.24	130	300.16
5	11.54	140	323.25
6	13.85	150	346.34
7	16.16	160	369.43
8	18.47	170	392.52
9	20.78	180	415.61
10	23.09	200	461.78
15	34.63	250	577.24
20	46.18	300	692.69
25	57.72	350	808.13
30	69.27	400	922.58
40	92.36	500	1154.48
50	115.45	600	1385.39
60	138.54	700	1616.30
70	161.63	800	1847.20
80	184.72	900	2078.10
90	207.81	1000	2309.00

Note: One pound of pressure per square inch of water equals 2.309 ft of water at 62°F. Therefore, to find the feet head of water for any pressure not given in the table above, multiply the pressure pounds per square inch by 2.309.

SOURCE: R. O. Parmley, *HVAC Field Manual,* McGraw-Hill, New York, 1988. Used with permission of publisher.

TABLE 1-15   Water Pressure Conversion: Pounds per Square Inch to Feet Head and Feet Head to Pounds per Square Inch (*Continued*)

Feet Head of Water to PSI

Feet head	Pounds per square inch	Feet head	Pounds per square inch
1	0.43	100	43.31
2	0.87	110	47.64
3	1.30	120	51.97
4	1.73	130	56.30
5	2.17	140	60.63
6	2.60	150	64.96
7	3.03	160	69.29
8	3.46	170	73.63
9	3.90	180	77.96
10	4.33	200	86.62
15	6.50	250	108.27
20	8.66	300	129.93
25	10.83	350	151.58
30	12.99	400	173.24
40	17.32	500	216.55
50	21.65	600	259.85
60	25.99	700	303.16
70	30.32	800	346.47
80	34.65	900	389.78
90	38.98	1000	433.00

Note: One foot of water at 62°F equals 0.433 lb pressure per square inch. To find the pressure per square inch for any feet head not given in the table above, multiply the feet head by 0.433.

TABLE 1-16   Conversion Factors for Length, Torque, Rotation, Moment of Inertia, Power, and Temperature, USCS and Metric

	MULTIPLY	BY	TO OBTAIN
Length	Meters	3.281	Feet
	Meters	39.37	Inches
	Inches	.0254	Meters
	Feet	.3048	Meters
	Millimeters	.0394	Inches
Torque	Newton-Meters	.7376	Lb-Ft
	Lb-Ft	1.3558	Newton-Meter
	Lb-In	.0833	Lb-Ft
	Lb-Ft	12.00	Lb-In
Rotation	RPM	6.00	Degrees/Sec.
	RPM	.1047	Rad./Sec.
	Degrees/Sec.	.1667	RPM
	Rad./Sec.	9.549	RPM
Moment of Inertia	Newton-Meters^2	2.42	Lb-Ft^2
	Oz-In^2	.000434	Lb-Ft
	Lb-In^2	.00694	Lb-Ft^2
	Slug-Ft^2	32.17	Lb-Ft^2
	Oz-In-Sec^2	.1675	Lb-Ft^2
	Lb-In-Sec^2	2.68	Lb-Ft^2
Power	Watts	.00134	HP
	Lb-Ft/Min	.0000303	HP
Temperature	Degree C = (Degree F -32) × 5/9 Degree F = (Degree C × 9/5) + 32		

SOURCE: Boston Gear, Quincy, MA.

TABLE 1-17   Conversion Factors for Pressure, Temperature, Weight of Liquid, Flow, and Work, USCS

Pressure	
1 lb/in²	= 2.31 ft water at 60°F
	= 2.04 inHg at 60°F
1 ft water at 60°F	= 0.433 lb/in²
	= 0.884 inHg at 60°F
1 inHg at 60°F	= 0.49 lb/in²
	= 1.13 ft water at 60°F
lb/in² absolute (psia)	= lb/in² gauge (psig) + 14.7

Temperature	
°C	= (°F − 32) × 5/9

Weight of Liquid	
1 gal (U.S.)	= 8.34 lb × sp. gr.
1 ft³	= 62.4 lb × sp. gr.
1 lb	= 0.12 U.S. gal ÷ sp. gr.
	= 0.016 ft³ ÷ sp. gr.

Flow	
1 gal/min (gpm)	= 0.134 ft³/min
	= 500 lb/h × sp. gr.
500 lb/h	= 1 gal/min ÷ sp. gr.
1 ft³/min (cfm)	= 448.8 gal/h (gph)

Work	
1 Btu (mean)	= 778 ft·lb
	= 0.293 Wh
	= 1/180 of heat required to change temp of 1 lb water from 32°F to 212°F
1 hp·h	= 2545 Btu (mean)
	= 0.746 kWh
1 kWh	= 3413 Btu (mean)
	= 1.34 hp·h

SOURCE: R. O. Parmley, *HVAC Field Manual,* McGraw-Hill, New York, 1988. Used with permission of publisher.

TABLE 1-18   DC and AC Power Circuits

POWER IN DC CIRCUITS:

$$\text{Watts} = \text{Volts} \times \text{Amperes} \qquad \text{Horsepower} = \frac{\text{Volts} \times \text{Amperes}}{746}$$

$$\text{Kilowatts} = \frac{\text{Volts} \times \text{Amperes}}{1000}$$

$$\text{Kilowatt-Hours} = \frac{\text{Volts} \times \text{Amperes} \times \text{Hours}}{1000}$$

POWER IN AC CIRCUITS:

Kilovolt - Amperes (KVA)

$$\text{KVA (Single-Phase)} = \frac{\text{Volts} \times \text{Amperes}}{1000}$$

$$\text{KVA (Three-Phase)} = \frac{\text{Volts} \times \text{Amperes} \times 1.73}{1000}$$

Kilowatt (Kw)

$$\text{Kw (Single-Phase)} = \frac{\text{Volts} \times \text{Amperes} \times \text{Power Factor}}{1000}$$

$$\text{Kw (Two-Phase)} = \frac{\text{Volts} \times \text{Amperes} \times \text{Power Factor} \times 1.42}{1000}$$

$$\text{Kw (Three-Phase)} \quad \frac{\text{Volts} \times \text{Amperes} \times \text{Power Factor} \times 1.73}{1000}$$

$$\text{Power Factor} = \frac{\text{Kilowatts}}{\text{Kilovolts} \times \text{Amperes}}$$

SOURCE: Boston Gear, Quincy, MA.

TABLE 1-19   Specific Gravity of Liquids

Liquid	Temp., °F	Specific gravity
Water (1 cu. ft. weighs 62.41 lb)	50	1.00
Brine (sodium chloride 25%)	32	1.20
Pennsylvania crude oil	80	0.85
Fuel oil Nos. 1 and 2	85	0.95
Gasoline	80	0.74
Kerosene	85	0.82
Lubricating oil SAE 10-20-30	115	0.94

TABLE 1-20   Specific Gravity of Gases (at 60°F and 29.9 inHg)

Gases	Specific gravity
Dry air (1 ft^3 at 60°F and 29.92 inHg weighs 0.07638 lb)	1.000
Acetylene (C_2H_2)	0.91
Ethane (C_2H_6)	1.05
Methane (CH_4)	0.554
Ammonia (NH_3)	0.596
Carbon dioxide (CO_2)	1.53
Carbon monoxide (CO)	0.967
Butane (C_4H_{10})	2.067
Butene (C_4H_8)	1.93
Chlorine (Cl_2)	2.486
Helium (He)	0.138
Hydrogen (H_2)	0.0696
Nitrogen (N_2)	0.9718
Oxygen (O_2)	1.1053

TABLE 1-21 Steam Table

Gauge press., lb	Temp., °F	Gauge press., lb	Temp., °F	Gauge press., lb	Temp., °F	Gauge press., lb	Temp., °F
5	228	55	302	110	344	210	391
6	230	56	303	112	345	212	392
7	233	57	304	114	346	214	393
8	235	58	305	116	348	216	394
9	237	59	306	118	349	218	394
10	240	60	307	120	350	220	395
11	242	61	308	122	351	222	396
12	244	62	309	124	352	224	397
13	246	63	310	126	353	226	397
14	248	64	311	128	354	228	398
15	250	65	312	130	355	230	399
16	252	66	312	132	356	232	400
17	254	67	313	134	357	234	400
18	255	68	314	136	359	235	401
19	257	69	315	138	360	237	401
20	259	70	316	140	361	239	402
21	261	71	317	142	362	241	403
22	262	72	317	144	363	243	403
23	264	73	318	146	364	245	404
24	265	74	319	148	365	247	405
25	267	75	320	150	366	249	405
26	268	76	321	152	367	251	406
27	270	77	321	154	368	253	407
28	271	78	322	156	368	255	407
29	273	79	323	158	369	257	408
30	274	80	324	160	370	259	409
31	275	81	324	162	371	261	409
32	277	82	325	164	372	263	410
33	278	83	326	166	373	265	411
34	279	84	327	168	374	267	411
35	281	85	327	170	375	269	412
36	282	86	328	172	376	271	413
37	283	87	329	174	377	273	413
38	284	88	330	176	378	275	414
39	285	89	330	178	378	277	415
40	287	90	331	180	379	279	415
41	288	91	332	182	380	281	416
42	289	92	332	184	381	283	416
43	290	93	333	186	382	285	417
44	291	94	334	188	383	295	420
45	292	95	334	190	384	305	423
46	293	96	335	192	384	355	437
47	294	97	336	194	385	375	442
48	295	98	336	196	386	385	444
49	297	99	337	198	387	405	449
50	298	100	338	200	388	455	460
51	299	102	339	202	388	510	472
52	300	104	340	204	389	560	481
53	301	106	341	206	390	585	486
54	302	108	343	208	391		

SOURCE: R. O. Parmley, *HVAC Field Manual,* McGraw-Hill, New York, 1988. Used with permission of publisher.

TABLE 1-22   Horsepower Required to Pump 1 Gallon of Water per Minute at 87% Pump Efficiency

Working pressure	HP per gal pumped	Working pressure	HP per gal pumped
100	0.067	1500	1.000
150	0.10	1600	.067
200	0.133	1800	1.20
300	0.20	2000	1.33
400	0.267	2250	1.50
500	0.33	2500	1.67
600	0.40	2750	1.83
700	0.467	3000	2.00
800	0.533	3500	2.33
900	0.60	4000	2.67
1000	0.667	4500	3.00
1200	0.80	5000	3.33
1300	0.867	5500	3.67
1400	0.933	6000	4.00

SOURCE: R. O. Parmley, *HVAC Field Manual,* McGraw-Hill, New York, 1988. Used with permission of publisher.

1-1   How to Calculate Horsepower, Accelerating Force for Linear Motion, Accelerating Torque for Rotary Motion, and Shaft Inertia*

Horsepower

For rotating objects:

$$HP = \frac{TN}{63,000}$$

where $T =$ torque (lb-in)
$N =$ speed (RPM)

or:

$$HP = \frac{TN}{5250}$$

---

* Text, art, and tables in this subsection adapted from material provided by Boston Gear, Quincy, MA. Used with permission.

where $T =$ torque (lb-ft)
       $N =$ speed (RPM)

For objects in linear motion:

$$HP = \frac{FV}{396,000}$$

where $F =$ force (lb)
       $V =$ velocity (IPM)

or:

$$HP = \frac{FV}{33,000}$$

where $F =$ force (lb)
       $V =$ velocity (FPM)

For pumps:

$$HP = \frac{(GPM) \times (\text{head in feet}) \times (\text{specific gravity})}{3950 \times (\text{efficiency of pump})}$$

For fans and blowers:

$$HP = \frac{CFM \times (\text{pressure in pounds/sq ft})}{33,000 \times \text{efficiency}}$$

When calculated horsepower falls between standard motor ratings, select the next higher rating.

Accelerating force for linear motion

The following formula can be used to calculate the approximate accelerating force required for linear motion. However, before sizing the drive, add the torque required to accelerate the motor armature, gears, pulleys, etc. to the linear-motion accelerating force converted to torque.

$$\text{Acceleration force } (F) = \frac{W (\Delta V)}{1933t}$$

where $W =$ weight (lb)
        $\Delta V =$ change in velocity (FPM)
        $t =$ time (seconds) to accelerate weight

Accelerating torque for rotary motion

When, in addition to the selection of a motor with proper torque capacity to start and maintain machine motion, a desired time for acceleration is involved and the required torque value may be affected, an additional formula must be considered. This formula makes it possible to calculate the average torque required over the complete range of speed change to accelerate a known inertia (WK^2).

On high inertia loads, accelerating torque may be the major factor in the drive selection.

The formula to calculate acceleration torque (torque required above load torque) of a rotating member:

$$T = \frac{(WK^2)(\Delta N)}{308t}$$

where $T =$ acceleration torque (lb-ft)
        $WK^2 =$ total system inertia (lb-ft^2) that the motor must accelerate. This value includes motor armature, reducer, and load.
        $\Delta N =$ change in speed required (RPM)
        $t =$ time to accelerate total system load (seconds)

The same formula can also be used to determine the minimum acceleration time of a given drive, or if it can accomplish the desired change in speed within the required time period.

$$t = \frac{(WK^2)(\Delta N)}{308T}$$

Inertia (WK^2)

The factor WK^2 is the weight (lb) of an object multiplied by the square of the radius of gyration (K). The unit measurement of the radius of gyration is expressed in feet.

For solid or hollow cylinders (Fig. 1-7), inertia may be calculated by the equations shown below.

FIGURE 1-7   Solid and hollow cylinders.

The inertia of solid steel shafting per inch of shaft length is given in Table 1-23. To calculate for hollow shafts, take the difference between the inertia values for the O.D. and I.D. as the value per inch. For shafts of materials other than steel, multiply the value for steel by the factors in Table 1-24.

Simplified, the WK^2 for a solid cylinder or disk $= W(r_2/2)$, where r = radius in feet and W is weight in pounds.

For a hollow cylinder:

$$WK^2 = W \times \frac{r_1{}^2 + r_2{}^2}{2}$$

where r_1 is ID and r_2 is OD.

TABLE 1-23   Inertia of Steel Shafting (per Inch of Length)

DIAM. (IN.)	WK² (=FT.²)	DIAM. (IN.)	WK² (=FT.²)
3/4	.00006	10-1/2	2.35
1	.0002	10-3/4	2.58
1-1/4	.0005	11	2.83
1-1/2	.001	11-1/4	3.09
1-3/4	.002	11-1/2	3.38
2	.003	11-3/4	3.68
2-1/4	.005	12	4.00
2-1/2	.008	12-1/4	4.35
2-3/4	.001	12-1/2	4.72
3	.016	12-3/4	5.11
3-1/2	0.029	13	5.58
3-3/4	0.038	13-1/4	5.96
4	0.049	13-1/2	6.42
4-1/4	0.063	13-3/4	6.91
4-1/2	0.079	14	7.42

TABLE 1-23   Inertia of Steel Shafting (per Inch of Length) (*Continued*)

DIAM. (IN.)	WK² (=FT.²)	DIAM. (IN.)	WK² (=FT.²)
5	0.120	14-1/4	7.97
5-1/2	0.177	14-1/2	8.54
6	0.250	14-3/4	9.15
6-1/4	0.296	15	9.75
6-1/2	0.345	16	12.59
6-3/4	0.402	17	16.04
7	0.464	18	20.16
7-1/4	0.535	19	25.03
7-1/2	0.611	20	30.72
7-3/4	0.699	21	37.35
8	0.791	22	44.99
8-1/4	0.895	23	53.74
8-1/2	1.00	24	63.71
8-3/4	1.13	25	75.02
9	1.27	26	87.76
9-1/4	1.41	27	102.06
9-1/2	1.55	28	118.04
9-3/4	1.75	29	135.83
10	1.93	30	155.55
10-1/4	2.13	–	–

The inertia of complex concentric rotating parts is calculated by breaking the part up into simple rotating cylinders, calculating their inertia, and summing their values, as shown in Fig. 1-8.

WK^2 of rotating elements

In practical mechanical systems, all the rotating parts do not operate at the same speed. The WK^2 of all moving parts

TABLE 1-24   Shaft Material Factors

SHAFT MATERIAL	FACTOR
Rubber	.121
Nylon	.181
Aluminum	.348
Bronze	1.135
Cast Iron	.922

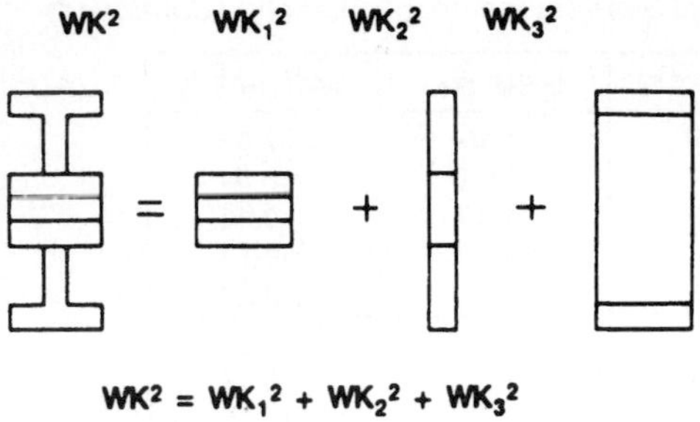

$$WK^2 = WK_1^2 + WK_2^2 + WK_3^2$$

FIGURE 1-8   Calculating the inertia of complex concentric rotating parts.

operating at each speed must be reduced to an equivalent WK^2 at the motor shaft, so that they can all be added together and treated as a unit, as follows:

$$\text{Equivalent } WK^2 = WK^2 \left[\frac{N}{N_M} \right]^2$$

where WK^2 = inertia of the moving part
N = speed of the moving part (RPM)
N_M = speed of the driving motor (RPM)

When using speed reducers, and the machine inertia is reflected back to the motor shaft, the equivalent inertia is equal to the machine inertia divided by the square of the drive reduction ratio:

$$\text{Equivalent } WK^2 = \frac{WK^2}{(\text{DR})^2}$$

where DR = drive reduction = $\dfrac{N_M}{N}$

WK^2 of linear motion

Not all driven systems involve rotating motion. The equivalent WK^2 of linearly moving parts can also be reduced to the motor shaft speed as follows:

$$\text{Equivalent } WK^2 = \frac{W(V)^2}{39.5\,(N_M)^2}$$

where W = weight of load (lbs)
          V = linear velocity of rack and load or conveyor
                and load (FPM)
          N_M = speed of the driving motor (RPM)

Note: This equation can only be used where the linear speed bears a continuous fixed relationship to the motor speed, such as with a conveyor.

Electrical formulas

Ohm's law:

$$\text{Amperes} = \frac{\text{volts}}{\text{ohms}} \qquad \text{Ohms} = \frac{\text{volts}}{\text{amperes}}$$

$$\text{Volts} = \text{amperes} \times \text{ohms}$$

1-2 Horsepower and Torque*

Power is the rate of doing work.

Work is the exerting of a *force* through a *distance*. One foot-pound is a unit of work. It is the work done in exerting a force of one pound through a distance of one foot.

The amount of work done (foot-pounds) is the force (pounds) exerted multiplied by the distance (feet) through which the force acts.

The amount of power used (foot-pounds per minute) is the work (foot-pounds) done divided by the time (minutes) required.

$$\text{Power (foot-pounds per minute)} = \frac{\text{work (ft-lb)}}{\text{time (minutes)}}$$

Power is usually expressed in terms of horsepower.

Horsepower is power (foot-pounds per minute) divided by 33,000 (Fig. 1-9).

---

* Text and unnumbered illustrations in this subsection adapted from material provided by Boston Gear, Quincy, MA. Used with permission.

TABLE 1-25   Horsepower and Torque Capacity of Shafting

Shaft Dia.	Shaft Horsepower Based on Pure Torsion at 10,000 PSI Maximum Shear Stress							Torque Capacity (Lb. Ins.) Based on 10,000 PSI Shear Stress
	Revolutions Per Minute							
	30	50	100	175	690	1150	1750	
3/8	0.049	0.082	0.164	0.287	1.13	1.88	2.87	103
7/16	0.078	0.130	0.261	0.456	1.79	2.99	4.56	164
1/2	0.117	0.194	0.389	0.681	2.68	4.47	6.80	245
9/16	0.166	0.277	0.554	0.969	3.82	6.36	9.69	349
5/8	0.228	0.380	0.760	1.32	5.24	8.73	13.2	479
11/16	0.303	0.506	1.01	1.76	6.97	11.6	17.6	637
3/4	0.394	0.656	1.31	2.29	9.05	15.0	22.9	827
13/16	0.501	0.834	1.66	2.92	11.5	19.1	29.2	1052
7/8	0.625	1.04	2.08	3.64	14.3	23.9	36.4	1314
15/16	0.769	1.28	2.56	4.48	17.6	29.4	44.3	1616
1	0.933	1.55	3.11	5.44	21.4	35.7	54.4	1961
1-1/16	1.12	1.86	3.73	6.53	25.7	42.9	65.3	2352
1-1/8	1.32	2.21	4.43	7.75	30.5	50.9	77.5	2792
1-3/16	1.56	2.60	5.21	9.11	35.9	59.9	91.1	3283
1-1/4	1.82	3.03	6.07	10.6	41.9	69.8	106.	3830
1-5/16	2.11	3.51	7.03	12.3	48.5	80.	123.	4433
1-3/8	2.42	4.04	8.08	11.1	55.8	93.	141.	5097
1-7/16	2.77	4.62	9.24	16.1	63.7	106.	161.	5824
1-1/2	3.15	5.25	10.5	18.3	72.4	120.	183.	6618
1-9/16	3.56	5.93	11.8	20.7	81.8	136.	207.	7480
1-5/8	4.00	6.67	13.3	23.3	92.1	153.	233.	8414
1-11/16	4.48	7.47	14.9	26.1	103.1	171.	261.	9422
1-3/4	5.00	8.33	16.6	29.1	115.0	191.	291.	10509
1-13/16	5.55	9.26	18.5	32.4	127.8	213.	324.	11675
1-7/8	6.15	10.2	20.5	35.8	141.5	235.	358.	12925
1-15/16	6.78	11.3	22.6	39.6	156.1	260.	396.	14261
2	7.46	12.4	24.8	43.5	171.7	286.	435.	15686
2-1/16	8.18	13.6	27.2	47.7	188.3	313.	477.	17203
2-1/8	8.95	14.9	29.8	52.2	206.0	343.	522.	18815
2-3/16	9.77	16.2	32.5	56.9	224.7	374.	569.	20525
2-1/4	10.6	17.7	35.4	62.0	244.5	407.	620.	22335
2-5/16	11.5	19.2	38.4	67.3	265.4	442.	673.	24248
2-3/8	12.5	20.8	41.6	72.9	287.6	479.	729.	26268
2-7/16	13.5	22.5	45.0	78.8	310.9	518.	788.	29396
2-1/2	14.5	24.3	48.6	85.0	335.1	559.	850.	30637
2-9/16	15.7	26.1	52.3	91.6	361.2	602.	916.	32993
2-5/8	16.8	28.1	56.2	98.4	388.3	647.	984.	35466

The above table is computed based on a torsional stress of 10,000 psi. For applications involving bending moments (gears, sprockets, etc.) the horsepower capacity must be reduced accordingly.

The stress level of 10,000 PSI is representative of medium carbon steel shafting. For other materials, a correction must be made accordingly.

SOURCE: Boston Gear, Quincy, MA.

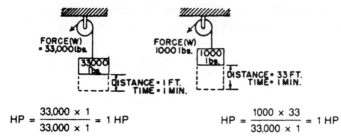

$$HP = \frac{33,000 \times 1}{33,000 \times 1} = 1\ HP \qquad\qquad HP = \frac{1000 \times 33}{33,000 \times 1} = 1\ HP$$

FIGURE 1-9 Illustration of horsepower.

Horsepower (HP)

$$= \frac{\text{power (ft-lb per minute)}}{33,000}$$

$$= \frac{\text{work (ft-pounds)}}{33,000 \times \text{time (min.)}}$$

$$= \frac{\text{force (lb)} \times \text{distance (feet)}}{33,000 \times \text{time (min.)}}$$

Horsepower (HP)

$$= \frac{\text{force (lb)} \times \text{distance (feet)}}{33,000 \times \text{time (min.)}}$$

Torque (T) is the product of a force (W) in pounds, times a radius (R) in inches from the center of shaft (lever arm) and is expressed in inch-pounds (Fig. 1-10).

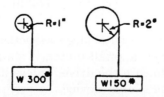

T=WR=300 × 1=300 In. Lbs.   T=WR=150 × 2=300 In. Lbs.

FIGURE 1-10 Illustration of torque.

If the shaft is revolved, the force (W) is moved through a distance, and work is done:

$$\text{Work (ft-pounds)} = W \times \frac{2\pi R}{12} \times \text{No. of revolutions of shaft}$$

When this work is done in a specified time, power is used:

$$\text{Power (ft-pounds per min.)} = W \times \frac{2\pi R}{12} \times \text{RPM}$$

Since horsepower = 33,000 foot-pounds per minute

$$\text{Horsepower (HP)} = W \times \frac{2\pi R}{12} \times \frac{\text{RPM}}{33,000} = \frac{W \times R \times \text{RPM}}{63,025}$$

but torque (inch-pounds) = force (W) × radius (R), therefore

$$\text{Horsepower (HP)} = \frac{\text{torque } (T) \times \text{RPM}}{63,025}$$

Where total reductions are small, 50 to 1 or less, HP figures are commonly used. Higher reductions require that torque figures be used to select drive components, because with large reductions, a small motor can produce extremely high torque at the final low speed. For example, $\frac{1}{12}$ HP reduced to 1 RPM using the formula below and neglecting friction:

$$\text{HP} = \frac{\text{torque} \times \text{RPM}}{63,025} \quad \text{or} \quad \text{Torque} = \frac{63,025 \times \text{HP}}{\text{RPM}}$$

$$\text{Torque} = \frac{63,025 \times 1/12}{1} = 5252 \text{ in-lb}$$

Therefore, motors for use with large reductions should be carefully selected. Even a small motor, if stalled, can produce enough torque to ruin the drive, unless it is protected by a shear pin or some similar device.

Neglecting frictional losses, Fig. 1-11 illustrates the manner in which torque increases as speed decreases.

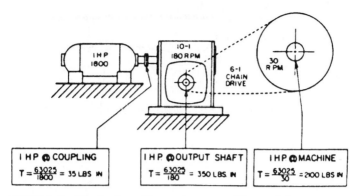

FIGURE 1-11 Illustration of torque increase with speed decrease.

1-3 Center-Distance Determination in Custom-Designed Belt Drives*

The following nomenclature and basic equations are involved (Fig. 1-12 illustrates the notation):

C = center distance, inches
L = belt length, inches = PNB
P = circumferential pitch of belt, inches
NB = number of teeth on belt = L/P
$N1$ = number of teeth (grooves) on larger pulley
$N2$ = number of teeth (grooves) on smaller pulley
ϕ = one half angle of wrap on smaller pulley, radians
$\theta = \dfrac{\pi}{2} - \phi$ = angle between straight portion of belt and line of centers, radians
R_1 = pitch radius of larger pulley, inches = $N1P/2\pi$
R_2 = pitch radius of smaller pulley, inches = $N2P/2\pi$
π = 3.14159 (ratio of circumference to diameter of circle)

---

* Text and illustration used in this subsection adapted from material provided by Stock Drive Products, New Hyde Park, NY. Used with permission.

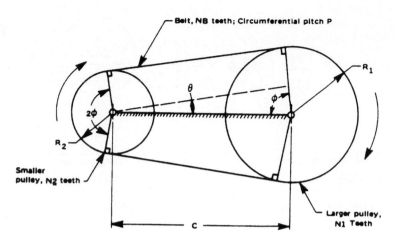

FIGURE 1-12 Belt geometry.

The basic equation for the determination of center distance is:

$$2C \sin \phi = L - \pi(R_1 + R_2) - (\pi - 2\phi)(R_1 - R_2)$$

where $C \cos \phi = R_1 - R_2$

These equations can be combined in different ways to yield various equations for the determination of center distance.

TABLE 1-26   Pulley Pitch and Outside Diameters

No. of Grooves	.080 PITCH FOR FIBREX BELTS		.0816 PITCH FOR FIBREX BELTS		.0816 PITCH FOR POLYURETHANE BELTS	
	Pitch Diameter	Outside Diameter	Pitch Diameter	Outside Diameter	Pitch Diameter	Outside Diameter
10	.255	.235	.260	.240	.260	.246
11	.280	.260	.286	.266	.286	.272
12	.306	.286	.312	.292	.312	.298
13	.331	.311	.338	.318	.338	.324
14	.357	.337	.364	.344	.364	.350
15	.382	.362	.390	.370	.390	.376
16	.407	.387	.415	.395	.416	.402
17	.433	.413	.442	.422	.442	.428
18	.458	.438	.467	.447	.468	.454
19	.484	.464	.494	.474	.494	.480
20	.509	.489	.519	.499	.520	.506
21	.535	.515	.546	.526	.546	.532
22	.560	.540	.571	.551	.572	.558
23	.580	.566	.598	.578	.598	.584
24	.611	.591	.623	.603	.624	.610
25	.637	.617	.650	.630	.650	.636
26	.662	.642	.675	.655	.676	.662
27	.687	.667	.701	.681	.702	.688
28	.713	.693	.727	.707	.728	.714
29	.738	.718	.753	.733	.754	.740
30	.764	.744	.779	.759	.780	.766
31	.789	.769	.805	.785	.806	.792
32	.815	.795	.831	.811	.832	.818
33	.840	.820	.857	.837	.858	.844
34	.866	.846	.883	.863	.884	.870
35	.891	.871	.909	.889	.910	.896
36	.917	.897	.935	.915	.936	.922
37	.942	.922	.961	.941	.962	.948
38	.968	.948	.987	.967	.988	.974
39	.993	.973	1.013	.993	1.014	1.000
40	1.019	.999	1.039	1.019	1.040	1.026
41	1.044	1.024	1.065	1.045	1.066	1.052
42	1.070	1.050	1.091	1.071	1.092	1.078
43	1.095	1.075	1.117	1.097	1.118	1.104
44	1.120	1.100	1.142	1.122	1.144	1.130
45	1.146	1.126	1.169	1.149	1.170	1.156
46	1.171	1.151	1.194	1.174	1.196	1.182
47	1.197	1.177	1.221	1.201	1.222	1.208
48	1.222	1.202	1.246	1.226	1.248	1.234
49	1.248	1.228	1.273	1.253	1.274	1.260
50	1.273	1.253	1.298	1.278	1.300	1.286
51	1.299	1.279	1.325	1.305	1.326	1.312
52	1.324	1.304	1.350	1.330	1.352	1.338
53	1.350	1.330	1.377	1.357	1.378	1.364
54	1.375	1.355	1.403	1.383	1.404	1.390
55	1.401	1.381	1.429	1.409	1.430	1.416
56	1.426	1.406	1.455	1.435	1.456	1.442

SOURCE: Stock Drive Products, New Hyde Park, NY.

TABLE 1-26   Pulley Pitch and Outside Diameters (*Continued*)

No. of Grooves	.080 PITCH FOR FIBREX BELTS		.0816 PITCH FOR FIBREX BELTS		FOR POLYURETHANE BELTS	
	Pitch Diameter	Outside Diameter	Pitch Diameter	Outside Diameter	Pitch Diameter	Outside Diameter
57	1.451	1.431	1.480	1.460	1.482	1.468
58	1.477	1.457	1.506	1.486	1.508	1.494
59	1.502	1.482	1.532	1.512	1.534	1.520
60	1.528	1.508	1.559	1.539	1.560	1.546
61	1.553	1.533	1.584	1.564	1.586	1.572
62	1.579	1.559	1.611	1.591	1.612	1.598
63	1.604	1.584	1.636	1.616	1.638	1.624
64	1.630	1.610	1.663	1.643	1.664	1.650
65	1.655	1.635	1.688	1.668	1.690	1.676
66	1.681	1.661	1.715	1.695	1.716	1.702
67	1.706	1.686	1.740	1.720	1.742	1.728
68	1.731	1.711	1.766	1.746	1.768	1.754
69	1.757	1.737	1.792	1.772	1.794	1.780
70	1.783	1.763	1.819	1.799	1.820	1.806
71	1.808	1.788	1.845	1.825	1.846	1.832
72	1.833	1.813	1.870	1.850	1.872	1.858
73	1.859	1.839	1.896	1.876	1.898	1.884
74	1.884	1.864	1.922	1.902	1.924	1.910
75	1.910	1.890	1.948	1.928	1.950	1.936
76	1.935	1.915	1.974	1.954	1.976	1.962
77	1.961	1.941	2.000	1.980	2.002	1.988
78	1.986	1.966	2.026	2.006	2.028	2.014
79	2.012	1.992	2.052	2.032	2.054	2.040
80	2.037	2.017	2.078	2.058	2.080	2.066
81	2.063	2.043	2.104	2.084	2.106	2.092
82	2.088	2.068	2.130	2.110	2.132	2.118
83	2.114	2.094	2.156	2.136	2.158	2.144
84	2.139	2.119	2.182	2.162	2.184	2.170
85	2.165	2.145	2.208	2.188	2.210	2.196
86	2.190	2.170	2.234	2.214	2.236	2.222
87	2.215	2.195	2.259	2.239	2.262	2.248
88	2.241	2.221	2.286	2.266	2.288	2.274
89	2.266	2.246	2.311	2.291	2.314	2.300
90	2.292	2.272	2.338	2.318	2.340	2.326
91	2.317	2.297	2.363	2.343	2.366	2.352
92	2.343	2.323	2.390	2.370	2.392	2.378
93	2.368	2.348	2.415	2.395	2.418	2.404
94	2.394	2.374	2.442	2.422	2.444	2.430
95	2.419	2.399	2.467	2.447	2.470	2.456
96	2.445	2.425	2.494	2.474	2.496	2.482
97	2.470	2.450	2.519	2.499	2.522	2.508
98	2.496	2.476	2.546	2.526	2.548	2.534
99	2.521	2.501	2.571	2.551	2.574	2.560
100	2.546	2.526	2.597	2.577	2.600	2.586
101	2.572	2.552	2.623	2.603	2.626	2.612
102	2.597	2.577	2.649	2.629	2.652	2.638
103	2.623	2.603	2.675	2.655	2.678	2.664

TABLE 1-26 Pulley Pitch and Outside Diameters (*Continued*)

No. of Grooves	.080 PITCH FOR FIBREX BELTS Pitch Diameter	.080 PITCH FOR FIBREX BELTS Outside Diameter	.0816 PITCH FOR FIBREX BELTS Pitch Diameter	.0816 PITCH FOR FIBREX BELTS Outside Diameter	.0816 PITCH FOR POLYURETHANE BELTS Pitch Diameter	.0816 PITCH FOR POLYURETHANE BELTS Outside Diameter
104	2.648	2.628	2.701	2.681	2.704	2.690
105	2.674	2.654	2.727	2.707	2.730	2.716
106	2.699	2.679	2.753	2.733	2.756	2.742
107	2.725	2.705	2.780	2.760	2.782	2.768
108	2.750	2.730	2.805	2.785	2.808	2.794
109	2.776	2.756	2.832	2.812	2.834	2.820
110	2.801	2.781	2.857	2.837	2.860	2.846
111	2.827	2.807	2.884	2.864	2.886	2.872
112	2.852	2.832	2.909	2.889	2.912	2.898
113	2.878	2.858	2.936	2.916	2.938	2.924
114	2.903	2.887	2.961	2.641	2.964	2.950
115	2.928	2.908	2.987	2.967	2.990	2.976
116	2.954	2.934	3.013	2.993	3.016	3.002
117	2.979	2.959	3.039	3.019	3.042	3.028
118	3.005	2.985	3.065	3.045	3.068	3.054
119	3.030	3.010	3.091	3.071	3.094	3.080
120	3.056	3.036	3.117	3.097	3.120	3.106
121	3.081	3.061	3.143	3.123	3.146	3.132
122	3.107	3.087	3.169	3.149	3.172	3.158
123	3.132	3.112	3.195	3.175	3.198	3.184
124	3.158	3.138	3.221	3.201	3.224	3.210
125	3.183	3.163	3.247	3.227	3.250	3.236
126	3.209	3.189	3.273	3.253	3.276	3.262
127	3.234	3.214	3.299	3.279	3.302	3.288
128	3.259	3.239	3.324	3.304	3.328	3.314
129	3.285	3.265	3.351	3.331	3.354	3.340
130	3.310	3.290	3.376	3.346	3.380	3.366
131	3.336	3.316	3.403	3.383	3.406	3.392
132	3.361	3.341	3.428	3.408	3.432	3.418
133	3.387	3.367	3.455	3.435	3.458	3.444
134	3.412	3.392	3.480	3.460	3.484	3.470
135	3.438	3.418	3.507	3.487	3.510	3.496
136	3.463	3.443	3.532	3.512	3.536	3.522
137	3.489	3.469	3.559	3.539	3.562	3.548
138	3.514	3.494	3.584	3.564	3.588	3.574
139	3.540	3.520	3.611	3.591	3.614	3.600
140	3.565	3.545	3.636	3.616	3.640	3.626
141	3.591	3.571	3.663	3.643	3.666	3.652
142	3.616	3.596	3.688	3.668	3.692	3.678
143	3.641	3.621	3.714	3.694	3.718	3.704
144	3.667	3.647	3.740	3.720	3.744	3.730
145	3.692	3.672	3.766	3.746	3.770	3.756
146	3.718	3.698	3.792	3.772	3.796	3.782
147	3.743	3.723	3.818	3.798	3.822	3.808
148	3.769	3.749	3.844	3.824	3.848	3.834
149	3.794	3.774	3.870	3.850	3.874	3.860
150	3.820	3.800	3.896	3.876	3.900	3.886

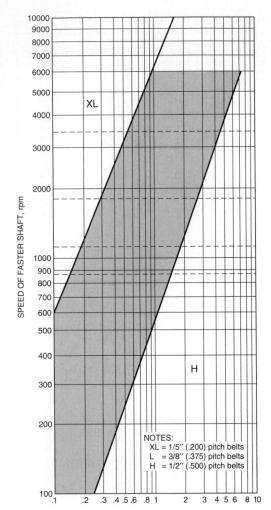

FIGURE 1-13 Belt pitch selection. (*Courtesy: Stock Drive Products, New Hyde Park, NY.*)

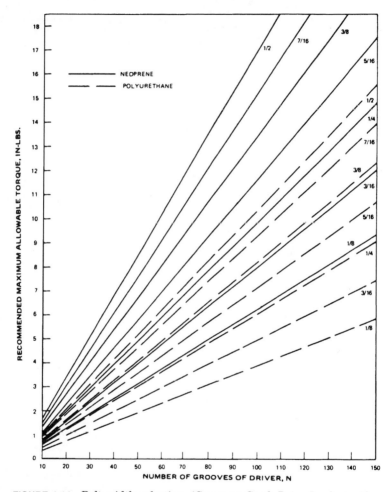

FIGURE 1-14 Belt width selection. (*Courtesy: Stock Drive Products, New Hyde Park, NY.*)

NOTES

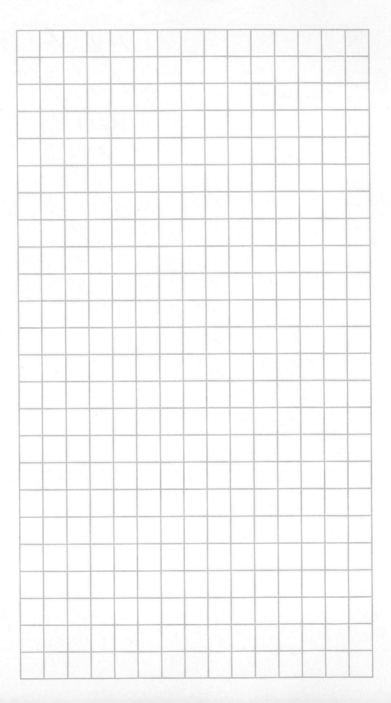

Heat Loss
Estimating

TABLE 2-1   Recommended Design for Indoor Winter Temperatures, by Type of Building

Type of building	Temp., °F
Schools:	
Classrooms	72
Assembly rooms, dining rooms	72
Playrooms, gymnasiums	65
Swimming pool	75
Locker rooms	70
Hospitals:	
Private rooms	72
Operating rooms	75
Wards	70
Toilets	70
Bathrooms	75
Kitchens and laundries	66
Theaters	72
Hotels:	
Bedrooms	70
Ballrooms	68
Residences	72
Stores	68
Offices	72
Factories	65

SOURCE: R. O. Parmley, *HVAC Field Manual,* McGraw-Hill, New York, 1988. Used with permission of publisher.

TABLE 2-2   Recommended Design for Outdoor Winter Temperatures, for Various Locations

State	City	Temp., °F
Ala.	Birmingham	10
Ariz.	Flagstaff	−10
Ariz.	Phoenix	25
Ark.	Little Rock	5
Calif.	Los Angeles	35
Calif.	San Francisco	35
Colo.	Denver	−10
Conn.	Hartford	0
D.C.	Washington	0
Fla.	Jacksonville	25
Fla.	Miami	35
Ga.	Atlanta	10
Idaho	Boise	−10
Ill.	Chicago	−10
Ind.	Indianapolis	−10
Iowa	Des Moines	−15
Kans.	Topeka	−10
Ky.	Louisville	0
La.	New Orleans	20
Maine	Portland	−5
Md.	Baltimore	0
Mass.	Boston	0
Mich.	Detroit	−10
Minn.	Minneapolis	−20
Miss.	Vicksburg	10
Mo.	St. Louis	0
Mont.	Helena	−20
Nebr.	Lincoln	−10
Nev.	Reno	−5
N.H.	Concord	−15
N.J.	Trenton	0
N.Mex.	Albuquerque	0
N.Y.	New York	0
N.C.	Greensboro	10
N.Dak.	Bismarck	−30
Ohio	Cincinnati	0
Okla.	Tulsa	0
Ore.	Portland	10
Pa.	Philadelphia	0
R.I.	Providence	0
S.C.	Charleston	15
S.Dak.	Rapid City	−20
Tenn.	Nashville	0
Tex.	Dallas	0
Tex.	Houston	20
Utah	Salt Lake City	−10
Vt.	Burlington	−10
Va.	Richmond	15

source: R. O. Parmley, *HVAC Field Manual,* McGraw-Hill, New York, 1988. Used with permission of publisher.

TABLE 2-3 Transmission Coefficients (U) for Wood and Steel Doors, Btu/h·ft²·°F

Nominal Door Thickness, in.	Description	No Storm Door	Wood Storm Door[c]	Metal Storm Door[d]
Wood Doors[a, b]				
1-3/8	Panel door with 7/16-in. panels[e]	0.57	0.33	0.37
1-3/8	Hollow core flush door	0.47	0.30	0.32
1-3/8	Solid core flush door	0.39	0.26	0.28
1-3/8	Panel door with 7/16-in. panels[e]	0.57	0.33	0.36
1-3/4	Hollow core flush door	0.46	0.29	0.32
1-3/4	Panel door with 1-1/8-in. panels[e]	0.39	0.26	0.28
1-3/4	Solid core flush door	0.33	0.28	0.25
2-1/4	Solid core flush door	0.27	0.20	0.21
Steel Doors[b]				
1-3/4	Fiberglass or mineral wool core with steel stiffeners, no thermal break[f]	0.60	—	—
1-3/4	Paper honeycomb core without thermal break[f]	0.56	—	—
1-3/4	Solid urethane foam core without thermal break[a]	0.40	—	—
1-3/4	Solid fire rated mineral fiberboard core without thermal break[f]	0.38	—	—
1-3/4	Polystyrene core without thermal break (18 gage commercial steel)[f]	0.35	—	—
1-3/4	Polystyrene core without thermal break (18 gage commercial steel)[f]	0.35	—	—
1-3/4	Polyurethane core without thermal break (18 gage commercial steel)[f]	0.29	—	—
1-3/4	Polyurethane core without thermal break (24 gage commercial steel)[f]	0.29	—	—
1-3/4	Polyurethane core with thermal break and wood perimeter (24 gage residential steel)[f]	0.20	—	—
1-3/4	Solid urethane foam core with thermal break[a]	0.19	0.16	0.17

Note: All U factors for exterior doors in this table are for doors with no glazing, except for the storm doors which are in addition to the main exterior door. Any glazing area in exterior doors should be included with the appropriate glass type and analyzed. Interpolation and moderate extrapolation are permitted for door thicknesses other than those specified.

[a] Values are based on a nominal 32 by 80 in. door size with no glazing.

[b] Outside air conditions: 15 mph wind speed, 0°F air temperature; inside air conditions: natural convection, 70°F air temperature.

[c] Values for wood storm door are for approximately 50% glass area.

[d] Values for metal storm door are for any percent glass area.

[e] 55% panel area.

[f] ASTM C 236 hotbox data on a nominal 3 by 7 ft door size with no glazing.

SOURCE: *ASHRAE Handbook—1989 Fundamentals.* Reprinted by permission.

TABLE 2-4   Average Heat Loss for Various Materials

Material	Density, lb per cu ft	Average Btu passing per hr through a plate of material 1 sq ft in area 1 in. thick per °F difference in temperature of the two faces
Seaweed including air.............	...	0.1788
Kapok (loose packed)..............	0.874	0.238
Wool felt.......................	20.6	0.363
Cork (ground)....................	4.867	0.3225
Corkboard.......................	4.74	0.305
Celotex 7/16 in. thick.............	13.5	0.33
Sawdust........................	11.86	0.406
Pine...........................	...	1.045
Oak (7/64 in. thick)..............	37.45	2.74
Mineral wool....................	12.50	0.2614
Asbestos paper..................	31.2	1.249
85%-Magnesia...................	13.48	0.5104
Sil-O-cel.......................	28.1	0.5805
Facing cement...................	...	1.010
Masonry (average)...............	128.000	4.10
Fire-felt.......................	26.000	0.596

SOURCE: S. Elonka and Q. W. Minich, *Standard Refrigeration and Air Conditioning Questions and Answers,* McGraw-Hill, New York, 1973. Used with permission.

TABLE 2-5   Below-Grade Heat Losses

Ground water temp., °F	Basement floor loss,* Btu/h/ft²	Below-grade wall loss, Btu/h/ft²
40	3.0	6.0
50	2.0	4.0
60	1.0	2.0

* Based on basement temperature of 70°F.
SOURCE: R. O. Parmley, *HVAC Field Manual,* McGraw-Hill, New York, 1988. Used with permission of publisher.

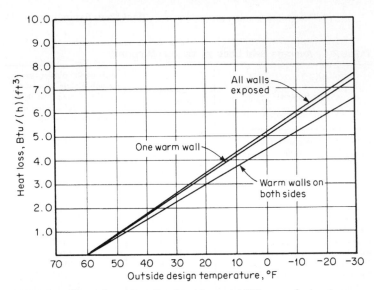

FIGURE 2-1 Chart for estimating heat loss at 60°F room design temperature for one-story building with flat roof (for skylight in roof add 7½ percent). (*Source: A. M. Khashab*, Heating, Ventilating and Air Conditioning Systems Estimating Manual, *2d ed., McGraw-Hill, New York, 1977. Used with permission.*)

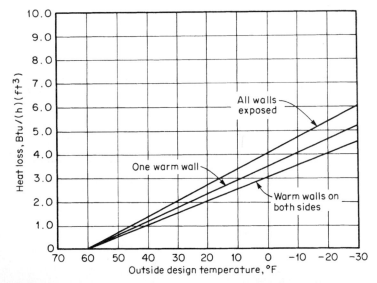

FIGURE 2-2 Chart for estimating heat loss at 60°F room design temperature for one-story building with heated space above. (*Source: A. M. Khashab,* Heating, Ventilating and Air Conditioning Systems Estimating Manual, *2d ed., McGraw-Hill, New York, 1977. Used with permission.*)

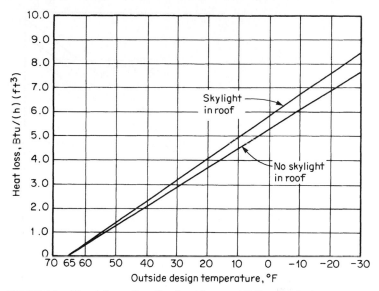

FIGURE 2-3 Chart for estimating heat loss at 65°F room design temperature for one-story building. (*Source: A. M. Khashab,* Heating, Ventilating and Air Conditioning Systems Estimating Manual, *2d ed., McGraw-Hill, New York, 1977. Used with permission.*)

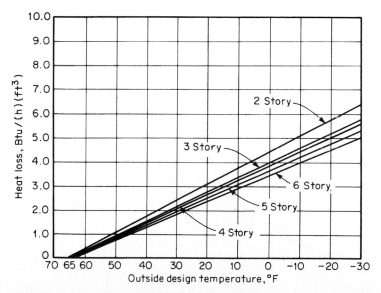

FIGURE 2-4 Chart for estimating heat loss at 65°F room design temperature for multistory building. (*Source: A. M. Khashab,* Heating, Ventilating and Air Conditioning Systems Estimating Manual, *2d ed., McGraw-Hill, New York, 1977. Used with permission.*)

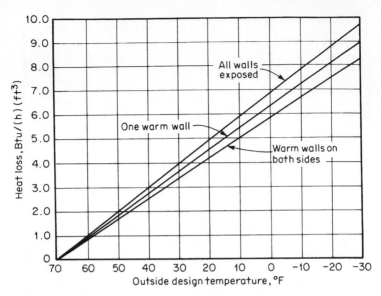

FIGURE 2-5 Chart for estimating heat loss at 70°F room design temperature for one-story building with flat roof. (*Source: A. M. Khashab,* Heating, Ventilating and Air Conditioning Systems Estimating Manual, *2d ed., McGraw-Hill, New York, 1977. Used with permission.*)

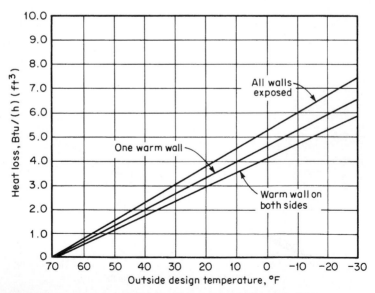

FIGURE 2-6 Chart for estimating heat loss at 70°F room design temperature for one-story heated space above. (*Source: A. M. Khashab,* Heating, Ventilating and Air Conditioning Systems Estimating Manual, *2d ed., McGraw-Hill, New York, 1977. Used with permission.*)

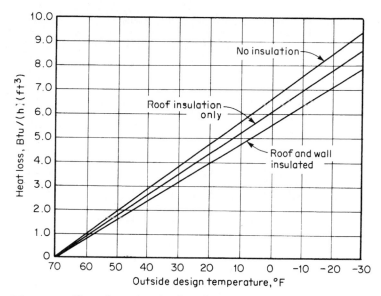

FIGURE 2-7   Chart for estimating heat loss at 70°F room design temperature for entire building (multistory) (add 10% for bad north and west exposures). (*Source: A. M. Khashab,* Heating, Ventilating and Air Conditioning
Systems Estimating Manual, *2d ed., McGraw-Hill, New York, 1977. Used
with permission.*)

DESIGN DATA: −25° TO 67° F.   **HEAT LOSS CALCULATIONS**   ⊢ UNHEATED ⊣

ITEM	"U"	1 AREA	BTU	2 AREA	BTU	3 AREA	BTU	4 AREA	BTU	5 AREA	BTU	6 AREA	BTU	7 AREA	BTU
A GROSS WALL	– – –	1131	✕	170	✕	270	✕	90	✕	233	✕	231	✕	✕	✕
B WINDOWS	0.61	96	59	16	10	8	5	0	✕	0	✕	0	✕		
C DOORS	0.07	40	3	0	✕	0	✕	0	✕	0	✕	60	4		
D WALLS	0.026	995	26	154	4	262	7	90	2	233	6	171	5	✕	✕
E COLD CEILING	0.025	1600	40	320	8	200	5	139	4	139	4	482	12		
F COLD FLOOR	0.04	1600	64	320	13	200	8	139	6	139	6	482	20		
G PERIMETER INFIL PER LIN. FT.	0.45	116	52	20	9	27	12	11	5	24	11	22	10		
H DRS. & WINDOWS INFIL / L.F.	0.45	169	76	23	10	11	5	0	✕	0	✕	42	19		
BTU LOSS PER 1° TEMP (F)		320		54		42		17		27		70		✕	
TOTAL ROOM LOSS 92°F TD		29,440		4,968		3,864		1,564		2,484		6,440			
CFM TOTAL EXHAUST PER ROOM		✕		250		✕		120		120		✕			
BTU / H / ROOM = 60TC ÷ 55		✕		N/A SUMMER		✕		12,044		12,044		✕			
NOTES															
TOTALS		29,440		4,968		3,864		13,608		14,528		6,440			

THERMAL PERFORMANCE:

MAX. = 12 BTU/H/FT^2 OF ABOVE GRADE EXTERIOR BLDG. ENVELOPE

$$TP = \frac{92(\text{BTU ROOMS})}{A + E \ (\text{AREA})}$$

$$= \frac{92 \times 413}{2125 + 2880}$$

$$= 7.59 \ \text{BTU/H/FT}^2$$

FIGURE 2-8   Typical heat loss calculation. (*Source: R. O. Parmley,* HVAC Field Manual, *McGraw-Hill, New York, 1988. Used with permission of publisher.*)

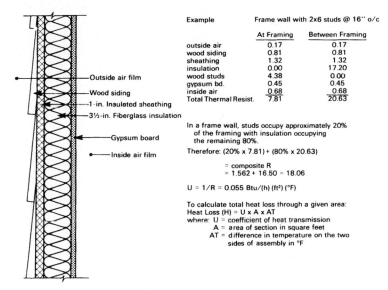

Outside air film
Wood siding
1-in. Insulated sheathing
3½-in. Fiberglass insulation
Gypsum board
Inside air film

Example   Frame wall with 2x6 studs @ 16″ o/c

	At Framing	Between Framing
outside air	0.17	0.17
wood siding	0.81	0.81
sheathing	1.32	1.32
insulation	0.00	17.20
wood studs	4.38	0.00
gypsum bd.	0.45	0.45
inside air	0.68	0.68
Total Thermal Resist.	7.81	20.63

In a frame wall, studs occupy approximately 20%
of the framing with insulation occupying
the remaining 80%.

Therefore: (20% x 7.81) + (80% x 20.63)

$$= \text{composite R}$$
$$= 1.562 + 16.50 = 18.06$$

$U = 1/R = 0.055$ Btu/(h) (ft²) (°F)

To calculate total heat loss through a given area:
Heat Loss (H) = U x A x AT
where: U = coefficient of heat transmission
       A = area of section in square feet
       AT = difference in temperature on the two
            sides of assembly in °F

FIGURE 2-9 Basic heat loss calculation for a frame wall. (*Source: R. O. Parmley,* HVAC Field Manual, *McGraw-Hill, New York, 1988. Used with permission of publisher.*)

NOTES

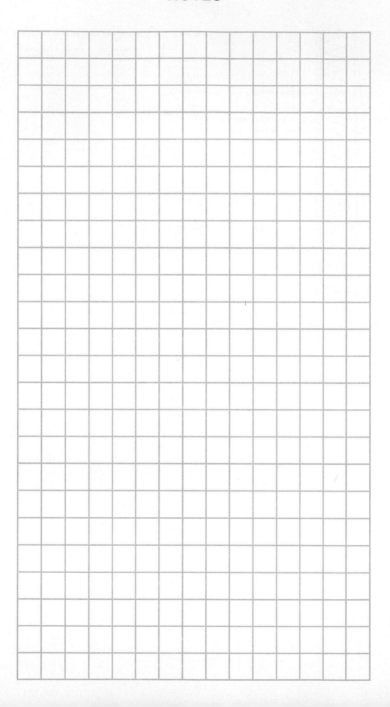

Building Material Properties

TABLE 3-1   Weights of Building Materials

Materials	Weight Lb. per Sq. Ft.	Materials	Weight Lb. per Sq. Ft.
CEILINGS		**PARTITIONS**	
Channel suspended		Clay Tile	
system	1	3 in.	17
Lathing and plastering	See Partitions	4 in.	18
Acoustical fiber tile	1	6 in.	28
		8 in.	34
FLOORS		10 in.	40
Steel Deck	See	Gypsum Block	
	Manufacturer	2 in.	9½
Concrete-Reinforced 1 in.		3 in.	10½
Stone	12½	4 in.	12½
Slag	11½	5 in.	14
Lightweight	6 to 10	6 in.	18½
		Wood Studs 2 × 4	
Concrete-Plain 1 in.		12–16 in. o.c.	2
Stone	12	Steel partitions	4
Slag	11	Plaster 1 inch	
Lightweight	3 to 9	Cement	10
		Gypsum	5
Fills 1 inch		Lathing	
Gypsum	6	Metal	½
Sand	8	Gypsum Board ½ in.	2
Cinders	4		
		WALLS	
Finishes		Brick	
Terrazzo 1 in.	13	4 in.	40
Ceramic or Quarry Tile ¾		8 in.	80
in.	10	12 in.	120
Linoleum ¼ in.	1	Hollow Concrete Block	
Mastic ¾ in.	9	(Heavy Aggregate)	
Hardwood ⅞ in.	4	4 in.	30
Softwood ¾ in.	2½	6 in.	43
		8 in.	55
ROOFS		12½ in.	80
Copper or tin	1	Hollow Concrete Block	
Corrugated steel	See p. 6-5, *Steel*	(Light Aggregate)	
	Construction	4 in.	21
	Manual, 8th ed.	6 in.	30
3-ply ready roofing	1	8 in.	38
3-ply felt and gravel	5½	12 in.	55
5-ply felt and gravel	6	Clay tile	
		(Load Bearing)	
Shingles		4 in.	25
Wood	2	6 in.	30
Asphalt	3	8 in.	33
Clay tile	9 to 14	12 in.	45
Slate ¼	10	Stone 4 in.	55
		Glass Block 4 in.	18
Sheathing		Windows, Glass, Frame	8
Wood ¾ in.	3	& Sash	
Gypsum 1 in.	4	Curtain Walls	See
			Manufacturer
Insulation 1 in.		Structural Glass 1 in.	15
Loose	½	Corrugated Cement As-	
Poured in place	2	bestos ¼ in.	3
Rigid	1½		

SOURCE: American Institute of Steel Construction.

TABLE 3-2 Weights and Specific Gravities

Substance	Weight Lb. per Cu. Ft.	Specific Gravity	Substance	Weight Lb. per Cu. Ft.	Specific Gravity
ASHLAR MASONRY			**MINERALS**		
Granite, syenite, gneiss	165	2.3-3.0	Asbestos	153	2.1-2.8
Limestone, marble	160	2.3-2.8	Barytes	281	4.50
Sandstone, bluestone	140	2.1-2.4	Basalt	184	2.7-3.2
			Bauxite	159	2.55
MORTAR RUBBLE			Borax	109	1.7-1.8
MASONRY			Chalk	137	1.8-2.6
Granite, syenite, gneiss	155	2.2-2.8	Clay, marl	137	1.8-2.6
Limestone, marble	150	2.2-2.6	Dolomite	181	2.9
Sandstone, bluestone	130	2.0-2.2	Feldspar, orthoclase	159	2.5-2.6
			Gneiss, serpentine	159	2.4-2.7
DRY RUBBLE MASONRY			Granite, syenite	175	2.5-3.1
Granite, syenite, gneiss	130	1.9-2.3	Greenstone, trap	187	2.8-3.2
Limestone, marble	125	1.9-2.1	Gypsum, alabaster	159	2.3-2.8
Sandstone, bluestone	110	1.8-1.9	Hornblende	187	3.0
			Limestone, marble	165	2.5-2.8
BRICK MASONRY			Magnesite	187	3.0
Pressed brick	140	2.2-2.3	Phosphate rock, apatite	200	3.2
Common brick	120	1.8-2.0	Porphyry	172	2.6-2.9
Soft brick	100	1.5-1.7	Pumice, natural	40	0.37-0.90
			Quartz, flint	165	2.5-2.8
CONCRETE MASONRY			Sandstone, bluestone	147	2.2-2.5
Cement, stone, sand	144	2.2-2.4	Shale, slate	175	2.7-2.9
Cement, slag, etc.	130	1.9-2.3	Soapstone, talc	169	2.6-2.8
Cement, cinder, etc.	100	1.5-1.7			
VARIOUS BUILDING					
MATERIALS			**STONE, QUARRIED, PILED**		
Ashes, cinders	40-45	....	Basalt, granite, gneiss	96	....
Cement, portland, loose	90	....	Limestone, marble, quartz	95	....
Cement, portland, set	183	2.7-3.2	Sandstone	82	....
Lime, gypsum, loose	53-64	....	Shale	92	....
Mortar, set	103	1.4-1.9	Greenstone, hornblende	107	....
Slags, bank slag	67-72	....			
Slags, bank screenings	98-117	....			
Slags, machine slag	96	....			
Slags, slag sand	49-55	....	**BITUMINOUS SUBSTANCES**		
			Asphaltum	81	1.1-1.5
EARTH, ETC., EXCAVATED			Coal, anthracite	97	1.4-1.7
Clay, dry	63	....	Coal, bituminous	84	1.2-1.5
Clay, damp, plastic	110	....	Coal, lignite	78	1.1-1.4
Clay and gravel, dry	100	....	Coal, peat, turf, dry	47	0.65-0.85
Earth, dry, loose	76	....	Coal, charcoal, pine	23	0.28-0.44
Earth, dry, packed	95	....	Coal, charcoal, oak	33	0.47-0.57
Earth, moist, loose	78	....	Coal, coke	75	1.0-1.4
Earth, moist, packed	96	....	Graphite	131	1.9-2.3
Earth, mud, flowing	108	....	Paraffine	56	0.87-0.91
Earth, mud, packed	115	....	Petroleum	54	0.87
Riprap, limestone	80-85	....	Petroleum, refined	50	0.79-0.82
Riprap, sandstone	90	....	Petroleum, benzine	46	0.73-0.75
Riprap, shale	105	....	Petroleum, gasoline	42	0.66-0.69
Sand, gravel, dry, loose	90-105	....	Pitch	69	1.07-1.15
Sand, gravel, dry, packed	100-120	....	Tar, bituminous	75	1.20
Sand, gravel, wet	118-120	....			
EXCAVATIONS IN WATER					
Sand or gravel	60	....	**COAL AND COKE, PILED**		
Sand or gravel and clay	65	....	Coal, anthracite	47-58	....
Clay	80	....	Coal, bituminous, lignite	40-54	....
River mud	90	....	Coal, peat, turf	20-26	....
Soil	70	....	Coal, charcoal	10-14	....
Stone riprap	65	....	Coal, coke	23-32	....

The specific gravities of solids and liquids refer to water at 4°C., those of gases to air at 0°C. and 760 mm. pressure. The weights per cubic foot are derived from average specific gravities, except where stated that weights are for bulk, heaped or loose material, etc.

TABLE 3-2   Weights and Specific Gravities (*Continued*)

Substance	Weight Lb. per Cu. Ft.	Specific Gravity	Substance	Weight Lb. per Cu. Ft.	Specific Gravity
METALS, ALLOYS, ORES			**TIMBER, U. S. SEASONED**		
Aluminum, cast,			**Moisture Content by**		
hammered	165	2.55-2.75	Weight:		
Brass, cast, rolled	534	8.4-8.7	Seasoned timber 15 to 20%		
Bronze, 7.9 to 14% Sn	509	7.4-8.9	Green timber up to 50%		
Bronze, aluminum	481	7.7	Ash, white, red	40	0.62-0.65
Copper, cast, rolled	556	8.8-9.0	Cedar, white, red	22	0.32-0.38
Copper ore, pyrites	262	4.1-4.3	Chestnut	41	0.66
Gold, cast, hammered	1205	19.25-19.3	Cypress	30	0.48
Iron, cast, pig	450	7.2	Fir, Douglas spruce	32	0.51
Iron, wrought	485	7.6-7.9	Fir, eastern	25	0.40
Iron, spiegel-eisen	468	7.5	Elm, white	45	0.72
Iron, ferro-silicon	437	6.7-7.3	Hemlock	29	0.42-0.52
Iron ore, hematite	325	5.2	Hickory	49	0.74-0.84
Iron ore, hematite in bank	160-180	....	Locust	46	0.73
Iron ore, hematite loose	130-160	....	Maple, hard	43	0.68
Iron ore, limonite	237	3.6-4.0	Maple, white	33	0.53
Iron ore, magnetite	315	4.9-5.2	Oak, chestnut	54	0.86
Iron slag	172	2.5-3.0	Oak, live	59	0.95
Lead	710	11.37	Oak, red, black	41	0.65
Lead ore, galena	465	7.3-7.6	Oak, white	46	0.74
Magnesium, alloys	112	1.74-1.83	Pine, Oregon	32	0.51
Manganese	475	7.2-8.0	Pine, red	30	0.48
Manganese ore, pyrolusite	259	3.7-4.6	Pine, white	26	0.41
Mercury	849	13.6	Pine, yellow, long-leaf	44	0.70
Monel Metal	556	8.8-9.0	Pine, yellow, short-leaf	38	0.61
Nickel	565	8.9-9.2	Poplar	30	0.48
Platinum, cast, hammered	1330	21.1-21.5	Redwood, California	26	0.42
Silver, cast, hammered	656	10.4-10.6	Spruce, white, black	27	0.40-0.46
Steel, rolled	490	7.85	Walnut, black	38	0.61
Tin, cast, hammered	459	7.2-7.5	Walnut, white	26	0.41
Tin ore, cassiterite	418	6.4-7.0			
Zinc, cast, rolled	440	6.9-7.2			
Zinc ore, blende	253	3.9-4.2	**VARIOUS LIQUIDS**		
			Alcohol, 100%	49	0.79
			Acids, muriatic 40%	75	1.20
VARIOUS SOLIDS			Acids, nitric 91%	94	1.50
			Acids, sulphuric 87%	112	1.80
Cereals, oats ....bulk	32	....	Lye, soda 66%	106	1.70
Cereals, barley ....bulk	39	....	Oils, vegetable	58	0.91-0.94
Cereals, corn, rye ..bulk	48	....	Oils, mineral, lubricants	57	0.90-0.93
Cereals, wheat ....bulk	48	....	Water, 4°C. max. density	62.428	1.0
Hay and Straw ....bales	20	....	Water, 100°C.	59.830	0.9584
Cotton, Flax, Hemp	93	1.47-1.50	Water, ice	56	0.88-0.92
Fats	58	0.90-0.97	Water, snow, fresh fallen	8	.125
Flour, loose	28	0.40-0.50	Water, sea water	64	1.02-1.03
Flour, pressed	47	0.70-0.80			
Glass, common	156	2.40-2.60			
Glass, plate or crown	161	2.45-2.72	**GASES**		
Glass, crystal	184	2.90-3.00			
Leather	59	0.86-1.02	Air, 0°C. 760 mm.	.08071	1.0
Paper	58	0.70-1.15	Ammonia	.0478	0.5920
Potatoes, piled	42	....	Carbon dioxide	.1234	1.5291
Rubber, caoutchouc	59	0.92-0.96	Carbon monoxide	.0781	0.9673
Rubber goods	94	1.0-2.0	Gas, illuminating	.028-.036	0.35-0.45
Salt, granulated, piled	48	....	Gas, natural	.038-.039	0.47-0.48
Saltpeter	67	....	Hydrogen	.00559	0.0693
Starch	96	1.53	Nitrogen	.0784	0.9714
Sulphur	125	1.93-2.07	Oxygen	.0892	1.1056
Wool	82	1.32			

The specific gravities of solids and liquids refer to water at 4°C., those of gases to air at 0°C. and 760 mm. pressure. The weights per cubic foot are derived from average specific gravities, except where stated that weights are for bulk, heaped or loose material, etc.

SOURCE: American Institute of Steel Construction.

TABLE 3-3   Thermal Properties of Typical Building and Insulation Materials—Design Values[a]

| Description | Density (lb/ft³) | Customary Unit | | | | | SI Unit | |
		Conductivity (k)	Conductance (C)	Resistance[b] (R) Per inch thickness (1/k)	Resistance[b] (R) For thickness listed (1/C)	Specific Heat, Btu/(lb·deg F)	Resistance[b] (R) (m·K)/W	Resistance[b] (R) (m²·K)/W
BUILDING BOARD								
Boards, Panels, Subflooring, Sheathing Woodboard Panel Products								
Asbestos-cement board .... 0.125 in.	120	4.0	—	0.25	—	0.24	1.73	0.005
Asbestos-cement board .... 0.125 in.	120	—	33.00	—	0.03			0.01
Asbestos-cement board .... 0.25 in.	120	—	16.50	—	0.06			0.06
Gypsum or plaster board .... 0.375 in.	50	—	3.10	—	0.32	0.26		0.06
Gypsum or plaster board .... 0.5 in.	50	—	2.22	—	0.45			0.08
Gypsum or plaster board .... 0.625 in.	50	—	1.78	—	0.56			0.10
Plywood (Douglas Fir)[c]	34	0.80	—	1.25	—	0.29	8.66	
Plywood (Douglas Fir) .... 0.25 in.	34	—	3.20	—	0.31			0.05
Plywood (Douglas Fir) .... 0.375 in.	34	—	2.13	—	0.47			0.08
Plywood (Douglas Fir) .... 0.5 in.	34	—	1.60	—	0.62			0.11
Plywood (Douglas Fir) .... 0.625 in.	34	—	1.29	—	0.77			0.19
Plywood or wood panels .... 0.75 in.	34	—	1.07	—	0.93	0.29		0.16
Vegetable Fiber Board								
Sheathing, regular density .... 0.5 in.	18	—	0.76	—	1.32	0.31		0.23
.... 0.78125 in.	18	—	0.49	—	2.06			0.36
Sheathing intermediate density .... 0.5 in.	22	—	0.82	—	1.22	0.31		0.21
Nail-base sheathing .... 0.5 in.	25	—	0.88	—	1.14	0.31		0.20
Shingle backer .... 0.375 in.	18	—	1.06	—	0.94			0.17
Shingle backer .... 0.3125 in.	18	—	1.28	—	0.78			0.14
Sound deadening board .... 0.5 in.	15	—	0.74	—	1.35	0.30		0.24
Tile and lay-in panels, plain or acoustic	18	0.40	—	2.50	—	0.14	17.33	
.... 0.5 in.	18	—	0.80	—	1.25			0.22
.... 0.75 in.	18	—	0.53	—	1.89			0.33
Laminated paperboard	30	0.50	—	2.00	—	0.33	13.86	
Homogeneous board from repulped paper	30	0.50	—	2.00	—	0.28	13.86	

TABLE 3-3 Thermal Properties of Typical Building and Insulation Materials—Design Values^a (Continued)

Description	Density (lb/ft³)	Conductivity (k)	Conductance (C)	Customary Unit Resistance^b (R) Per inch thickness (1/k)	For thickness listed (1/C)	Specific Heat, Btu/(lb)(deg F)	SI Unit Resistance^b (R) (m·K)/W	(m²·K)/W
Hardboard								
Medium density	50	0.73	—	1.37	—	0.31	9.49	
High density, service temp. service								
underlay	55	0.82	—	1.22	—	0.32	8.46	
High density, std. tempered	63	1.00	—	1.00	—	0.32	6.93	
Particleboard								
Low density	37	0.54	—	1.85	—	0.31	12.82	
Medium density	50	0.94	—	1.06	—	0.31	7.35	
High density	62.5	1.18	—	0.85	—	0.31	5.89	
Underlayment ....0.625 in.	40	—	1.22	—	0.82	0.29		0.14
Wood subfloor ....0.75 in.		—	1.06	—	0.94	0.33		0.17
BUILDING MEMBRANE								
Vapor—permeable felt.	—	—	16.70	—	0.06			0.01
Vapor—seal, 2 layers of mopped								
15-lb felt.	—	—	8.35	—	0.12			0.02
Vapor—seal, plastic film	—	—	—	—	Negl.			
FINISH FLOORING MATERIALS								
Carpet and fibrous pad	—	—	0.48	—	2.08	0.34		0.37
Carpet and rubber pad	—	—	0.81	—	1.23	0.33		0.22
Cork tile ....0.125 in.	—	—	3.60	—	0.28	0.48		0.05
Terrazzo ....1 in.	—	—	12.50	—	0.08	0.19		0.01
Tile—asphalt, linoleum, vinyl, rubber	—	—	20.00	—	0.05	0.30		0.01
vinyl asbestos						0.24		
ceramic.								
Wood, hardwood finish ....0.75 in.		—	1.47	—	0.68	0.19		0.12
INSULATING MATERIALS								
BLANKET AND BATT^d								
Mineral Fiber, fibrous form processed								
from rock, slag, or glass								
approx.^c 3–3.5 in.	0.3–2.0	—	—	—	11^d			1.94
approx.^c 5.50–6.5	0.3–2.0	—	—	—	19^d			3.35
approx.^c 6–7 in.	0.3–2.0	—	—	—	22^d			3.87
approx.^c 8.5–9 in.	0.3–2.0	—	—	—	30^d			5.28
approx. 12 in.	0.3–2.0	—	—	—	38^d			6.69

Description	Density	k	C	1/k	1/C	Sp. Ht.		
Cellular glass	8.5	0.35	—	2.86	—	0.18	*19.81*	
Glass fiber, organic bonded	4-9	0.25	—	4.00	—	0.23	*27.72*	
Expanded perlite, organic bonded	1.0	0.36	—	2.78	—	0.30	*19.26*	
Expanded rubber (rigid)	4.5	0.22	—	4.55	—	0.40	*31.53*	
Expanded polystyrene extruded								
Cut cell surface	1.8	0.25	—	4.00	—	0.29	*27.72*	
Smooth skin surface	1.8-3.5	0.20	—	5.00	—	0.29	*34.65*	
Expanded polystyrene, molded beads	1.0	0.26	—	—	—		*26.3*	*3.8*
	1.25	0.25	—	—	—		*27.8*	*4.0*
	1.5	0.24	—	—	—		*29.1*	*4.2*
	1.75	0.24	—	—	—		*29.1*	*4.2*
	2.0	0.23	—	—	—		*29.8*	*4.3*
Cellular polyurethane[f] (R-11 exp.)(unfaced) (Thickness 1 in. or greater)	1.5	0.16	—	6.25	—	0.38	*43.82*	
(Thickness 1 in. or greater—high resistance to gas permeation facing)	2.5							
Foil-faced, glass fiber-reinforced cellular	1.5	0.14						
Polyisocyanurate (R-11 exp.)[b]	2	0.14	—	7.04	—	0.22	*48.79*	
Nominal 0.5 in.			0.278		*3.6*			*0.63*
Nominal 1.0 in.			0.139		*7.2*			*1.27*
Nominal 2.0 in.			0.069		*14.4*			*2.53*
Mineral fiber with resin binder	15	0.29	—	3.45	—	0.17	*23.91*	
Mineral fiberboard, wet felted								
Core or roof insulation	16-17	0.34	—	2.94	—		*20.38*	
Acoustical tile	18	0.35	—	2.86	—	0.19	*19.82*	
Acoustical tile	21	0.37	—	2.70	—		*18.71*	
Mineral fiberboard, wet molded								
Acoustical tile[g]	23	0.42	—	2.38	—	0.14	*16.49*	
Wood or cane fiberboard								
Acoustical tile[g] ... 0.5 in.			0.80		*1.25*	0.31		*0.22*
Acoustical tile[g] ... 0.75 in.			0.53		*1.89*			*0.33*
Interior finish (plank, tile)	15	0.35	—	2.86	—	0.32	*19.82*	
Cement fiber slabs (shredded wood with Portland cement binder)	25-27	0.50-0.53	—	2.0-1.89	—		*13.87*	
Cement fiber slabs (shredded wood with magnesia oxysulfide binder)	22	0.57	—	1.75	—	0.31	*12.16*	

TABLE 3-3 Thermal Properties of Typical Building and Insulation Materials—Design Values^a (Continued)

Description	Density (lb/ft³)	Conductivity (k)	Conductance (C)	Resistance^b (R) Per inch thickness (1/k)	Resistance^b (R) For thickness listed (1/C)	Specific Heat, Btu/(lb)(deg F)	SI Unit Resistance^b (R) (m·K)/W	SI Unit Resistance^b (R) (m²·K)/W
LOOSE FILL								
Cellulosic insulation (milled paper or wood pulp)	2.3–3.2	0.27–0.32	—	3.13–3.70	—	0.33	21.69–25.64	
Sawdust or shavings	8.0–15.0	0.45	—	2.22	—	0.33	15.39	
Wood fiber, softwoods	2.0–3.5	0.30	—	3.33	—	0.33	23.08	
Perlite, expanded	2.0–4.1	0.27–0.31	3.7–3.3	2.70	—	0.26	18.71	
	4.1–7.4	0.31–0.36	3.3–2.8					
	7.4–11.0	0.36–0.42	2.8–2.4					
Mineral fiber (rock, slag or glass)								
approx.^c 3.75–5 in.	0.6–2.0	—	—		11	0.17		1.94
approx.^c 6.5–8.75 in.	0.6–2.0	—	—		19			3.35
approx.^c 7.5–10 in.	0.6–2.0	—	—		22			3.87
approx.^c 10.25–13.75 in.	0.6–2.0	—	—		30			5.28
Vermiculite, exfoliated	7.0–8.2	0.47	—	2.13	—	3.20	14.76	
	4.0–6.0	0.44	—	2.27	—		15.73	
ROOF INSULATION^h								
Preformed, for use above deck								
Different roof insulations are available in different thicknesses to provide the design C values listed.^h								
Consult individual manufacturers for actual thickness of their material		0.36 to 0.05			2.7 to 20			0.49 to 3.52
MASONRY MATERIALS								
CONCRETES								
Cement mortar	116	5.0	—	0.20	—	—	1.39	
Gypsum-fiber concrete 87.5% gypsum, 12.5% wood chips	51	1.66	—	0.60	—	0.21	4.16	
Lightweight aggregates including expanded shale, clay or slate; expanded slags; cinders; pumice; vermiculite; also cellular concretes	120	5.2	—	0.19	—	—	1.32	
	100	3.6	—	0.28	—		1.94	
	80	2.5	—	0.40	—		2.77	
	60	1.7	—	0.59	—		4.09	
	40	1.15	—	0.86	—		5.96	
	30	0.90	—	1.11	—		7.69	
	20	0.70	—	1.43	—		9.91	
Perlite, expanded	40	0.93		1.08		0.32	7.48	
	30	0.71		1.41			9.77	
	20	0.50		2.00			13.86	

Material								
Sand and gravel or stone aggregate (oven dried)	140	9.0	—	0.11	—	0.22	0.76	
Sand and gravel or stone aggregate (not dried)	140	12.0	—	0.08	—		0.55	
Stucco	116	5.0	—	0.20	—		1.39	
MASONRY UNITS								
Brick, common[i]	120	5.0	—	0.20	—	0.19	1.39	
Brick, face[i]	130	9.0	—	0.11	—	0.21	0.76	
Clay tile, hollow:								
1 cell deep .... 3 in.			1.25		0.80			0.14
1 cell deep .... 4 in.			0.90		1.11			0.20
2 cells deep .... 6 in.			0.66		1.52			0.27
2 cells deep .... 8 in.			0.54		1.85			0.33
2 cells deep .... 10 in.			0.45		2.22			0.39
3 cells deep .... 12 in.			0.40		2.50			0.44
Concrete blocks, three oval core:								
Sand and gravel aggregate .... 4 in.			1.40		0.71	0.22		0.13
.... 8 in.			0.90		1.11			0.20
.... 12 in.			0.78		1.28			0.23
Cinder aggregate .... 3 in.			1.16		0.86	0.21		0.15
.... 4 in.			0.90		1.11			0.20
.... 8 in.			0.58		1.72			0.30
.... 12 in.			0.53		1.89			0.33
Lightweight aggregate .... 3 in.			0.79		1.27	0.21		0.22
(expanded shale, clay, slate .. 4 in.			0.67		1.50			0.26
or slag; pumice) .... 8 in.			0.50		2.00			0.35
.... 12 in.			0.44		2.27			0.40
Concrete blocks, rectangular core *[j]								
Sand and gravel aggregate								
2 core, 8 in. 36 lb.[k]*			0.96		1.04	0.22		0.18
Same with filled cores[l]*			0.52		1.93	0.22		0.34
Lightweight aggregate (expanded shale, clay, slate or slag; pumice):								
3 core, 6 in. 19 lb.[k]*			0.61		1.65	0.21		0.29
Same with filled cores[l]*			0.33		2.99			0.53
2 core, 8 in. 24 lb.[k]*			0.46		2.18			0.38
Same with filled cores[l]*			0.20		5.03			0.89
3 core, 12 in. 38 lb.[k]*			0.40		2.48			0.44
Same with filled cores[l]*			0.17		5.82			1.02
Stone, lime or sand.		12.50	—	0.08	—	0.19	0.55	
Gypsum partition tile:								
3 × 12 × 30 in. solid			0.79		1.26	0.19		0.22
3 × 12 × 30 in. 4-cell			0.74		1.35			0.24
4 × 12 × 30 in. 3-cell			0.60		1.67			0.29

METALS   See Table 3-4.

TABLE 3-3 Thermal Properties of Typical Building and Insulation Materials—Design Values[a] (Continued)

Description	Density (lb/ft³)	Conductivity (k)	Conductance (C)	Resistance[b] (R) Per inch thickness (1/k)	Resistance[b] (R) For thickness listed (1/C)	Specific Heat, Btu/(lb)(deg F)	SI Unit Resistance[b] (R) (m·K)/W	SI Unit Resistance[b] (R) (m²·K)/W
PLASTERING MATERIALS								
Cement plaster, sand aggregate	116	5.0	—	0.20	—	0.20	1.39	
Sand aggregate . . . . 0.375 in.	—	—	13.3	—	0.08	0.20		0.01
Sand aggregate . . . . 0.75 in.	—	—	6.66	—	0.15	0.20		0.03
Gypsum plaster:								
Lightweight aggregate . . . . . . 0.5 in.	45	—	3.12	—	0.32	—		0.06
Lightweight aggregate . . . . . . 0.625 in.	45	—	2.67	—	0.39	—		0.07
Lightweight agg. on metal lath . . . 0.75 in.	—	—	2.13	—	0.47	—		0.08
Perlite aggregate	45	1.5	—	0.67	—	0.32	4.64	
Sand aggregate	105	5.6	—	0.18	—	0.20	1.25	
Sand aggregate . . . . 0.5 in.	105	—	11.10	—	0.09	—		0.02
Sand aggregate . . . . 0.625 in.	105	—	9.10	—	0.11	—		0.02
Sand aggregate on metal lath . . . 0.75 in.	—	—	7.70	—	0.13	—		0.02
Vermiculite aggregate	45	1.7	—	0.59	—	0.31	4.09	
ROOFING								
Asbestos-cement shingles	120	—	4.76	—	0.21	0.24		0.04
Asphalt roll roofing	70	—	6.50	—	0.15	0.36		0.03
Asphalt shingles	70	—	2.27	—	0.44	0.30		0.08
Built-up roofing . . . . 0.375 in.	70	—	3.00	—	0.33	0.35		0.06
Slate . . . . 0.5 in.	—	—	20.00	—	0.05	0.30		0.01
Wood shingles, plain and plastic film faced	—	—	1.06	—	0.94	0.31		0.17
SIDING MATERIALS (On Flat Surface)								
Shingles								
Asbestos-cement	120	—	4.75	—	0.21	0.24		0.04
Wood, 16 in., 7.5 exposure	—	—	1.15	—	0.87	0.31		0.15
Wood, double, 16-in., 12-in. exposure	—	—	0.84	—	1.19	0.28		0.21
Wood, plus insul. backer board, 0.3125 in.	—	—	0.71	—	1.40	0.31		0.25
Siding								
Asbestos-cement, 0.25 in., lapped	—	—	4.76	—	0.21	0.24		0.04
Asphalt roll siding	—	—	6.50	—	0.15	0.35		0.03
Asphalt insulating siding (0.5 in. bed.)	—	—	0.69	—	1.46	0.35		0.26
Hardboard siding, 0.4375 in.	40	1.49	—	0.67	—	0.28	4.65	
Wood, drop, 1 × 8 in.	—	—	1.27	—	0.79	0.28		0.14
Wood, bevel, 0.5 × 8 in., lapped	—	—	1.23	—	0.81	0.28		0.14
Wood, bevel, 0.75 × 10 in., lapped	—	—	0.95	—	1.05	0.28		0.18
Wood, plywood, 0.375 in., lapped	—	—	1.59	—	0.59	0.29		0.10

Description	Density, lb/ft³	k	C	1/k	1/C		
Aluminum or Steel[m], over sheathing							
Hollow-backed	—	—	1.61	—	0.61	0.29	0.11
Insulating-board backed nominal 0.375 in.	—	—	0.55	—	1.82	0.32	0.32
Insulating-board backed nominal 0.375 in., foil backed	—	—	0.34	—	2.96	0.20	0.52
Architectural glass	—	—	10.00	—	0.10		0.02
WOODS[o,p]							
Maple, oak, and similar hardwoods	45	1.10	—	0.91	—	6.31	0.30
Fir, pine, etc.	32	0.80	—	1.25	—	8.66	0.33
	32						
0.75 in.			1.06		0.94		0.17
1.5 in.			0.53		1.88		0.33
2.5 in.			0.32		3.12		0.55
3.5 in.			0.23		4.38		0.77
5.5 in.			0.14		7.14		1.26
7.25 in.			0.11		9.09		1.60
7.25 in.			0.09		11.11		1.96
3.25 in.			0.07		14.28		2.15
1.25 in.							

a Representative values for dry materials were selected by ASHRAE TC 4.4, Thermal Insulation and Moisture Retarders (Total Thermal Performance Design Criteria). They are intended as design (not specification) values for materials in normal use. Insulation materials in actual service may have thermal values which vary from design values depending on their in-situ properties such as density and moisture content. For properties of a particular product, use the value supplied by the manufacturer or by unbiased tests.

b Resistance values are the reciprocals of C before rounding off C to two decimal places.

c Also see Insulating Materials, Board.

d Does not include paper backing and facing, if any. Where insulation forms a boundary (reflective or otherwise) of an air space, see Tables 1 and 2 for the insulating value of air space for the appropriate effective emittance and temperature conditions of the space.

c Conductivity varies with fiber diameter. (See Chapter 21, Thermal Conductivity section, and Fig. 1) Insulation is produced by different densities; therefore, there is a wide variation in thickness for the same R-value among manufacturers. No effort should be made to relate any specific R-value to any specific thickness. Commercial thicknesses generally available range from 2 to 8.5.

f Values are for aged, unfaced, board stock. For change in conductivity with age of expanded urethane, see Chapter 20, Factors Affecting Thermal Conductivity.

g Insulating values of acoustical tile vary, depending on density of the board and on type, size, and depth of perforations.

h ASTM C-855-77 recognizes the specification of roof insulation on the basis of the C-values shown. Roof insulation is made in thicknesses to meet these values.

i Face brick and common brick do not always have these specific densities. When density is different from that shown, there will be a change in thermal conductivity.

j Data on rectangular core concrete blocks differ from the above data on oval core blocks, due to core configuration, different mean temperatures, and possibly differences in unit weights. Weight data on the oval core blocks tested are not available.

k Weights of units approximately 7.625 in. high and 15.75 in. long. Where units are given as a means of describing the blocks tested, but conductance values are all for 1 ft² of area.

l Vermiculite, perlite, or mineral wool insulation. Where insulation is used, vapor barriers or other precautions must be considered to keep insulation dry.

m Values for metal siding applied over flat surfaces vary widely, depending on amount of ventilation of air space beneath the siding; whether air space is reflective or nonreflective; and on thickness, type, and application of insulating backing-board used. Values given are averages for use as design guides, and were obtained from several guarded hotbox tests (ASTM C236) or calibrated hotbox (BSS 77) on hollow-backed types and types made using backing-boards of wood fiber, foamed plastic, and glass fiber. Departures of ±50% or more from the values given may occur.

n Time-aged values for board stock with gas-barrier quality (0.001 in. thickness or greater) aluminum foil facers on two major surfaces.

o Forest Products Laboratory Wood Handbook, U.S. Dept. of Agriculture #72, 1974, Tables 3 and 4.

p L. Adams: Supporting cryogenic equipment with wood (Chemical Engineering, May 17, 1971).

SOURCE: *ASHRAE Handbook—1981 Fundamentals*. Reprinted by permission.

TABLE 3-4 Properties of Solids

Material Description	Specific Heat J/kg·K	Density kg/m³	Thermal Conductivity W/m·K	Ratio	Emissivity Surface Condition
Aluminum (alloy 1100)	896[b]	2 740[u]	221[u]	0.09[n] / 0.20[n]	commercial sheet / heavily oxidized
Aluminum Bronze (76%Cu, 22%Zn, 2% Al)	400[u]	8 280[u]	100[u]		
Alundum (aluminum oxide)	779[b]	2 400[u]			
Asbestos: fiber insulation	1050[b]	580[b]	0.170[u]		
Ashes, wood	800[i]	640[b]	0.16[b]		
Asphalt	800[i]	2 110[b]	0.071[b] [323]	0.93[b]	"paper"
Bakelite	920[b]	1 300[u]	0.74[b]		
Bell Metal	1500[b]		17[u]		
Bismuth Tin	360[i] [323]		65.0*		
Brick, building	170[u]	1 970[u]	0.7[b]	0.93*	
Brass: red (85% Cu, 15% Zn)	800[b]	8 780[u]	150[u]	0.030[b]	highly polished
yellow (65% Cu, 35% Zn)	400[u]	8 310[u]	120[u]	0.033[b]	highly polished
Bronze	400[u]	8 490[i]	29[d] [273]	0.02[d]	
Cadmium	435[i]	8 650[i]	92.9[b] [256]		
Carbon (gas retort)	230[b]		0.35[b] [256]	0.81[a]	
Cardboard	710[a]		0.07[b]		
Cellulose	1300[b]	54[i]	0.057[i]		
Cement (Portland clinker)	670[b]	1 920[i]	0.029[i]	0.34*	about 394 K
Chalk	900[i]	2 290[i]	0.83[b] [473]		
Charcoal (wood)	840[i]	240[a]	0.05[a] [473]		
Chrome Brick	710[b]	3 200[b]	1.2[b]		
Clay	920[b]	1 000[i]	0.17[f] [273]		
Coal	1000[b]	1 400[i]	0.1[b]		
Coal Tars	1500[b] [313]	1 200[b]			
Coke (petroleum, powdered)	1500[b] [673]	990[b]	0.95[b] [673]		
Concrete (stone)	653[b] [473]	2 300[b]	0.93[b]		
Copper (electrolytic)	390[u]	8 910[u]	393[u]	0.072[n]	commercial, shiny
Cork (granulated)	2030[i]	86[i]	0.048[i] [268]		
Cotton (fiber)	1340[i]	1 500[u]	0.042[u]		
Cryolite (AlF₃ · 3NaF)	1060[b]	2 900[b]			
Diamond	616[b]	2 420[i]	47[i]	0.41*	
Earth (dry and packed)		1 500[i]	0.064*		
Felt		330[b]	0.05[b]		
Fireclay Brick	829[b] [373]	1 790[i]	1[b] [473]	0.75[n]	at 1273 K
Fluorspar (CaF₂)	880[b]	3 190[u]	1.1[b] [473]		
German Silver (nickel silver)	400[u]	8 730[u]	33[u]	0.135[n]	polished

Material					Remarks
Glass:					
crown (soda-lime)	750[b]	2 470[u]	1.0[t] [366]	0.94[n]	smooth
flint (lead)	490[b]	4 280[u]	1.4[t]		
pyrex	840[b]	2 230[t]	1.0[t] [366]		
"wool"	657[b]	52.0[u]	0.038[t]		
Gold	131[u]	19 350[u]	297[t]	0.02[n]	highly polished
Graphite:					
powder	691*	1 870[u]	0.183*		
"Karbate" (impervious)	670[u]	1 200[b]	130[u]	0.75[n]	on a smooth plate
Gypsum	1080[b]	1 500[u]	0.43[b]	0.903[b]	on a smooth plate
Hemp (fiber)	1352.3[u]	921[b]			
Ice: [0°C]	2040[t]		2.24[b]	0.95*	
[−20°C]	1950[t]		2.44*		
Iron:					
cast	500[v] [373]	7 210[t]	47.7[b] [327]	0.435[b]	freshly turned
wrought		7 770[b]	60.4[b]	0.94[b]	dull, oxidized
Lead	129[u]	11 300[u]	34.8[b]	0.28[n]	gray, oxidized
Leather (sole)		1 000[b]	0.16[b]	0.36*	at 336 to 467 K
Limestone	909[b]	1 650[b]	0.93[b]	to 0.90	
Linen			0.09[b]		
Litharge (lead monoxide)	230[b]	7 850[b]			
Magnesia:					
powdered	980[b] [373]	796[b]	0.61[b] [320]		
light carbonate		210[b]	0.059[b]		
Magnesite Brick	930[b] [373]	2 530[b]	3.8[b] [478]		
Magnesium	1000[b]	1 730[b]	160[u]	0.55[n]	oxidized
Marble	880[b]	2 600[b]	2.6[b]	0.931[b]	light gray, polished
Nickel	440[u]	8 890[u]	59.5[u]	0.045[n]	electroplated, polished
Paints:					
White lacquer				0.80[n]	on rough plate
White enamel				0.91[n]	
Black lacquer				0.80[n]	
Black shellac				0.91[n]	"matte" finish
Flat black lacquer		1 000[u]	0.26[u]	0.96[n]	
Aluminum lacquer				0.39[n]	on rough plate
Paper	1300*	930[b]	0.13[b]	0.92[b]	pasted on tinned plate
Paraffin	2900[t]	900[t]	0.24[b] [273]	0.91[b]	rough
Plaster		2 110[b]	0.74[b] [348]	0.054[b]	polished
Platinum	130[u]	21 470[u]		0.92[b]	glazed
Porcelain	750	260[u]	69.0[u]		
Pyrites (Copper)	549[b]	4 200[b]	2.2[u]		
Pyrites (Iron)	569[b] [342]	4 970[v]			

*Data source not determined.

TABLE 3-4 Properties of Solids (Continued)

Material Description	Specific Heat J/kg · K	Density kg/m³	Thermal Conductivity W/m · K	Emissivity Ratio	Emissivity Surface Condition
Rock Salt	917[u]	2 180[u]			
Rubber:					
Vulcanized (soft)	2000[a]	1 100[t]	0.1[t]	0.86[b]	rough
(hard)		1 190[t]	0.16[t]	0.95[b]	glossy
Sand	800[b]	1 520[b]	0.33[b]		
Sawdust		190[b]	0.05[b]		
Silica	1320[b]	2 240[v]	1.4[t] [366]		
Silver	235[u]	10 500[u]	424[u]	0.02[t]	polished and at 500 K
Snow (freshly fallen) (at 32 F)		100[t]	0.598[t]		
Steel (mild)	500[t]	500[t]	2.2[t]		
Stone (quarried)	800[b]	7 830[t]	45.3[b]	0.12[t]	cleaned
Tar:		1 500[t]			
pitch	2500[v]	1 100[u]	0.88[v]		
bituminous		1 200[t]	0.71[u]		
Tin	233[u]	7 290[u]	64.9[u]	0.06[h]	bright and at 323 K
Tungsten	130[u]	19 400[u]	201[u]	0.032[n]	filament at 300 K
Wood:					
Hardwoods:	1900/2700[b]	370/1 100[z]	0.11/0.255[z]		
Ash, white		690[z]	0.172[z]		
Elm, American		580[z]	0.153[z]		
Hickory		800[z]			
Mahogany		550[u]	0.13[u]		
Maple, sugar	2390[b]	720[z]	0.187[z]		
Oak, white		750[z]	0.176[z]	0.90[n]	planed
Walnut, black		630[z]			
Softwoods:	See Table 3A, Chap. 23, source text	350/740[z]	0.11/0.16[z]		
Fir, white		430[z]	0.12[z]		
Pine, white		430[z]	0.11[z]		
Spruce		420[z]	0.11[z]		
Wool:					
Fiber	1360[u]	1 300[u]	0.036/0.063[u]		
Fabric		110/330[u]			
Zinc:					
Cast	390[u]	7 130[u]	110[u]	0.05[n]	polished
Hot-rolled	390[b]	7 130[b]	110[b]		
Galvanizing				0.23[n]	fairly bright

*Source unknown.

[a] *Handbook of Chemistry & Physics* (Chemical Rubber Publishing Co., Cleveland, OH, 47th ed., 1966, 49th ed., 1969).

[b] J. H. Perry: *Chemical Engineers' Handbook* (McGraw-Hill Book Co., Inc., New York, NY, 2nd ed., 1941, 4th ed., 1963).

[c] *Tables of Thermodynamic and Transport Properties of Air, Argon, Carbon Dioxide, Carbon Monoxide, Hydrogen, Nitrogen, Oxygen and Steam* (Pergamon Press, Elmsford, NY, 1960).

[d] *American Institute of Physics Handbook* (McGraw-Hill Book Co., Inc., New York, NY, 2nd ed., 1963).

[e] Organick and Studhalter: Thermodynamic properties of benzene (*Chemical Engineering Progress,* November 1948, p. 847).

[f] Lange: *Handbook of Chemistry* (McGraw-Hill Book Co., Inc., New York, NY, revised 10th ed., 1967).

[g] ASHRAE *Thermodynamic Properties of Refrigerants* (ASHRAE, 1969).

[h] Reid and Sherwood: The Properties of Gases and Liquids (McGraw-Hill Book Co., Inc., New York, NY, 2nd ed., 1966).

[i] ASHRAE *Handbook of Fundamentals* (ASHRAE, 1967).

[j] *T.P.R.C. Data Book* (Thermophysical Properties Research Center, W. Lafayette, IN, 1966).

[k] Estimated.

[l] L. N. Canjar, Max Goldman, and Henry Marchman: Thermodynamic properties of propylene *(Industrial and Engineering Chemistry,* May 1951, p. 1183).

[m] *ASME Steam Tables* (American Society of Mechanical Engineers, New York, NY, 1967).

[n] W. H. McAdams: *Heat Transmission* (McGraw-Hill Book Co., Inc., New York, NY, 3rd ed., 1954).

[o] D. R. Stull: Vapor pressure of pure substances (organic compounds) *(Industrial and Engineering Chemistry,* April 1947, p. 517).

[p] *JANAF Thermochemical Tables* (PB 168 370, National Technical Information Service, Springfield, VA, 1965).

[q] *Physical Properties of Chemical Compounds* (American Chemical Society, Washington, DC, 1955-1961).

[r] *International Critical Tables of Numerical Data* (National Research Council of U. S. A., published by McGraw-Hill Book Co., Inc., New York, NY, 1928).

[s] *Matheson Gas Data Book* (Matheson Company, Inc., East Rutherford, NJ, 4th ed., 1966).

[t] Baumeister and Marks: *Standard Handbook for Mechanical Engineers* (McGraw-Hill Book Co., Inc., New York, NY, 1967).

[u] Miner and Seastone: *Handbook of Engineering Materials* (John Wiley and Sons, New York, NY, 1955).

[v] Kirk and Othmer: *Encyclopedia of Chemical Technology,* (Interscience Division of John Wiley and Sons, New York, NY, 1966).

[w] Gouse and Stevens: *Chemical Technology of Petroleum* (McGraw-Hill Book Co., Inc., New York, NY, 3rd ed., 1960).

[x] *Saline Water Conversion Engineering Data Book* (M. W. Kellogg Co. for U. S., Dept. of Interior, 1955).

[y] J. Timmermans: *Physicochemical Constants of Pure Organic Compounds* (American Elsevier, New York, NY, Vol. 2, 2nd ed., 1965).

[z] *Wood Handbook* (Handbook No. 72, Forest Products Laboratory, U.S. Dept. of Agriculture, 1955).

[aa] Thermophysical Properties, of Refrigerants (ASHRAE 1975).

SOURCE: *ASHRAE Handbook—1981 Fundamentals.* Reprinted by permission.

TABLE 3-5 Coefficients of Transmission (U) of Frame Walls[a]

These coefficients are expressed in Btu per (hour) (square foot) (degree Fahrenheit difference in temperature between the air on the two sides), and are based on an outside wind velocity of 15 mph

Replace Air Space with 3.5-in. R-11 Blanket Insulation (New Item 4)

| Construction | Resistance (R) | | | |
| | 1 | | 2 | |
	Between Framing	At Framing	Between Framing	At Framing
1. Outside surface (15 mph wind)	0.17	0.17	0.17	0.17
2. Siding, wood, 0.5 in.×8 in. lapped (average)	0.81	0.81	0.81	0.81
3. Sheathing, 0.5-in. vegetable fiber board	1.32	1.32	1.32	1.32
4. Nonreflective air space, 3.5 in. (50 F mean; 10 deg F temperature difference)	1.01	—	11.00	—
5. Nominal 2-in. ×4-in. wood stud	—	4.35	—	4.35
6. Gypsum wallboard, 0.5 in.	0.45	0.45	0.45	0.45
7. Inside surface (still air)	0.68	0.68	0.68	0.68
Total Thermal Resistance (R)	$R_i=4.44$	$R_s=7.78$	$R_i=14.43$	$R_s=7.78$

Construction No. 1: $U_i = 1/4.44 = 0.225$; $U_s = 1/7.81 = 0.128$. With 20% framing (typical of 2-in. × 4-in. studs @ 16-in. o.c.), $U_{av} = 0.8$ $(0.225) + 0.20 (0.128) = 0.199$ (See Eq 9).

Construction No. 2: $U_i = 1/14.43 = 0.069$; $U_s = 0.128$. With framing unchanged, $U_{av} = 0.8(0.069) + 0.2(0.128) = 0.081$.

[a]See section Calculating Overall Coefficients in Chapter 23 of source text for basis of calculations.

SOURCE: *ASHRAE Handbook—1981 Fundamentals.* Reprinted by permission.

TABLE 3-6 Coefficients of Transmission (U) of Frame Partitions of Interior Walls[a]

Coefficients are expressed in Btu per (hour) (square foot) (degree Fahrenheit difference in temperature between the air on the two sides), and are based on still air (no wind) conditions on both sides

Replace Air Space with 3.5-in. R-11 Blanket Insulation (New Item 3)

| | Resistance (R) | | | |
| | 1 | | 2 | |
Construction	Between Framing	At Framing	Between Framing	At Framing
1. Inside surface (still air)	0.68	0.68	0.68	0.68
2. Gypsum wallboard, 0.5 in.	0.45	0.45	0.45	0.45
3. Nonreflective air space, 3.5 in. (50 F mean; 10 deg F temperature difference)	1.01	—	11.00	—
4. Nominal 2-in. × 4-in. wood stud	—	4.38	—	4.38
5. Gypsum wallboard, 0.5 in.	0.45	0.45	0.45	0.45
6. Inside surface (still air)	0.68	0.68	0.68	0.68
Total Thermal Resistance (R)	$R_i = 3.27$	$R_s = 6.64$	$R_i = 13.26$	$R_s = 6.64$

Construction No. 1: $U_i = 1/3.27 = 0.306$; $U_s = 1/6.64 = 0.151$. With 12% framing (typical of 2-in. × 4-in. studs @ 24-in. o.c.), $U_{av} = 0.9(0.306) + 0.12(0.151) = 0.293$

Construction No. 2: $U_i = 1/13.26 = 0.076$ $U_s = 1/6.64 = 0.151$. With framing unchanged, $U_{av} = 0.9(0.075) + 0.1(0.151) = 0.083$

[a]See section Calculating Overall Coefficients in Chapter 23 of source text for basis of calculations.

SOURCE: ASHRAE Handbook—1981 Fundamentals. Reprinted by permission.

TABLE 3-7 Coefficients of Transmission (U) of Solid Masonry Walls[a]

Coefficients are expressed in Btu per (hour) (square foot) (degree Fahrenheit difference in temperature between the air on the two sides), and are based on an outside wind velocity of 15 mph

Replace Furring Strips and Air Space with 1-in. Expanded Polystyrene Extruded, Smooth Skin Surface, 2.2 lb/ft³ (New Item 4)

	1		2
	Resistance (R)		
Construction	Between Furring	At Furring	
1. Outside surface (15 mph wind)	0.17	0.17	0.17
2. Common brick, 8 in.	1.60	1.60	1.60
3. Nominal 1-in. ×3-in. vertical furring	—	0.94	—
4. Nonreflective air space, 0.75 in. (50 F mean; 10 deg F temperature difference)	1.01	—	5.00
5. Gypsum wallboard, 0.5 in.	0.45	0.45	0.45
6. Inside surface (still air)	0.68	0.68	0.68
Total Thermal Resistance (R)	$R_i = 3.91$	$R_s = 3.84$	$R_i = 7.90 = R_s$

Construction No. 1: $U_i = 1/3.91 = 0.256$; $U_s = 1/3.84 = 0.260$. With 20% framing (typical of 1-in. × 3-in. vertical furring on masonry @ 16-in. o.c.)
$U_{av} = 0.8(0.256) + 0.2(0.260) = 0.257$
Construction No. 2: $U_i = U_s = U_{av} = 1/7.90 = 0.127$

See section Calculating Overall Coefficients in Chapter 23 of source text for basis of calculations.

[a]
SOURCE: *ASHRAE Handbook—1981 Fundamentals*. Reprinted by permission.

TABLE 3-8 Coefficients of Transmission (U) of Masonry Walls

Coefficients are expressed in Btu per (hour) (square foot) (degree Fahrenheit difference in temperature between the air on the two sides), and are based on an outside wind velocity of 15 mph

Construction	Replace Cinder Aggregate Block with 6-in. Light-weight Aggregate Block with Cores Filled (New Item 4)			
	Resistance (R)			
	1		2	
	Between Furring	At Furring	Between Furring	At Furring
1. Outside surface (15 mph wind)	0.17	0.17	0.17	0.17
2. Face brick, 4 in.	0.44	0.44	0.44	0.44
3. Cement mortar, 0.5 in.	0.10	0.10	0.10	0.10
4. Concrete block, cinder aggregate, 8 in.	1.72	1.72	2.99	2.99
5. Reflective air space, 0.75 in. (50 F mean; 30 deg F temperature difference) E = $0.05^{b,c}$	2.77	—	2.77	—
6. Nominal 1-in. × 3-in. vertical furring	—	0.94	—	0.94
7. Gypsum wallboard, 0.5 in.	0.45	0.45	0.45	0.45
8. Inside surface (still air)	0.68	0.68	0.68	0.68
Total Thermal Resistance (R)	$R_i = 6.33$	$R_s = 4.50$	$R_i = 7.60$	$R_s = 5.77$

Construction No. 1: $U_i = 1/6.33 = 0.158$; $U_s = 1/4.50 = 0.222$. With 20% framing (typical of 1-in. × 3-in. vertical furring on masonry @ 16-in. o.c.), $U_{av} = 0.8(0.158) + 0.2(0.222) = 0.171$

Construction No. 2: $U_i = 1/7.60 = 0.132$, $U_s = 1/5.77 = 0.173$. With framing unchanged, $U_{av} = 0.8(0.132) + 0.2(0.173) = 0.140$

aSee section Calculating Overall Coefficients in Chapter 23 of source text for basis of calculations.

bBased on E =0.05 for bright, polished surface. Due to surface oxidation, and other factors, actual value of the effective emissivity as it ages is expected to be E = 0.10 to 20.

cSee "Caution" p. 23.5 para. 2—re: air leakage and adjusted U-values.

SOURCE: *ASHRAE Handbook—1981 Fundamentals.* Reprinted by permission.

TABLE 3-9 Coefficients of Transmission (U) of Masonry Cavity Walls[a]

Coefficients are expressed in Btu per (hour) (square foot) (degree Fahrenheit difference in temperature between the air on the two sides), and are based on an outside wind velocity of 15 mph

| | Resistance (R) | | |
| Construction | Replace Furring Strips and Gypsum Wallboard with 0.625-in. Plaster (Sand Aggregate) Applied Directly to Concrete Block-Fill 2.5-in. Air Space with Vermiculite Insulation, 7-8.2 lb/ft³ (New Items 3 and 7) | | |
	Between Furring (1)	At Furring (1)	(2)
1. Outside surface (15 mph wind)	0.17	0.17	0.17
2. Common brick, 4 in.	0.80	0.80	0.80
3. Nonreflective air space, 2.5 in. (30 F mean; 10 deg F temperature difference)	1.10*	1.10*	5.32**
4. Concrete block, three-oval core, stone and gravel aggregate, 4 in.	0.71	0.71	0.71
5. Nonreflective air space 0.75 in. (50 F mean; 10 deg F temperature difference)	1.01	—	—
6. Nominal 1-in. × 3-in. vertical furring	—	0.94	—
7. Gypsum wallboard, 0.5 in.	0.45	0.45	0.11
8. Inside surface (still air)	0.68	0.68	0.68
Total Thermal Resistance (R)	$R_i = 4.92$	$R_s = 4.85$	$R_l = R_s = 7.79$

Construction No. 1: $U_i = 1/4.92 = 0.203; U_s = 1/4.85 = 0.206; U_{av} = 0.8(0.203) + 0.2(0.206) = 0.204$. With 20% framing (typical of 1-in. × 3-in. vertical furring on masonry @16-in. o.c.),

Construction No. 2: $U_i = U_s = U_{av} = 1.79 = 0.128$

[a]See section Calculating Overall Coefficients in Chapter 23 of source text for basis of calculations.

*Interpolated value from Table 2 in Chapter 23 of source text.

**Calculated value from Table 3 in Chapter 23 of source text.

SOURCE: *ASHRAE Handbook—1981 Fundamentals.* Reprinted by permission.

TABLE 3-10　Coefficients of Transmission (U) of Masonry Partitions[a]

Coefficients are expressed in Btu per (hour) (square foot) (degree Fahrenheit difference in temperature between the air on the two sides), and are based on still air (no wind) conditions on both sides

Replace Concrete Block with 4-in. Gypsum Tile (New Item 3) Construction	1	2
1. Inside surface (still air)	0.68	0.68
2. Plaster, lightweight aggregate, 0.625 in.	0.39	0.39
3. Concrete block, cinder aggregate, 4 in.	1.11	1.67
4. Plaster, lightweight aggregate, 0.625 in.	0.39	0.39
5. Inside surface (still air)	0.68	0.68
Total Thermal Resistance(R)	3.25	3.81

Construction No. 1: $U = 1/3.25 = 0.308$
Construction No. 2: $U = 1/3.81 = 0.262$

[a]See section Calculating Overall Coefficients in Chapter 23 of source text for basis of calculations.

SOURCE: *ASHRAE Handbook—1981 Fundamentals*. Reprinted by permission.

TABLE 3-11 Coefficients of Transmission (U) of Frame Construction Ceilings and Floors[a]

Coefficients are expressed in Btu per (hour) (square foot) (degree Fahrenheit difference between the air on the two sides), and are based on still air (no wind) on both sides

Assume Unheated Attic Space above Heated Room with Heat Flow Up—Remove Tile, Felt, Plywood, Subfloor and Air Space—Replace with R-19 Blanket Insulation (New Item 4)

Construction (Heat Flow Up)	1		2	
	Between Floor Joists	At Floor Joists	Between Floor Joists	At Floor Joists
1. Bottom surface (still air)	0.61	0.61	0.61	0.61
2. Metal lath and lightweight aggregate, plaster, 0.75 in.	0.47	0.47	0.47	0.47
3. Nominal 2-in. × 8-in. floor joist	—	9.06	—	9.06
4. Nonreflective airspace, 7.25-in. (50 F mean; 10 deg F temperature difference)	0.93*	—	19.00	—
5. Wood subfloor, 0.75 in.	0.94	0.94	—	—
6. Plywood, 0.625 in.	0.77	0.77	—	—
7. Felt building membrane	0.06	0.06	—	—
8. Tile	0.05	0.05	—	—
9. Top surface (still air)	0.61	0.61	0.61	0.61
Total Thermal Resistance (R)	$R =$ 4.44	$R_s =$ 12.57	$R_i =$ 20.69	$R_s =$ 10.75

Construction No. 1: $U_i = 1/4.45 = 0.225$; $U_s = 1/12.58 = 0.079$. With 10% framing (typical of 2-in. joists @ 16-in. o.c.), $U_{av} = 0.9(0.225) + 0.1 (0.079) = 0.210$

Construction No. 2: $U_i = 1/20.69 = 0.048$; $U_s = 1/10.75 = 0.093$. With framing unchanged, $U_{av} = 0.9(0.048) + 0.1 (0.093) = 0.053$

[a]See section Calculating Overall Coefficients in Chapter 23 of source text for basis of calculations.

*Use largest air space (3.5-in.) value shown in Table 2, Chapter 23, of source text.

SOURCE: *ASHRAE Handbook—1981 Fundamentals.* Reprinted by permission.

TABLE 3-12 Coefficients of Transmission (U) of Flat Masonry Roofs with Built-Up Roofing, with and without Suspended Ceilings[a,b] (Winter Conditions, Upward Flow)

These Coefficients are expressed in Btu per (hour) (square foot) (degree Fahrenheit difference in temperature between the air on the two sides), and are based upon an outside wind velocity of 15 mph

Add **Rigid Roof Deck Insulation,** $C = 0.24$ ($R = 1/C = 4.17$) (New Item 7) Construction (Heat Flow Up)	1	2
1. Inside surface (still air)	0.61	0.61
1. Metal lath and lightweight aggregate plaster, 0.75 in.	0.47	0.47
3. Nonreflective air space, greater than 3.5 in. (50 F mean; 10 deg F temperature difference)	0.93*	0.93*
4. Metal ceiling suspension system with metal hanger rods	0**	0**
5. Corrugated metal deck	0	0
6. Concrete slab, lightweight aggregate, 2 in. (30 lb/ft³)	2.22	2.22
7. Rigid roof deck insulation (none)	—	4.17
8. Built-up roofing, 0.375 in.	0.33	0.33
9. Outside surface (15 mph wind)	0.17	0.17
Total Thermal Resistance (R)	4.73	8.90

Construction No. 1: U_{av} = 1/4.73 = 0.211
Construction No. 2: U_{av} = 1/8.90 = 0.112

[a]See section Calculating Overall Coefficients in Chapter 23 of source text for basis of calculations.
[b]To adjust U values for the effect of added insulation between framing members, see Table 5 or 6 of source text.
*Use largest air space (3.5-in.) value shown in Table 2, Chapter 23. of source text.
**Area of hanger rods is negligible in relation to ceiling area.

SOURCE: *ASHRAE Handbook—1981 Fundamentals*. Reprinted by permission.

TABLE 3-13 Coefficients of Transmission (U) of Wood Construction Flat Roofs and Ceilings[a] (Winter Conditions, Upward Flow)

Coefficients are expressed in Btu per (hour)(square foot)(degree Fahrenheit difference in temperature between the air on the two sides), and are based upon an outside wind velocity of 15 mph

Replace Roof Deck Insulation and Partially Fill the 7.25-in. Air Space with 6-in. R-19 Blanket Insulation and 1.25-in. Air Space (New Items 5 and 7)

Construction (Heat Flow Up)	Resistance (R) 1		2	
	Between Joists	At Joists	Between Joists	At Joists
1. Inside surface (still air)	0.61	0.61	0.61	0.61
2. Acoustical tile, fiberboard, 0.5 in.	1.25	1.25	1.25	1.25
3. Gypsum wallboard, 0.5 in.	0.45	0.45	0.45	0.45
4. Nominal 2-in. × 8-in. ceiling joists	—	9.06	—	9.06
5. Nonreflective air space, 7.25 in. (50 F mean; 10 deg F temperature difference)	0.93*	—	1.05**	—
6. Plywood deck, 0.625 in.	0.78	0.78	0.78	0.78
7. Rigid roof deck insulation, c = 0.72, (R = 1/C)	1.39	1.39	19.00	—
8. Built-up roof	0.33	0.33	0.33	0.33
9. Outside surface (15 mph wind)	0.17	0.17	0.17	0.17
Total Thermal Resistance (R)	R_i=5.91	R_s=14.04	R_i=23.64	R_s=12.65

Construction No. 1: $U_i = 1/5.91 = 0.169$; $U_s = 1/14.04 = 0.071$. With 10% framing (typical of 2-in. joists @ 16-in. o.c.), $U_{av} = 0.9(0.169) + 0.1(0.071) = 0.159$

Construction No. 2: $U_i = 1/23.64 = 0.042$; $U_s = 1/12.65 = 0.079$. With framing unchanged, $U_{av} = 0.9(0.042) + 0.1(0.079) = 0.046$

[a]See section Calculating Overall Coefficients in Chapter 23 of source text for basis of calculations.
*Use largest air space (3.5-in.) value shown in Table 2, Chapter 23, of source text.
**Interpolated value (0°F mean; 10°F temperature difference).

SOURCE: *ASHRAE Handbook—1981 Fundamentals.* Reprinted by permission.

TABLE 3-14 Coefficients of Transmission (U) of Metal Construction Flat Roofs and Ceilings[a] (Winter Conditions, Upward Flow)

Coefficients are expressed in Btu per (hour) (square foot) (degree Fahrenheit difference in temperature between the air on the two sides), and are based on upon outside wind velocity of 15 mph

Construction (Heat Flow Up)	Replace Rigid Roof Deck Insulation ($C = 0.24$) and Sand Aggregate Plaster with Rigid Roof Deck Insulation, $C = 0.36$ and Lightweight Aggregate Plaster 0.75 in. on Metal Lath (New Items 2 and 6)	
	1	2
1. Inside surface (still air)	0.61	0.61
2. Metal lath and sand aggregate plaster, 0.75 in	0.13	0.47
3. Structural beam	0.00*	0.00*
4. Nonreflective air space (50 F mean; 10 deg F temperature difference) thickness	0.93**	0.93**
5. Metal deck	0.00*	0.00*
6. Rigid roof deck insulation, $C = 0.24$ ($R = 1/c$)	4.17	2.78
7. Built-up roofing, 0.375 in.	0.33	0.33
8. Outside surface (15 mph wind)	0.17	0.17
Total Thermal Resistance (R)	6.34	5.29

Construction No. 1: $U = 1/6.34 = 0.158$
Construction No. 2: $U = 1/5.29 = 0.189$

[a]See section Calculating Overall Coefficients in Chapter 23 of source text for basis of calculations.
*If structural beams and metal deck are to be considered, the technique shown in Examples 1 and 2 and Fig. 3 of Chapter 23 of the source text may be used to estimate total R. Full scale testing of a suitable portion of the construction is, however, preferable.
**Use largest air space (3.5-in.) value shown in Table 2, Chapter 23, of source text.

SOURCE: *ASHRAE Handbook—1981 Fundamentals*. Reprinted by permission.

TABLE 3-15 Coefficients of Transmission (U) of Pitched Roofs[a,b]

Coefficients are expressed in Btu per (hour) (square foot) (degree Fahrenheit difference in temperature between the air on the two sides), and are based on an outside wind velocity of 15 mph for heat flow upward and 7.5 mph forheat flow downward

Find U_{av} for same Construction 2 with Heat Flow Down (Summer Conditions)

Construction 1 (Heat Flow Up) (Reflective Air Space)	1 Between Rafters	At Rafters	2 Between Rafters	At Rafters
1. Inside surface (still air)	0.62	0.62	0.76	0.76
2. Gypsum wallboard 0.5 in., foil backed	0.45	0.45	0.45	0.45
3. Nominal 2-in. × 4-in. ceiling rafter	—	4.35	—	4.35
4. 45 deg slope reflective air space, 3.5 in. (50 F mean, 30 deg F temperature difference) E = 0.05	2.17	—	4.33	—
5. Plywood sheathing, 0.625 in.	0.77	0.77	0.77	0.77
6. Permeable felt building membrane	0.06	0.06	0.06	0.06
7. Asphalt shingle roofing	0.44	0.44	0.44	0.44
8. Outside surface (15 mph wind)	0.17	0.17	0.25**	0.25**
Total Thermal Resistance (R)	R_l =4.68	R_s =6.86	R_l =7.06	R_s =7.08

Construction No. 1: U_l=1/4.69= 0.213; U_s = 1/6.90 = 0.145. With 10% framing (typical of 2-in. rafters @16-in. o.c.), U_{av} = 0.9 (0.213) + 0.1 (0.145) = 0.206

Construction No. 2: U_l=1/7.07 = 0.141; U_s = 1/7.12 = 0.140. With framing unchanged, U_{av} = 0.9 (0.141) + 0.1 (0.140) = 0.141

Find U_{av} for same Construction 2 with Heat Flow Down (Summer Conditions)

Construction 1 (Heat Flow Up) (Non-Reflective Air Space)	3 Between Rafters	At Rafters	4 Between Rafters	At Rafters
1. Inside surface (still air)	0.62	0.62	0.76	0.76
2. Gypsum wallboard, 0.5 in.	0.45	0.45	0.45	0.45
3. Nominal 2-in. × 4-in. ceiling rafter	—	4.35	—	4.35
4. 45 deg slope, nonreflective air space, 3.5 in. (50 F mean, 10 deg F temperature difference)	0.96	—	0.90*	—
5. Plywood sheathing, 0.625 in.	0.77	0.77	0.77	0.77
6. Permeable felt building membrane	0.06	0.06	0.06	0.06
7. Asphalt shingle roofing	0.44	0.44	0.44	0.44
8. Outside surface (15-mph wind)	0.17	0.17	0.25**	0.25**
Total Thermal Resistance (R)	R_l =3.47	R_s =6.86	R_l =3.63	R_s =7.08

Construction No. 3: $U = 1/3.48 = 0.287$; $U_s = 1/6.90 = 0.145$. With 10% framing (typical of 2-in. rafters @ 16-in. o.c.), $U_{av} = 0.9\,(0.287) + 0.1\,(0.145) = 0.273$

Construction No. 4: $U_i = 1/3.64 = 0.275$; $U_s = 1/7.12 = 0.140$. With framing unchanged, $U_{av} = 0.9\,(0.275) + 0.1\,(0.140) = 0.262$

[a]See section Calculating Overall Coefficients in Chapter 23 of source text for basis of calculation.

[b]Pitch of roof—45 deg.

* Air space value at 90°F, 10°F temperature difference.

** 7.5-mi/h wind.

SOURCE: *ASHRAE Handbook—1981 Fundamentals*. Reprinted by permission.

TABLE 3-16 Coefficients of Transmission (U) of Windows, Skylights, and Light Transmitting Partitions

These values are for heat transfer from air to air, Btu/hr · ft² · F). To calculate total heat gain including solar transmission, see Chapter 28.

PART A—VERTICAL PANELS (EXTERIOR WINDOWS, SLIDING PATIO DOORS, AND PARTITIONS)—FLAT GLASS, GLASS BLOCK, AND PLASTIC SHEET

Description	Exterior[a]		Interior
	Winter	Summer	
Flat Glass[b]			
single glass	1.10	1.04	0.73
insulating glass—double[c]			
0.1875-in. air space[d]	0.62	0.65	0.51
0.25-in. air space[d]	0.58	0.61	0.49
0.5-in. air space[e]	0.49	0.56	0.46
0.5-in. air space, low emittance coating[f]			
e = 0.20	0.32	0.38	0.32
e = 0.40	0.38	0.45	0.38
e = 0.60	0.43	0.51	0.42
insulating glass—triple[c]			
0.25-in. air spaces[d]	0.39	0.44	0.38
0.5-in. air spaces[g]	0.31	0.39	0.30
storm windows			
1-in. to 4-in. air space[d]	0.50	0.50	0.44
Plastic Sheet			
single glazed			
0.125-in. thick	1.06	0.98	—

PART B—HORIZONTAL PANELS (SKYLIGHTS)—FLAT GLASS, GLASS BLOCK, AND PLASTIC DOMES

Description	Exterior[a]		Interior[f]
	Winter[i]	Summer[j]	
Flat Glass[e]			
single glass	1.23	0.83	0.96
insulating glass—double[c]			
0.1875-in. air space[d]	0.70	0.57	0.62
0.25-in. air space[d]	0.65	0.54	0.59
0.5-in. air space[e]	0.59	0.49	0.56
0.5-in. air space, low emittance coating[f]			
e = 0.20	0.48	0.36	0.39
e = 0.40	0.52	0.42	0.45
e = 0.60	0.56	0.46	0.50
Glass Block[h]			
11 × 11 × 3 in. thick with cavity divider	0.53	0.35	0.44
12 × 12 × 4 in. thick with cavity divider	0.51	0.34	0.42
Plastic Domes[k]			
single-walled	1.15	0.80	—
double-walled	0.70	0.46	—

0.25-in. thick	0.96	0.89	—
0.5-in. thick	0.81	0.76	—
insulating unit—double[c]			
0.25-in. air space[c]	0.55	0.56	—
0.5-in. air space[c]	0.43	0.45	—
Glass Block[h]			
6 × 6 × 4 in. thick	0.60	0.57	0.46
8 × 8 × 4 in. thick	0.56	0.54	0.44
—with cavity divider	0.48	0.46	0.38
12 × 12 × 4 in. thick	0.52	0.50	0.41
—with cavity divider	0.44	0.42	0.36
12 × 12 × 2 in. thick	0.60	0.57	0.46

PART C—ADJUSTMENT FACTORS FOR VARIOUS WINDOW AND SLIDING PATIO DOOR TYPES (MULTIPLY U VALUES IN PARTS A AND B BY THESE FACTORS)

Description	Single Glass	Double or Triple Glass	Storm Windows
Windows			
All Glass[l]	1.00	1.00	1.00
Wood Sash—80% Glass	0.90	0.95	0.90
Wood Sash—60% Glass	0.80	0.85	0.80
Metal Sash—80% Glass	1.00	1.20[m]	1.20[m]
Sliding Patio Doors			
Wood Frame	0.95	1.00	—
Metal Frame	1.00	1.10[n]	—

[a]See Part C of this table for adjustment for various window and sliding patio door types.
[b]Emittance of uncooled glass surface = 0.84.
[c]Double and triple refer to the number of lights of glass.
[d]0.125-in. glass.
[e]0.25-in. glass.
[f]Coating on either glass surface facing air space; all other glass surfaces uncoated.
[g]Window design: 0.25-in. glass—0.125-in. glass—0.25-in. glass.
[h]Dimensions are nominal.
[i]For heat flow up.
[j]For heat flow down.
[k]Based on area of opening, not total surface area.
[l]Refers to windows with negligible opaque area.
[m]Values will be less than these when metal sash and frame incorporate thermal break designs, U-values will be equal to or less than those for the glass. Window manufacturers should be consulted for specific data.

TABLE 3-17   Coefficients of Transmission (U) for Slab Doors

Thickness[a]	Btu per (hr · ft² · F)			
	Winter			Summer
	Solid Wood, No Storm Door	Storm Door[b]		No Storm Door
		Wood	Metal	
1-in.	0.64	0.30	0.39	0.61
1.25-in.	0.55	0.28	0.34	0.53
1.5-in.	0.49	0.27	0.33	0.47
2-in.	0.43	0.24	0.29	0.42
Steel Door				
1.75 in.				
A[c]	0.59	—	—	0.58
B[d]	0.19	—	—	0.18
C[e]	0.47	—	—	0.46

[a]Nominal thickness.

[b]Values for wood storm doors are for approximately 50% glass; for metal storm door values apply for any percent of glass.

[c]A = Mineral fiber core (2 lb/ft³).

[d]B = Solid urethane foam core with thermal break.

[e]C = Solid polystyrene core with thermal break.

SOURCE: *ASHRAE Handbook and Product Directory—1977 Fundamentals.* Reprinted by permission.

NOTES

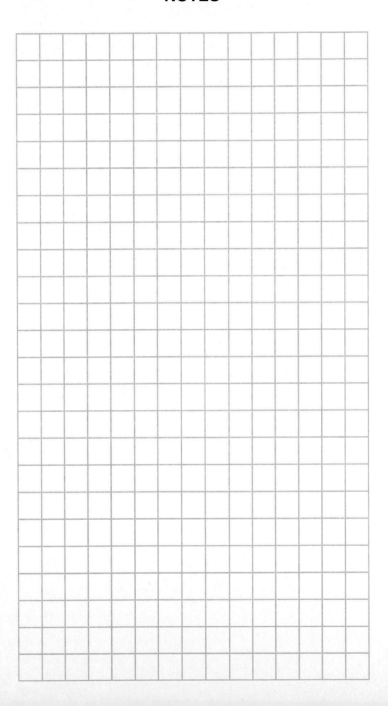

NOTES

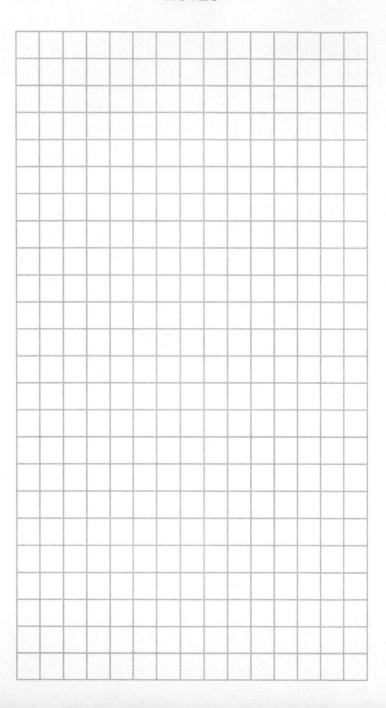

Heating Fuel Data

TABLE 4-1   Heat Values of Wood Fuel

Type of wood	Average wt/cord,* lb	Btu/cord†	Btu/lb	Comments
Hickory	3595	30,600,000	8510	Highest heat value
Hard maple	3075	29,000,000	9430 ⎤	
Beech	3240	27,800,000	8580 ⎟	
White oak	3750	27,700,000	7380 ⎬ High heat value
Red oak	3240	26,300,000	8110 ⎟	
Birch	3000	26,200,000	8730 ⎦	
Elm	2750	24,500,000	8900 ⎤ Hard to split
Tamarack	2500	24,010,000	9600 ⎦	
Soft Maple	2500	24,000,000	9600	Acceptable
Cherry	2550	23,500,000	9210	Difficult to find
Ash	2950	22,600,000	7660 ⎤	
Spruce	2100	18,100,000	8610 ⎟	
Hemlock	2100	17,910,000	8520 ⎟	
White Pine	1800	17,900,000	9940 ⎬ Marginal heat value
Aspen	1900	17,700,000	9310 ⎟	
Basswood	1900	17,001,000	8940 ⎦	

*Wood cut for fuel or pulpwood (128 ft³) as arranged in a stack 4 ft × 4 ft × 8 ft. "Fireplace" or "face" cords are only about one-third of a true cord.

†Btu: The British thermal unit (Btu) is the quantity of heat required to raise the temperature of 1 lb of water 1°F.

SOURCE: R. O. Parmley, *HVAC Field Manual,* McGraw-Hill, New York, 1988. Used with permission of publisher.

TABLE 4-2   Typical Btu Values of Fuels

ASTM rank solids	Btu values per lb
Anthracite Class I	11,230
Bituminous Class II Group 1	14,100
Bituminous Class II Group 3	13,080
Subbituminous Class III Group 1	10,810
Subbituminous Class III Group 2	9,670

Liquids	Btu values per gal
Fuel oil No. 1	138,870
Fuel oil No. 2	143,390
Fuel oil No. 4	144,130
Fuel oil No. 5	142,720
Fuel oil No. 6	137,275

Gases	Btu values per ft^3
Natural gas	1030 to 1132
Producers gas	163
Illuminating gas	534
Mixed (coke oven and water gas)	545

Coal, lignite, peat, etc.	Btu values per lb
Eastern coal	13,250
Western coal	9,000
Lignite	7,000
Peat (sods or milled) (@ 30% M.C.)*	6,000
Agri fuel pellets or briquettes (@ 8% M.C.)*	8,000
Wood chips (@ 45% M.C.)*	4,700

*M.C. = Moisture Content

SOURCE: R. O. Parmley, *HVAC Field Manual,* McGraw-Hill, New York, 1988. Used with permission of publisher.

TABLE 4-3   Heat Values of Petrofuels

Type	Btu/lb	Btu/gal or ft^3
Fuel oil No. 2	20,571	144,000/gal
Natural gas	18,000	1,030/ft^3
Propane	21,564	2,572/ft^3
Butane	21,440	3,200/ft^3

SOURCE: R. O. Parmley, *HVAC Field Manual,* McGraw-Hill, New York, 1988. Used with permission of publisher.

TABLE 4-4   Physical Properties of Propane

Formula	C_3H_8
Btu/gal	91,500
Btu/ft³ of gas at 60°F, atmospheric pressure	2,520
Btu/lb of gas	21,560
Range of inflammability: percent of gas in gas-air mixture	2.15 to 9.6%
Vapor pressure, psig at 60°F	92
Vapor pressure, psig at 100°F	172
Pounds per gal of liquid at 60°F, storage tank pressure	4.23
Specific gravity of liquid at 60°F (water = 1)	0.51
Boiling point of liquid at atmospheric pressure	−44°F
Cubic feet of gas per lb of liquid (at 60°F, atmospheric pressure)	8.59
Cubic feet of gas per gal of liquid (at 60°F, atmospheric pressure)	36.5
Specific gravity of gas (air = 1)	1.53

SOURCE: R. O. Parmley, *HVAC Field Manual,* McGraw-Hill, New York, 1988. Used with permission of publisher.

TABLE 4-5   Heating Efficiency Comparison

Fuel	Heating equipment efficiency percentage*
Electricity	95
No. 2 oil	80
Propane	78
Natural gas	80
Firewood (air-dried)	55
No. 5 & No. 6 oil (low sulfur)	80
Agri fuel pellets or briquettes (@ 8% M.C.)	78
Eastern coal	78
Western coal	75
Wood chips (@ 45% M.C.)	65
Lignite	75
Peat (@ 30% M.C.)	68

*These percentages are general. For specific equipment, see manufacturers' literature and certified test data.

SOURCE: R. O. Parmley, *HVAC Field Manual,* McGraw-Hill, New York, 1988. Used with permission of publisher.

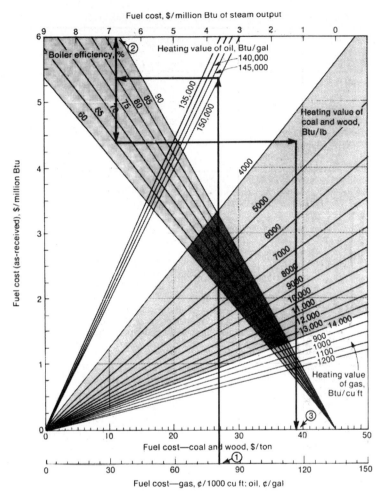

FIGURE 4-1 Fuel cost, dollars per million Btu of steam output. (*Source: T. C. Elliott et al.,* Standard Handbook of Powerplant Engineering, *McGraw-Hill, New York, 1989. Used with permission of publisher.*)

TABLE 4-6　Tank Sizing (in Gallons) for Vapor Withdrawal

Maximum gas needed to vaporize*	Tank size (in gallons) required if lowest outdoor temperature (average for 24 h) reaches:						
	32°F	20°F	10°F	0°F	−10°F	−20°F	−30°F
125,000 Btu/h (50 ft³/h)	115	115	115	250	250	400	600
250,000 Btu/h (100 ft³/h)	250	250	250	400	500	1000	1500
375,000 Btu/h (150 ft³/h)	300	400	500	500	1000	1500	2500
500,000 Btu/h (200 ft³/h)	400	500	750	1000	1200	2000	3500
750,000 Btu/h (300 ft³/h)	750	1000	1500	2000	2500	4000	5000

*Average rate of withdrawal during an 8-h phase.

SOURCE: R. O. Parmley, *HVAC Field Manual,* McGraw-Hill, New York, 1988. Used with permission of publisher.

TABLE 4-7 Combustion Reactions of Common Fuel Constituents

Constituent	Molecular Symbol	Combustion Reactions	Stoichiometric Oxygen and Air Requirements			
			lb/lb Fuel[a]		ft³/ft³ Fuel	
			O_2	Air	O_2	Air
Carbon (to CO)	C	$C+0.5\alpha_2 \rightarrow CO$	1.33	5.75	—	—
Carbon (to CO₂)	C	$C+O_2 \rightarrow CO_2$	2.66	11.51	—	—
Carbon Monoxide	CO	$CO+0.5\alpha_2 \rightarrow CO_2$	0.57	2.47	0.50	2.39
Hydrogen	H_2	$H_2+0.5\alpha_2 \rightarrow H_2O$	7.94	34.28	0.50	2.39
Methane	CH_4	$CH_4+2O_2 \rightarrow CO_2+2H_2O$	3.99	17.24	2.00	9.57
Ethane	C_2H_6	$C_2H_6+3.5O_2 \rightarrow 2CO_2+3H_2O$	3.72	16.09	3.50	16.75
Propane	C_3H_8	$C_3H_8+5O_2 = 3CO_2+4H_2O$	3.63	15.68	5.00	23.95
Butane	C_4H_{10}	$C_4H_{10}+6.5O_2 = 4CO_2+5H_2O$	3.58	15.47	6.50	31.14
—	C_nH_{2n+2}	$C_nH_{2n+2}+(1.5n+0.5)O_2 \rightarrow nCO_2+(n+1)H_2O$	—	—	$1.5n+0.5$	$7.18n+2.39$
Ethylene	C_2H_4	$C_2H_4+3O_2 \rightarrow 2CO_2+2H_2O$	3.42	14.78	3.00	14.38
Propylene	C_3H_6	$C_3H_6+4.5O_2 \rightarrow 3CO_2+3H_2O$	3.42	14.78	4.50	21.53
—	C_nH_{2n}	$C_nH_{2n}+1.5nO_2 \rightarrow nCO_2+nH_2O$	3.42	14.78	$1.50n$	$7.18n$
Acetylene	C_2H_2	$C_2H_2+2.5\alpha_2 \rightarrow 2CO_2+H_2O$	3.07	13.27	2.50	11.96
—	C_nH_{2m}	$C_nH_{2m}+(n+0.5m)O_2 \rightarrow nCO_2+mH_2O$	—	—	$n+0.5m$	$4.78n+2.39m$
Sulfur (to SO₂)	S	$S+O_2 \rightarrow SO_2$	1.00	4.31	—	—
Sulfur (to SO₃)	S	$S+1.5O_2 \rightarrow SO_3$	1.50	6.47	—	—
Hydrogen Sulfide	H_2S	$H_2S+1.5O_2 \rightarrow SO_2+H_2O$	1.41	6.08	1.50	7.18

[a] Atomic masses: H = 1.008; C = 12.01; O = 16.00; S = 32.06.

SOURCE: *ASHRAE Handbook—1989 Fundamentals.* Reprinted by permission.

TABLE 4-8 ASTM Classification for U.S. Coals

Coal rank			Coal analysis, inherent equilibrium basis						Rank (FC)
Class	Group	Designation*	M, %	VM, %	FC, %	A, %	S, %	GCV, Btu/lb†	GCV, Btu/lb†
I	1	ma	4.5	1.7	84.1	9.7	0.77	12,745	(99.2)
	2	an	2.5	6.2	79.4	11.9	0.60	12,925	(94.1)
	3	sa	2.0	10.6	67.2	20.2	0.62	11,925	(88.7)
II	1	lvb	1.0	16.6	77.3	5.1	0.74	14,715	(82.8)
	2	mvb	1.5	20.8	67.5	10.2	1.68	13,720	(77.5)
	3	hvAb	1.5	30.7	56.6	11.2	1.82	13,325	(65.8)/15,230
	4	hvBb	5.8			11.7	2.70	11,910	13,710
	5	hvCb	12.2			9.0	3.20	11,340	12,630
III	1	subA	14.1			7.0	0.43	11,140	12,075
	2	subB	25.0			3.7	0.30	9,345	9,745
	3	subC	31.0			4.8	0.55	8,320	8,790
IV	1	ligA	37.0			4.2	0.40	7,255	7,610

*ma = meta-anthracite;   an = anthracite;   sa = semianthracite;   lvb = low-volatile   bituminous   coal;
mvb = medium-volatile bituminous coal; hvAb = high-volatile A bituminous coal; hvBb = high-volatile B bituminous coal; hvCb = high-volatile C bituminous coal; subA = subbituminous A coal; subB = subbituminous B coal;
subC = subbituminous C coal; ligA = lignite A.

†kJ/kg = Btu/lb × 2.33.

SOURCE: T. C. Elliott et al., *Standard Handbook of Powerplant Engineering*, McGraw-Hill,
New York, 1989. Used with permission of publisher.

TABLE 4-9 Classification of Coals by Rank*

Class	Group	Fixed-carbon limits (dry, mineral-matter-free basis), % — Equal to or greater than	Less than	Volatile-matter limits (dry, mineral-matter-free basis), % — Greater than	Equal to or less than	Calorific value limits (moist,† mineral-matter-free basis), Btu/lb — Equal to or greater than	Less than	Agglomerating character
I. Anthracitic	1. Meta-anthracite	98	—	—	2	—	—	Nonagglomerating
	2. Anthracite	92	98	2	8	—	—	
	3. Semianthracite‡	86	92	8	14	—	—	
II. Bituminous	1. Low-volatile bituminous coal	78	86	14	22	—	—	Commonly agglomerating¶
	2. Medium-volatile bituminous coal	69	78	22	31	—	—	
	3. High-volatile A bituminous coal	—	69	31	—	14,000§	—	
	4. High-volatile B bituminous coal	—	—	—	—	13,000§	14,000	
	5. High-volatile C bituminous coal	—	—	—	—	11,500	13,000	
		—	—	—	—	10,500	11,500	Agglomerating
III. Subbituminous	1. Subbituminous A coal	—	—	—	—	10,500	11,500	Nonagglomerating
	2. Subbituminous B coal	—	—	—	—	9,500	10,500	
	3. Subbituminous C coal	—	—	—	—	8,300	9,500	
IV. Lignitic	1. Lignite A	—	—	—	—	6,300	8,300	
	2. Lignite B	—	—	—	—	—	6,300	

*This classification does not include a few coals, principally nonbanded varieties, which have unusual physical and chemical properties and which come within the limits of fixed carbon or calorific value of the high-volatile bituminous and subbituminous ranks. All these coals either contain less than 48 percent dry, mineral-matter-free fixed carbon or have more than 15,500 moist, mineral-matter-free Btu/lb. To obtain kilojoules per kilogram, multiply by 2.33.

†*Moist* refers to coal containing its natural inherent moisture but not including visible water on the surface of the coal.

‡If agglomerating, classify in low-volatile group of the bituminous class.

§Coals having 69 percent or more fixed carbon on the dry, mineral-matter-free basis shall be classified according to fixed carbon, regardless of calorific value.

¶It is recognized that there may be nonagglomerating varieties in these groups of the bituminous class and that there are notable exceptions in high-volatile C bituminous group.

SOURCE: T. C. Elliott et al., *Standard Handbook of Powerplant Engineering*, McGraw-Hill, New York, 1989. Used with permission of publisher.

109

TABLE 4-10 Typical Ultimate Analyses for Coals

Rank	Btu/lb As Received	Constituents, Percent by Weight					
		Oxygen	Hydrogen	Carbon	Nitrogen	Sulfur	Ash
Anthracite	12,700	5.0	2.9	80.0	0.9	0.7	10.5
Semianthracite	13,600	5.0	3.9	80.4	1.1	1.1	8.5
Low-Volatile Bituminous	14,350	5.0	4.7	81.7	1.4	1.2	6.0
Medium-Volatile Bituminous	14 000	5.0	5.0	81.4	1.4	1.5	6.0
High-Volatile Bituminous A	13,800	9.3	5.3	75.9	1.5	1.5	6.5
High-Volatile Bituminous B	12,500	13.8	5.5	67.8	1.4	3.0	8.5
High-Volatile Bituminous C	11,000	20.6	5.8	59.6	1.1	3.5	9.4
Subbituminous B	9000	29.5	6.2	52.5	1.0	1.0	9.8
Subbituminous C	8500	35.7	6.5	46.4	0.8	1.0	9.6
Lignite	6900	44.0	6.9	40.1	0.7	1.0	7.3

SOURCE: *ASHRAE Handbook—1989 Fundamentals.* Reprinted by permission.

TABLE 4-11   ASTM Specifications Classify Oils According to Grade

Grade	Flash point, °F	Pour point, °F	Water and sediment, vol %	Max ash, wt %	Carbon residue on 10% bottoms, %	Saybolt viscosity, s				Kinematic viscosity, cSt				Minimum gravity, deg API	Sulfur, %
						Universal at 100 °F		Furol at 122 °F		at 100 °F		at 122 °F			
						Min	Max	Min	Max	Min	Max	Min	Max		
1	100ᵃ	0ᶜ	Trace	—	0.15	—	—	—	—	1.4	2.2	—	—	35	0.5ᵃ,ᵉ
2	100ᵃ	20ᶜ	0.05	—	0.35	(32.6)ᵇ	(37.9)—	—	—	2.0ᶜ	3.6	—	—	30	0.5ᵃ,ᵉ
4	130ᵃ	20ᶜ	0.5	0.1	—	45	125	—	—	(5.8)	(26.4)ᶠ	—	—	—	Legal
5 light	130ᵃ	—	1	0.1	—	150	300	—	—	(32)	(65)ᶠ	—	—	—	Legal
5 heavy	130ᵃ	—	1	0.1	—	350	750	(23)	(40)	(75)	(162)ᶠ	(42)	(81)	—	Legal
6	150	—ᵈ	2ᵍ	—	—	(900)	(9000)	45	300	—	—	(92)	(638)ᶠ	—	Legal

ᵃOr legal.

ᵇViscosity values in parentheses are for information only and not necessarily limiting.

ᶜLower or higher pour points may be specified whenever required by conditions of storage or use. When pour point less than 0°F is specified, the minimum viscosity for grade no. 2 shall be 1.8 cSt (32.0 s Saybolt Universal), and the minimum 90% point shall be waived.

ᵈWhere low-sulfur fuel oil is required, grade no. 6 fuel oil will be classified as low-pour (60°F max) or high-pour (no max). Low-pour fuel oil should be used unless all tanks and lines are heated.

ᵉIn countries outside the United States, other sulfur limits may apply.

ᶠWhere low-sulfur fuel oil is required, fuel oil falling in the viscosity range of a lower-numbered grade, down to and including no. 4, may be supplied by agreement between purchaser and supplier. The viscosity range of the initial shipment shall be identified, and advance notice shall be required when changing from one viscosity range to another. This notice shall be in sufficient time to permit the user to make the necessary adjustments.

ᵍThe amount of water by distillation plus the sediment by extraction shall not exceed 2%. The amount of sediment by extraction shall not exceed 0.5%. A deduction in quantity shall be made for all water and sediment in excess of 1%. Note: Data outlined in this table come from the American Society for Testing & Materials' Standard D 396, Standard Specification for Fuel Oils. When using this table, bear in mind that the failure to meet any requirement of a given grade does not automatically place an oil in the next lower grade unless it meets all requirements of the lower grade.

°C = (°F − 32)/1.8.

SOURCE: T. C. Elliott et al., *Standard Handbook of Powerplant Engineering*, McGraw-Hill, New York, 1989. Used with permission of publisher.

TABLE 4-12   Heating Values of Fuel Oils vs. API Gravity at 60°F (15.6°C)

Deg API	Heating value			
	Gross		Net	
	Btu/lb*	Btu/gal†	Btu/lb*	Btu/gal†
5	17,980	155,470	16,990	146,860
10	18,260	152,280	17,270	144,000
15	18,500	149,030	17,480	140,750
20	18,740	145,880	17,660	137,510
25	18,940	142,820	17,830	134,350
30	19,130	139,660	17,980	131,300
35	19,300	136,720	18,110	128,350
40	19,450	133,760	18,230	125,390
45	19,590	130,910	18,340	122,530

*kJ/kg = Btu/lb × 2.33.

†kJ/L = Btu/gal × 0.279.

SOURCE: T. C. Elliott et al., *Standard Handbook of Powerplant Engineering*, McGraw-Hill, New York, 1989. Used with permission of publisher.

TABLE 4-13   Heat Gain from Typical Electric Motors

Motor Name-plate or Rated Horse-power	Motor Type	Nom-inal rpm	Full Load Motor Effici-ency in Percent	Location of Motor and Driven Equipment with Respect Conditioned Space or Airstream		
				A	B	C
				Motor in, Driven Equip-ment in Btu/h	Motor out, Driven Equip-ment in Btu/h	Motor Driven Equip-ment out Btu/h
0.05	Shaded Pole	1500	35	360	130	240
0.08	Shaded Pole	1500	35	580	200	380
0.125	Shaded Pole	1500	35	900	320	590
0.16	Shaded Pole	1500	35	1160	400	760
0.25	Split Phase	1750	54	1180	640	540
0.33	Split Phase	1750	56	1500	840	660
0.50	Split Phase	1750	60	2120	1270	850
0.75	3-Phase	1750	72	2650	1900	740
1	3-Phase	1750	75	3390	2550	850
1	3-Phase	1750	77	4960	3820	1140
2	3-Phase	1750	79	6440	5090	1350
3	3-Phase	1750	81	9430	7640	1790
5	3-Phase	1750	82	15,500	12,700	2790
7.5	3-Phase	1750	84	22,700	19,100	3640
10	3-Phase	1750	85	29,900	24,500	4490
15	3-Phase	1750	86	44,400	38,200	6210
20	3-Phase	1750	87	58,500	50,900	7610
25	3-Phase	1750	88	72,300	63,600	8680
30	3-Phase	1750	89	85,700	76,300	9440
40	3-Phase	1750	89	114,000	102,000	12,600
50	3-Phase	1750	89	143,000	127,000	15,700
60	3-Phase	1750	89	172,000	153,000	18,900
75	3-Phase	1750	90	212,000	191,000	21,200
100	3-Phase	1750	90	283,000	255,000	28,300
125	3-Phase	1750	90	353,000	318,000	35,300
150	3-Phase	1750	91	420,000	382,000	37,800
200	3-Phase	1750	91	569,000	509,000	50,300
250	3-Phase	1750	91	699,000	636,000	62,900

SOURCE: *ASHRAE Handbook—1989 Fundamentals.* Reprinted by permission.

TABLE 4-14 Rates of Heat Gain from Occupants of Conditioned spaces[a,b,c]

Degree of Activity	Typical Application	Total Heat Adults, Male, Btu/h	Total Heat Adjusted,[d] Btu/h	Sensible Heat, Btu/h	Latent Heat, Btu/h
Seated at theater	Theater—Matinee	390	330	225	105
Seated at theater	Theater—Evening	390	350	245	105
Seated, very light work	Offices, hotels, apartments	450	400	245	155
Moderately active office work	Offices, hotels, apartments	475	450	250	200
Standing, light work; walking	Department store, retail store	550	450	250	200
Walking; standing	Drug store, bank	550	500	250	250
Sedentary work	Restaurant[e]	490	550	275	275
Light bench work	Factory	800	750	275	475
Moderate dancing	Dance hall	900	850	305	545
Walking 3 mph; light machine work	Factory	1000	1000	375	625
Bowling[e]	Bowling alley	1500	1450	580	870
Heavy work	Factory	1500	1450	580	870
Heavy machine work; lifting	Factory	1600	1600	635	965
Athletics	Gymnasium	2000	1800	710	1090

[a] Tabulated values are based on 75°F room dry-bulb temperature. For 80°F room dry-bulb, the total heat remains the same, but the sensible heat values should be decreased by approximately 20%, and the latent heat values increased accordingly.

[b] Also refer to Table 1, Chapter 8, of source text.

[c] All values are rounded to nearest 5 Btu/h.

[d] *Adjusted heat gain* is based on normal percentage of men, women, and children for the application listed, with the postulate that the gain from an adult female is 85% of that for an adult male, and that the gain from a child is 75% of that for an adult male.

[e] Adjusted total heat gain for *Sedentary work, Restaurant,* includes 60 Btu/h for food per individual (30 Btu/h sensible and 30 Btu/h latent).

[f] For *Bowling,* figure one person per alley actually bowling, and all others as sitting (400 Btu/h) or standing and walking slowly (550 Btu/h).

SOURCE: *ASHRAE Handbook—1989 Fundamentals.* Reprinted by permission.

TABLE 4-15 Analyses of Refuse Components, Percentage by Weight

Refuse item	Component wgt, %	C	H	O	N	Cl	S	Water	Ash
Corrugated boxboard	5	36.79	5.08	35.41	0.11	0.12	0.23	20.00	2.26
Newspaper	12	36.62	4.66	31.76	0.11	0.11	0.19	25.00	1.55
Magazines	3	32.93	4.64	32.85	0.11	0.13	0.21	16.00	13.13
All other paper	23	32.41	4.51	29.91	0.31	0.61	0.19	23.00	9.06
Plastics	3	56.43	7.79	8.05	0.85	3.00	0.29	15.00	8.59
Rubber, leather	2	43.09	5.37	11.57	1.34	4.97	1.17	10.00	22.49
Wood	3	41.22	5.03	34.55	0.24	0.09	0.07	16.00	2.82
Textiles	3	37.23	5.02	27.11	3.11	0.27	0.28	25.00	1.98
Yard trimmings	10	23.29	2.93	17.54	0.89	0.13	0.15	45.00	10.07
Food wastes	10	17.93	2.55	12.85	1.13	0.38	0.06	60.00	5.10
Fines, −1 in (−25 mm)	10	15.03	1.91	12.15	0.50	0.36	0.15	25.00	44.90
Metallic	7								
Glass, ceramics, etc.	9								
	100								

SOURCE: T. C. Elliott et al., *Standard Handbook of Powerplant Engineering*, McGraw-Hill, New York, 1989. Used with permission of publisher.

TABLE 4-16 Ordinary Ranges of Overall Coefficients of Heat Transfer

Type of heat exchanger	State of controlling resistance		Typical fluid	Typical apparatus
	Free convection, U	Forced convection, U		
Liquid to liquid	25–60 [141.9–340.7]	150–300 [851.7–1703.4]	Water	Liquid-to-liquid heat exchangers
Liquid to liquid	5–10 [28.4–56.8]	20–50 [113.6–283.9]	Oil	
Liquid to gas*	1–3 [5.7–17.0]	2–10 [11.4–56.8]	—	Hot-water radiators
Liquid to boiling liquid	20–60 [113.6–340.7]	50–150 [283.9–851.7]	Water	Brine coolers
Liquid to boiling liquid	5–20 [28.4–113.6]	25–60 [141.9–340.7]	Oil	
Gas* to liquid	1–3 [5.7–17.0]	2–10 [11.4–56.8]	—	Air coolers, economizers
Gas* to gas	0.6–2 [3.4–11.4]	2–6 [11.4–34.1]	—	Steam superheaters
Gas* to boiling liquid	1–3 [5.7–17.0]	2–10 [11.4–56.8]	—	Steam boilers
Condensing vapor to liquid	50–200 [283.9–1136]	150–800 [851.7–4542.4]	Steam to water	Liquid heaters and condensers
Condensing vapor to liquid	10–30 [56.8–170.3]	20–60 [113.6–340.7]	Steam to oil	
Condensing vapor to liquid	40–80 [227.1–454.2]	60–150 [340.7–851.7]	Organic vapor to water	
Condensing vapor to liquid	—	15–300 [85.2–1703.4]	Steam-gas mixture	
Condensing vapor to gas*	1–2 [5.7–11.4]	2–10 [11.4–56.8]	—	Steam pipes in air, air heaters
Condensing vapor to boiling liquid	40–100 [227.1–567.8]	—	Steam to water	Scale-forming evaporators
Condensing vapor to boiling liquid	300–800 [1703.4–4542.4]	—	—	
Condensing vapor to boiling liquid	50–150 [283.9–851.7]	—	Steam to oil	

*At atmospheric pressure.

Note: $U = $ Btu/(h·ft^2·°F) [W/(m^2·°C)]. Under many conditions, either higher or lower values may be realized.

SOURCE: T. G. Hicks, *Power Generation Calculations Reference Guide*, McGraw-Hill, New York, 1987. Used with permission of publisher.

TABLE 4-17   Heat Gain Factors of Typical Appliances under Hoods

Appliance	Usage Factor F_{UA}	Radiation Factor F_{RA}	Load Factor $F_L = F_u F_R$ Elec or Steam	Load Factor $F_L = F_u F_R/F_{FL}$ Gas
Griddle	0.10	0.35	0.04	0.025
Hot-top range				
Without oven	0.79	0.47	0.37	0.231
With oven	0.59	0.48	0.28	0.175
Convection oven	0.14	0.30	0.04	0.025
Broiler	0.55	0.36	0.20	0.125
Charbroiler	0.50	0.15	0.08	0.050
Fryer	0.03	0.30	0.01	0.056
Steam cooker	0.13	0.30	0.04	0.025

SOURCE: *ASHRAE Handbook—1989 Fundamentals.* Reprinted by permission.

4-1   Solar Angles and Sun-Path Diagrams*

Solar angles

The position of the sun in relation to specific geographic locations, seasons, and times of day can be determined by several different methods. Model measurements, by means of sun machines or shade dials, have the advantage of direct visual observations. Tabulative and calculative methods have the advantage of exactness. However, graphic projection methods are usually preferred by architects as they are easily understood and can be correlated to both radiant energy and shading calculations.

Sun-path diagrams

The most practical graphic projection is the sun-path diagram method. Such diagrams depict the path of the sun within the skyvault as projected onto a horizontal plane. The horizon is represented as a circle with the observation point in the center. The sun's position at any date and hour

---

* Text, table, and illustrations in this subsection adapted from Joseph N. Boaz (ed.), Charles G. Ramsey's and Harold R. Sleeper's *Architectural Graphic Standards,* 6th ed., Wiley, New York, 1970. Reprinted by permission of the publisher.

can be determined from the diagram in terms of its altitude (α) and bearing angle (β). (See Fig. 4-2.) The graphs are constructed in equidistant projection. The altitude angles are represented at 10° intervals by equally spaced concentric circles; they range from 0° at the outer circle (horizon) to 90° at the center point. These intervals are graduated along the south meridian. Bearing angles are represented at 10° intervals by equally spaced radii; they range from 0° at the south meridian to 180° at the north meridian. These intervals are graduated along the periphery. The sun's bearing will be to the east during morning hours, and to the west during afternoon hours.

The earth's axis is inclined 23°27′ to its orbit around the sun and rotates 15° hourly. Thus, from all points on the earth, the sun appears to move across the skyvault on various parallel circular paths with maximum declinations of ±23°27′. The declination of the sun's path changes in a cycle between the extremes of the summer solstice and winter solstice. (See Table 4-18.) Thus, the sun follows the same path on two corresponding dates each year. Due to irregularities between the calendar year and the astronomical data, here a unified calibration is adapted. The differences, as they do not exceed 41′, are negligible for architectural purposes.

The ellipitical curves in Fig. 4-3 represent the horizontal projections of the sun's path. They are given on the 21st day

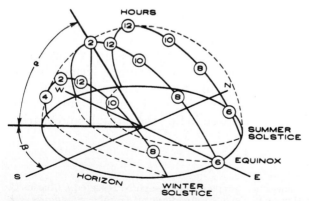

FIGURE 4-2   Sun-path diagram method.

TABLE 4-18   Declination of the Sun

Date	Declination	Corresp. Date	Declination	Unified Calibr.
June 21	+23°27'			+23°27'
May 21	+20°09'	July 21	+20°31'	+20°20'
Apr. 21	+11°48'	Aug. 21	+12°12'	+12°00'
Mar. 21	+ 0°10'	Sep. 21	+ 0°47'	+ 0°28'
Feb. 21	−10°37'	Oct. 21	−10°38'	−10°38'
Jan. 21	−19°57'	Nov. 21	−19°53'	−19°55'
Dec. 21	−23°27'			−23°27'

of each month. Roman numerals designate the months. A cross grid of curves graduates the hours indicated in arabic numerals. Eight sun-path diagrams are shown at 4° intervals from 24° N to 52° N latitude.

Example. Find the sun's position in Columbus, Ohio, on February 21st at 2 P.M.

Step 1. Locate Columbus on the map. The latitude is 40° N.

Step 2. In the 40° sun-path diagram select the February path (marked with II), and locate the 2-hour line. Where the two lines cross is the position of the sun.

Step 3. Read the altitude on the concentric circles (32°) and the bearing angle along the outer circle (35°30'W).

4-2   Calculation of Solar Position*

One can calculate accurately the solar position to any locale and time by relating the spherical triangle formed by the observer's celestial meridian, the meridian of the sun, and the great circle passing through zenith and the sun. The following formulas can be used to find the altitude and bearing angles:

---

* Text in this subsection adapted from Joseph N. Boaz (ed.), Charles G. Ramsey's and Harold R. Sleeper's *Architectural Graphic Standards,* 6th ed., Wiley, New York, 1970. Reprinted by permission of the publisher.

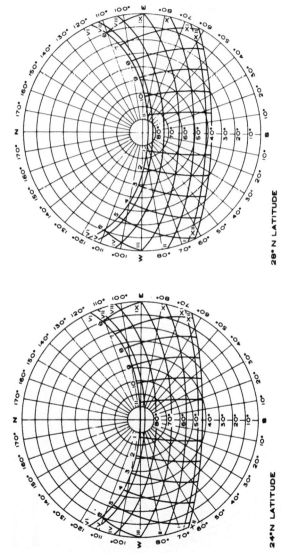

FIGURE 4-3 Sun-path diagrams.

119

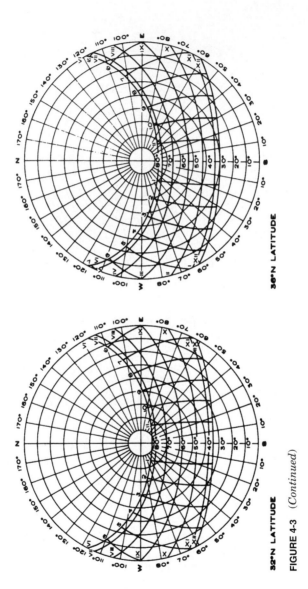

36°N LATITUDE

32°N LATITUDE

FIGURE 4-3 (*Continued*)

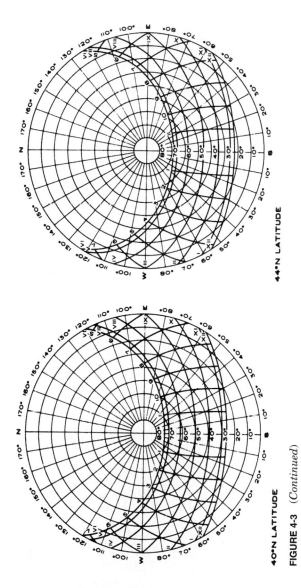

40°N LATITUDE

44°N LATITUDE

FIGURE 4-3 (*Continued*)

121

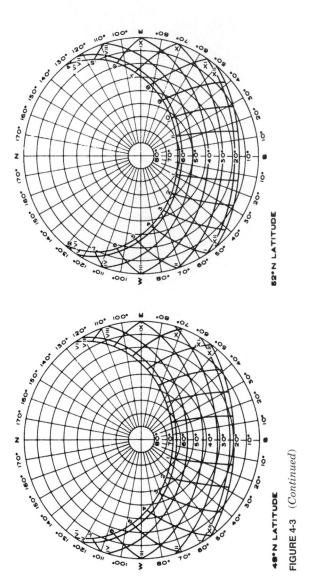

52°N LATITUDE

48°N LATITUDE (*Continued*)

FIGURE 4-3

122

$$\sin A = \sin d \, \sin l + \cos d \, \cos l \, \cos t \qquad (1)$$

$$\cos B = \frac{\sin d \, \cos l - \cos d \, \sin l \, \cos t}{\cos A} \qquad (2)$$

where A = altitude of the sun in degrees, measured from the horizontal.

d = declination of the sun (see page on solar angles) at the desired date. (North declinations are conventionally positive; south declinations negative.)

l = latitude of the locale; conventionally negative in the southern hemisphere.

t = hour angle of the sun in degrees, measured counter-clockwise from north towards east. At solar noon it is zero and changes 15° per hour.

B = bearing angle of the sun in degrees; here measured clockwise from north towards east.

4-3   Calculation of Solar Radiation*

To evaluate the importance of solar shading one has to know the amount of solar energy falling on the exposed surface. As the primary protection of the shading devices is from the direct solar radiation, only these energy calculations are described here.

The magnitude of solar radiation depends, first of all, on the sun's altitude. The tabulated values indicate direct radiation energies received under clear atmospheric conditions at normal incidence (ID):

Solar altitude (deg)	5	10	15	20	25	30	35	40	45	50	60	70	80	90
Btu/ft^2/h	67	123	166	197	218	235	248	258	266	273	283	289	292	294

The energy received on a surface depends also on the cosine of the angle of incidence. As this is a spacial angle,

---

* Text and tables in this subsection adapted from Joseph N. Boaz (ed.), Charles G. Ramsey's and Harold R. Sleeper's *Architectural Graphic Standards,* 6th ed., Wiley, New York, 1970. Reprinted by permission of the publisher.

it is conventional on vertical surfaces to substitute it with the functions of the altitude ($\angle A$) and the azimuth ($\angle a$) angles related to the normal of the surface in question. Thus, the direct radiation on vertical surfaces (R) can be defined as:

$$R = I_D \times \cos A \cos a$$

For horizontal surfaces the received direct radiation energy will be:

$$R = I_D \times \sin A$$

In Table 4-19 calculated values of solar position in degrees, and direct radiation energies in Btu/sq ft/hour values, are shown at different orientations. Table 4-19 indicates from 24°N to 46°N latitude at 4° intervals.

4-4   Radiation Calculator*

Radiation calculations can be performed by graphical means. The upper half of the direct radiation calculator Fig. 4-4 charts the energies falling on a horizontal plain under clear sky conditions. The equi-intensity radiation lines are indicated at 25-Btu/sq ft/hour intervals. The lower half circle shows the amount of direct radiation falling on a vertical surface. The calculator can be used at any latitude and at any orientation. The calculator is in the same scale and projection as the sun-path diagrams shown in the material on solar angles. Transfer the calculator diagram to a transparent overlay, and superimpose it on a sun-path diagram in the desired orientation: the radiation values can be read directly.

---

* Text and illustration in this subsection adapted from Joseph N. Boaz (ed.), Charles G. Ramsey's and Harold R. Sleeper's *Architectural Graphic Standards,* 6th ed., Wiley, New York, 1970. Reprinted by permission of the publisher.

TABLE 4-19 Solar Position Values and Direct Radiation Energies

26° N. LATITUDE

JUNE 22

AM →	ALT.	BEAR.	S	SE	E	NE	N	SW	HOR.
6 a.m.	10.05	111.30		49	113	111	44		22
7	22.82	105.97		93	185	168	53		81
8	35.93	101.15		113	199	168	39		147
9	49.24	96.45		113	176	139	20		206
10	62.69	88.83	3	94	131	90			253
11	76.15	82.61	9	55	69	43			282
12	87.45	0.00	13	9				9	293
PM →	α	β	S	SW	W	NW	N	SE	HOR.

MARCH 21, SEPT. 24

AM →	ALT.	BEAR.	S	SE	E	NE	N	SW	HOR.
6 a.m.	0.00	90.00							
7	13.45	83.30	17	117	147	92			36
8	26.70	75.80	49	172	194	102			101
9	39.46	66.33	80	185	182	72			163
10	51.11	52.79	104	171	137	23			213
11	60.25	31.44	120	137	73				246
12	64.00	0.00	125	88					257
PM →	α	β	S	SW	W	NW	N	SE	HOR.

DECEMBER 22

AM →	ALT.	BEAR.	S	SE	E	NE	N	SW	HOR.
7 a.m.	2.23	62.48	14	29	27	9			1
8	13.76	54.88	87	149	123	26			37
9	24.12	45.30	138	196	139	1			88
10	32.66	33.01	171	199	111				131
11	38.46	17.65	190	177	61				159
12	45.55	0.00	197	139					168
PM →	α	β	S	SW	W	NW	N	SE	HOR.

30° N. LATITUDE

JUNE 22

AM →	ALT.	BEAR.	S	SE	E	NE	N	SW	HOR.
6 a.m.	11.48	110.59		55	124	121	47		27
7	23.87	104.30		100	189	168	48		86
8	36.60	98.24		121	200	162	29		150
9	49.53	91.79		121	177	129	6		207
10	62.50	83.46	15	103	131	82			252
11	75.11	67.48	29	69	69	29			281
12	83.45	0.00	33	24				24	291
PM →	α	β	S	SW	W	NW	N	SE	HOR.

MARCH 21, SEPT. 24

AM →	ALT.	BEAR.	S	SE	E	NE	N	SW	HOR.
6 a.m.	0.00	90.00							
7	12.95	82.37	19	115	143	88			33
8	25.66	73.90	55	174	181	96			95
9	37.76	63.44	90	190	176	63			155
10	48.59	49.11	117	179	136	13			203
11	56.77	28.19	135	147	72			44	234
12	60.00	0.00	142	100				100	245
PM →	α	β	S	SW	W	NW	N	SE	HOR.

DECEMBER 22

AM →	ALT.	BEAR.	S	SE	E	NE	N	SW	HOR.
7 a.m.	0.38	62.40	2	5	5	2			1
8	11.44	54.15	78	131	133	21			27
9	21.27	44.12	135	189	131	3			73
10	29.28	31.73	173	197	137			47	114
11	34.64	16.77	195	179	59			96	140
12	36.55	0.00	202	143				143	150
PM →	α	β	S	SW	W	NW	N	SE	HOR.

TABLE 4-19 Solar Position Values and Direct Radiation Energies (Continued)

34° N. LATITUDE

JUNE 22

AM	ALT.	BEAR.	S	SE	E	NE	N	SW	HOR.
5 a.m. 7 p.m.	1.47	117.57		6	17	19	9		1
6 6	12.86	109.78		61	135	130	49		33
7 5	24.80	102.54		106	192	166	43		91
8 4	37.07	95.28		129	200	155	19		152
9 3	49.49	87.10	9	131	177	119			207
10 2	61.79	76.00	32	115	130	69			250
11 1	73.17	55.11	48	83	69	15			278
12	79.45	0.00	53	38				38	287
PM	α	β	S	SE	W	NW	N	SE	HOR.

MARCH 21, SEPT. 24

AM	ALT.	BEAR.	S	SE	E	NE	N	SW	HOR.
6 a.m. 6 p.m.	0.00	90.00							
7 5	12.39	81.48	21	113	139	83			31
8 4	24.49	72.11	60	175	187	90			89
9 3	35.89	60.79	99	195	177	55			146
10 2	45.89	45.92	129	186	134	3			192
11 1	53.21	25.60	149	156	71			55	221
12	56.00	0.00	156	110				110	231
PM	α	β	S	SE	W	NW	N	SE	HOR.

DECEMBER 22

AM	ALT.	BEAR.	S	SE	E	NE	N	SW	HOR.
8 a.m. 4 p.m.	9.08	53.57	66	110	90	17			18
9 3	18.38	43.12	129	177	121			6	59
10 2	25.86	30.65	171	193	101			49	96
11 1	30.81	16.05	196	178	56			99	121
12	32.55	0.00	204	144				144	130
PM	α	β	S	SE	W	NW	N	SE	HOR.

38° N. LATITUDE

JUNE 22

AM	ALT.	BEAR.	S	SE	E	NE	N	SW	HOR.
5 a.m. 7 p.m.	3.32	118.42		13	39	42	20		3
6 6	14.18	108.87		68	146	138	50		39
7 5	25.60	100.70		112	195	164	37		95
8 4	37.33	92.25		136	201	148	8		153
9 3	49.13	82.47	23	141	176	108			206
10 2	60.58	69.06	50	127	130	57			246
11 1	70.61	45.67	67	96	69	1			273
12	75.45	0.00	73	52				52	281
PM	α	β	S	SE	W	NW	N	SE	HOR.

MARCH 21, SEPT. 24

AM	ALT.	BEAR.	S	SE	E	NE	N	SW	HOR.
6 a.m. 6 p.m.	0.00	90.00							
7 5	11.77	80.63	22	110	134	79			28
8 4	23.20	70.43	65	175	182	83			83
9 3	33.86	58.38	107	198	173	47			137
10 2	43.03	42.16	140	192	131			6	179
11 1	49.57	23.52	162	164	71			65	207
12	52.00	0.00	169	120				120	217
PM	α	β	S	SE	W	NW	N	SE	HOR.

DECEMBER 22

AM	ALT.	BEAR.	S	SE	E	NE	N	SW	HOR.
8 a.m. 4 p.m.	6.69	53.12	51	84	68	12			10
9 3	15.44	42.30	120	162	109			8	45
10 2	22.40	29.75	166	185	95			50	79
11 1	26.96	15.45	193	174	53			99	102
12	28.55	0.00	202	143				143	110
PM	α	β	S	SE	W	NW	N	SE	HOR.

TABLE 4-19 Solar Position Values and Direct Radiation Energies (Continued)

42° N. LATITUDE

JUNE 22

AM	PM	ALT.	BEAR.	BTU/Sq. ft./Hour S	SE	E	NE	N	SW	HOR.
5 a.m.	7 p.m.	5.15	117.16		21	61	65	31		6
6	6	15.44	107.87		74	155	145	50		45
7	5	26.28	98.78		118	197	161	30		98
8	4	37.38	89.19	3	144	201	140			153
9	3	48.45	77.96	37	151	176	98			203
10	2	58.95	62.79	67	138	129	44			242
11	1	67.64	38.62	85	109	68			12	266
12		71.45	0.00	92	65				65	274
	PM	α	β	S	SW	W	NW	N	SE	HOR.

MARCH 21, SEPT. 24

AM	PM	ALT.	BEAR.	BTU/Sq. ft./Hour S	SE	E	NE	N	SW	HOR.
6 a.m.	6 p.m.	0.00	90.00							
7	5	11.09	79.84	23	107	128	74		25	25
8	4	21.81	68.88	68	174	177	77		76	76
9	3	31.70	57.81	113	200	169	40		126	126
10	2	40.06	40.79	150	197	129			15	166
11	1	45.88	21.82	173	171	69			73	192
12		48.00	0.00	181	128				128	201
	PM	α	β	S	SW	W	NW	N	SE	HOR.

DECEMBER 22

AM	PM	ALT.	BEAR.	BTU/Sq. ft./Hour S	SE	E	NE	N	SW	HOR.
8 a.m.	4 p.m.	4.28	52.82	35	57	46	8		4	4
9	3	12.46	41.63	105	141	94			8	31
10	2	18.91	29.01	157	173	87			50	62
11	1	23.09	14.96	187	167	50			97	82
12		24.55	0.00	197	139				139	90
	PM	α	β	S	SW	W	NW	N	SE	HOR.

46° N. LATITUDE

JUNE 22

AM	PM	ALT.	BEAR.	BTU/Sq. ft./Hour S	SE	E	NE	N	SW	HOR.
5 a.m.	7 p.m.	6.97	116.78		28	79	84	40		11
6	6	16.63	106.77		80	162	149	49		50
7	5	26.82	96.80		124	199	157	24		101
8	4	37.22	86.15	14	151	201	132			153
9	3	47.47	73.66	51	160	175	87			199
10	2	56.95	57.25	83	149	128	32			235
11	1	64.40	33.33	103	121	68			25	258
12		67.45	0.00	110	78				78	265
	PM	α	β	S	SW	W	NW	N	SE	HOR.

MARCH 21, SEPT. 24

AM	PM	ALT.	BEAR.	BTU/Sq. ft./Hour S	SE	E	NE	N	SW	HOR.
6 a.m.	6 p.m.	0.00	90.00							
7	5	10.36	79.09	23	103	122	70		23	23
8	4	20.32	67.45	71	172	172	71		69	69
9	3	29.42	54.27	119	200	165	33		114	114
10	2	36.98	38.75	157	200	126			22	152
11	1	42.14	20.43	182	176	68			81	175
12		44.00	0.00	190	134				134	184
	PM	α	β	S	SW	W	NW	N	SE	HOR.

DECEMBER 22

AM	PM	ALT.	BEAR.	BTU/Sq. ft./Hour S	SE	E	NE	N	SW	HOR.
8 a.m.	4 p.m.	1.36	52.65	15	25	20	3		1	1
9	3	9.46	41.12	87	115	76			8	19
10	2	15.41	28.41	143	156	77			46	45
11	1	19.23	14.56	176	156	46			92	63
12		20.55	0.00	187	132				132	70
	PM	α	β	S	SW	W	NW	N	SE	HOR.

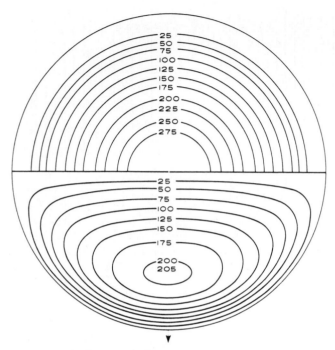

FIGURE 4-4   Radiation calculator.

4-5   Shading: Devices, Masks, Protractor*

Shading devices

The effect of shading devices can be plotted in the same manner as the sun-path was projected. The diagrams show which part of the skyvault will be obstructed by the devices and are projections of the surface covered on the skyvault as seen from an observation point at the center of the diagram. These projections also represent those parts of the skyvault from which no sunlight will reach the observation point; if the sun passes through such an area, the observation point will be shaded.

* Text and illustration in this subsection adapted from Joseph N. Boaz (ed.), Charles G. Ramsey's and Harold R. Sleeper's *Architectural Graphic Standards,* 6th ed., Wiley, New York, 1970. Reprinted by permission of the publisher.

Shading masks

Any building element will define a characteristic form in these projection diagrams, known as "shading masks." Masks of horizontal devices (overhangs) will create a segmental pattern; vertical intercepting elements (fins) produce a radial pattern; shading devices with horizontal and vertical members (eggcrate type) will make a combinative pattern. A shading mask can be drawn for any shading device, even for very complex ones, by geometric plotting. As the shading masks are geometric projections, they are independent of latitude and exposed directions; therefore they can be used in any location and at any orientation. By overlaying a shading mask in the proper orientation on the sun-path diagram, one can read off the times when the sun rays will be intercepted. Masks can be drawn for full shade (100% mask) when the observation point is at the lowest point of the surface needing shading; or for 50% shading when the observation point is placed at the halfway mark on the surface. It is customary to design a shading device in such a way that as soon as shading is needed on a surface the masking angle should exceed 50%. Solar calculations should be used to check the specific loads. Basic shading devices are shown in Fig. 4-5, with their obstruction effect on the sky-vault and with their projected shading masks.

Shading mask protractor

In Fig. 4-6 the half of the protractor showing segmental lines is used to plot lines parallel and normal to the observed vertical surface. The half showing bearing and altitude lines is used to plot shading masks of vertical fins or any other obstruction objects. The protractor is in the same projection and scale as the sun-path diagrams (see material on solar angles); therefore it is useful to transfer the protractor to a transparent overlay to read the obstruction effect.

Examples of various types
of shading devices

The illustrations in Fig. 4-7 show a number of basic types of devices, classified as horizontal, vertical, and eggcrate

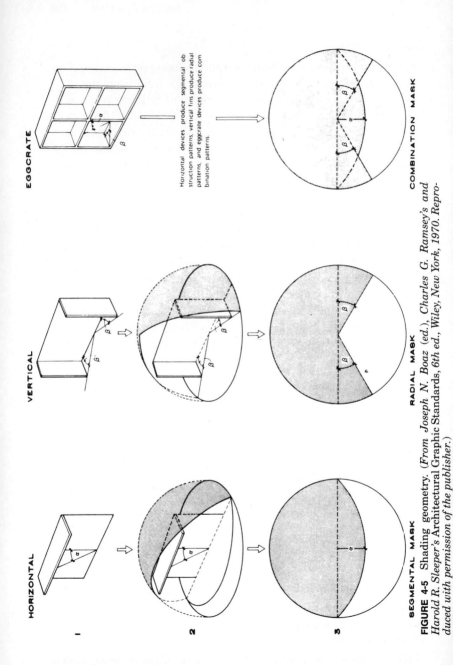

HORIZONTAL   VERTICAL   EGGCRATE

Horizontal devices produce segmental obstruction patterns, vertical fins produce radial patterns, and eggcrate devices produce combination patterns.

SEGMENTAL MASK   RADIAL MASK   COMBINATION MASK

FIGURE 4-5 Shading geometry. (*From Joseph N. Boaz (ed.), Charles G. Ramsey's and Harold R. Sleeper's* Architectural Graphic Standards, *6th ed., Wiley, New York, 1970. Reproduced with permission of the publisher.*)

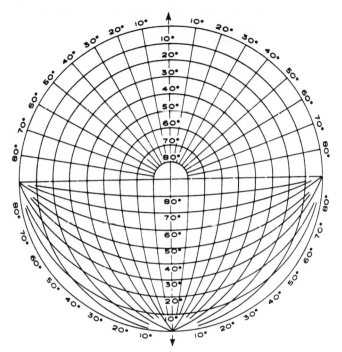

FIGURE 4-6 Shading mask protractor.

types. The dash lines shown in the section diagram in each case indicate the sun angle at the time of 100% shading. The shading mask for each device is also shown, the extent of 100% shading being indicated by the gray area.

General rules can be deduced for the types of shading devices to be used for different orientations. Southerly orientations call for shading devices with segmental mask characteristics, and horizontal devices work in these directions efficiently. For easterly and westerly orientations vertical devices serve well, having radial shading masks. If slanted, they should incline toward the north, to give more protection from the southern positions of the sun. The eggcrate type of shading device works well on walls facing southeast, and is particularly effective for southwest orientations. Because of this type's high shading ratio and low winter heat admission, its best use is in hot climate regions.

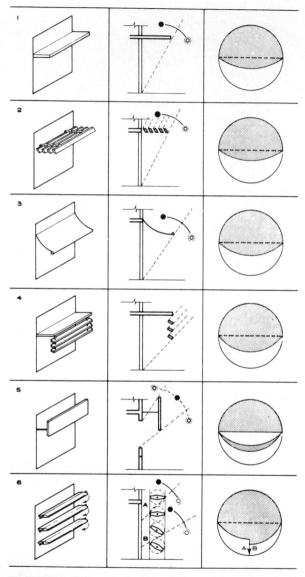

FIGURE 4-7 Types of shading devices. [*From Joseph N. Boaz (ed.), Charles G. Ramsey's and Harold R. Sleeper's* Architectural Graphic Standards, *6th ed., Wiley, New York, 1970. Reproduced with permission of the publisher.*]

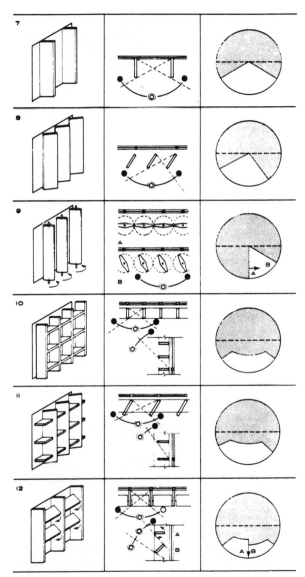

FIGURE 4-7 (*Continued*)

For north walls, fixed vertical devices are recommended; however, their use is needed only for large glass surfaces, or in hot regions. At low latitudes on both south and north exposures eggcrate devices work efficiently.

Whether the shading devices be fixed or movable, the same recommendations apply in respect to the different orientations. The movable types can be most efficiently utilized where the sun's altitude and bearing angles change rapidly: on the east, southeast, and especially, because of the afternoon heat, on the southwest and west.

Horizontal types. (1) Horizontal overhangs are most efficient toward south, or around southern orientations. Their mask characteristics are segmental. (2) Louvers parallel to wall have the advantage of permitting air circulation near the elevation. (3) Slanted louvers will have the same characteristics as solid overhangs, and can be made retractable. (4) When protection is needed for low sun angles, louvers hung from solid horizontal overhangs are efficient. (5) A solid, or perforated screen strip parallel to wall cuts out the lower rays of the sun. (6) Movable horizontal louvers change their segmental mask characteristics according to their positioning.

Vertical types. (7) Vertical fins serve well toward the near east and near west orientations. Their mask characteristics are radial. (8) Vertical fins oblique to wall will result in asymmetrical mask. Separation from wall will prevent heat transmission. (9) Movable fins can shade the whole wall, or open up in different directions according to the sun's position.

Eggcrate types. (10) Eggcrate types are combinations of horizontal and vertical types, and their masks are superimposed diagrams of the two masks. (11) Solid eggcrate with slanting vertical fins results in asymmetrical mask. (12) Eggcrate device with movable horizontal elements shows flexible mask characteristics. Because of their high shading ratio, eggcrates are efficient in hot climates.

4-6   Example of Shading Device Calculation*

The structure in this example is located in New York, N.Y. Two sides of the building (here called north and south) are fully glazed. The two other sides are closed.

Step 1.   Position structure to true orientation (see material on orientation). The long axis of the building lies 15° north of east (see Fig. 4-8).

Step 2.   To evaluate the need for solar control, the amount of solar energy falling on the exposed glass surfaces should be calculated. New York lies nearest to the 40°N latitude. The most penetrating sun angles occur at June 21st. Superimpose over the 40° sun-path diagram (see material on solar angles) the radiation calculator (see material on solar energy calculation), and turning the calculator 15° east of south, read along the June 21st sun-path the hourly radiation impacts. Figure 4-9 shows the Btu/sq ft/hour sun energy values falling on the south side. One can see from it that this surface receives an eight-hour insolation, with energies over 90 Btu/sq ft/hour around 11 A.M. Figure 4-10 shows the sun

---

* Text and illustrations in this subsection adapted from Joseph N. Boaz (ed.), Charles G. Ramsey's and Harold R. Sleeper's *Architectural Graphic Standards,* 6th ed., Wiley, New York, 1970. Reprinted by permission of the publisher.

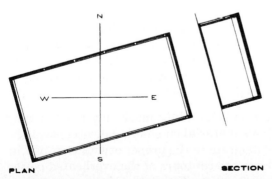

PLAN   **SECTION**

FIGURE 4-8   Example of shading device calculation (step 1).

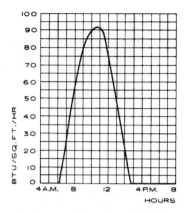

FIGURE 4-9 Example of shading device calculation (step 2).

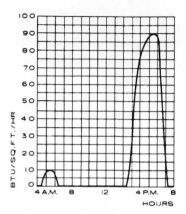

FIGURE 4-10 Example of shading device calculation (step 2 continued).

energies impinging on the north side. One can see from it that the early morning impact is negligible, but around 6 P.M. a considerable amount of energy falls on the surface. Conclusion: both exposed sides should be protected by shading devices.

Step 3. To determine the times when shading is needed: during cool times of the year (called "underheated period") the warming effect of the sun is desirable. During the warm times (called "overheated period") shading is needed to approach comfort conditions. For practical use the 70°F temperature can be accepted as a dividing line between these two conditions.

Figure 4-11 illustrates the New York (40°N latitude) sunpath diagram on which are charted all conditions throughout the year when the temperatures equal or exceed 70°F. In these overheated times, illustrated with the shaded area on the graph, shading will be needed.

Step 4. Construction of shading mask: lay the "shading mask protractor" (see material on shading devices) over the overheated period diagram in the proper orientation, as Fig. 4-12 illustrates. Here the contours of the overheated period are shown by the dotted line. From the shading mask lines one can see that towards the south, devices with segmental

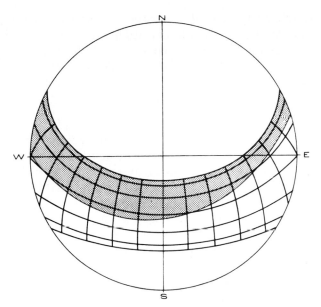

FIGURE 4-11   Example of shading device calculation (step 3).

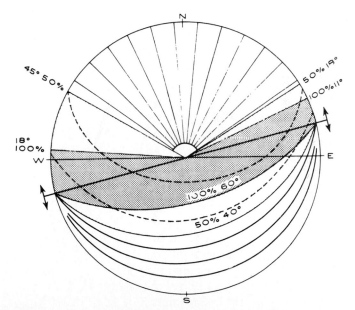

FIGURE 4-12   Example of shading device calculation (step 4).

character (overhang types) will cover conveniently the over-heated period. At the north side the application of devices having radial mask patterns (fin types) will be effective.

In Fig. 4-11 in darker tone is shown the 100% shading effect when the total wall surface is in shadow. In lighter tone is shown the 50% shading effect when only half of the surface will be in shade.

Step 5. Design of shading devices from shading masks; the mask defines the type and the angles of the devices only, and possibilities remain for various design arrangements. In the example at the north side vertical fins will serve effectively. As the required angles in the shading mask are different towards the west (18° for 100% and 45° for 50% shading effect) than for the easterly direction (11° for 100% and 18° for 50% shading effect), the device shall be oblique to the wall surface. Figure 4-13 illustrates an application for the north side. The necessary shading angle is measured in the plan from the middle of the shading fins; the full shade giving angles are measured from the inside corners of the shading elements.

On the south side one could apply a 60° solid overhang. However, this might be too long to cantilever. Instead, the solution here is a combination of horizontal and vertical ele-

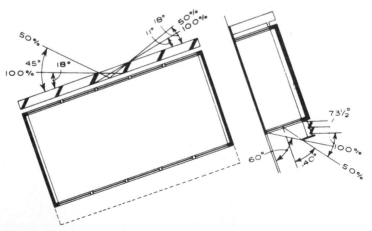

FIGURE 4-13 Example of shading device calculation (step 5).

ments, which corresponds to the same shading mask (see material on Shading Devices). In the section the 50% shading effect is measured from the middle of the glass pane, the 100% shading effect from the bottom. The horizontal part of the shading device could be solid; however, it is constructed here with louver elements to secure ventilation. The critical angle for the louvers is 73½°, to correspond to the sun's highest altitude angle at this latitude. (See Fig. 4-13.)

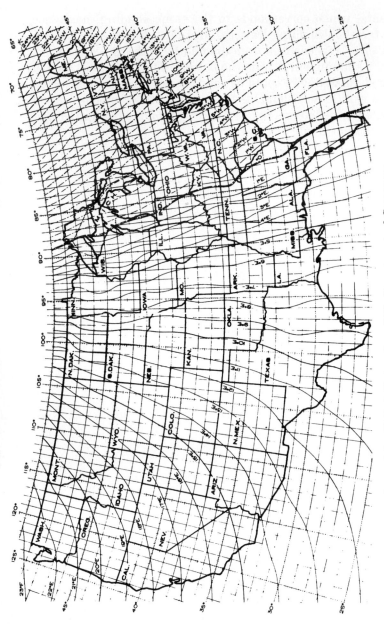

FIGURE 4-14 Isogonic chart of the United States, from U.S. Department of Commerce, Coast and Geodetic Survey, 1965. [From Joseph N. Boaz (ed.), Charles G. Ramsey's and Harold R. Sleeper's Architectural Graphic Standards, 6th ed., Wiley, New York, 1970. Reproduced with permission of the publisher.]

NOTES

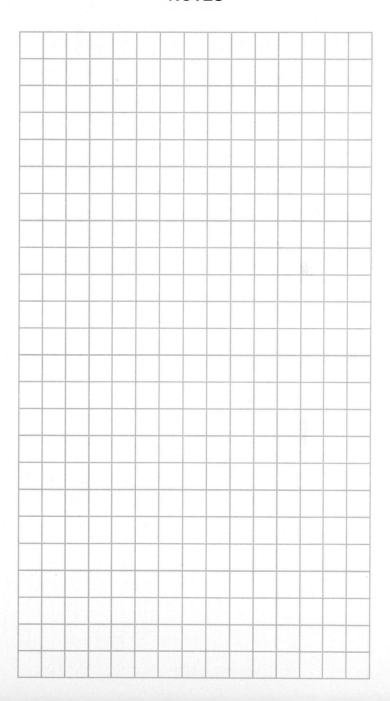

NOTES

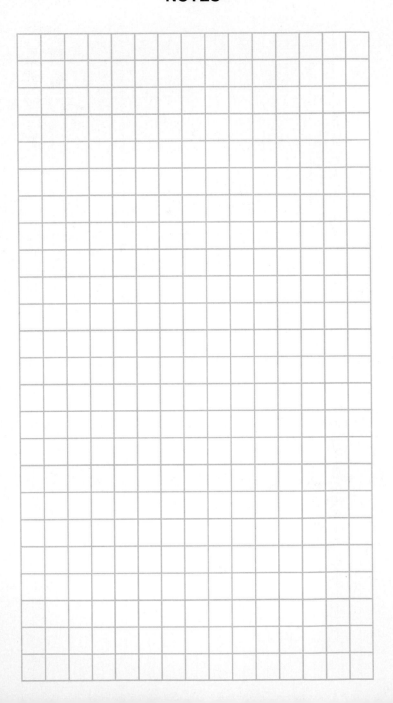

Heating Systems

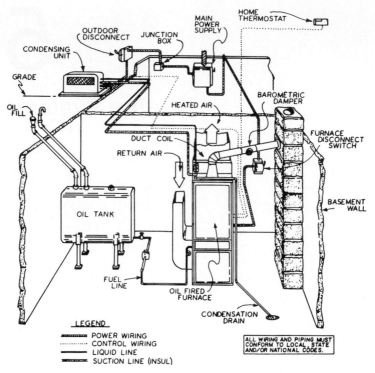

FIGURE 5-1 Typical oil-fired forced-air heating system. (*Source: R. O. Parmley,* HVAC Field Manual, *McGraw-Hill, New York, 1988. Used with permission of publisher.*)

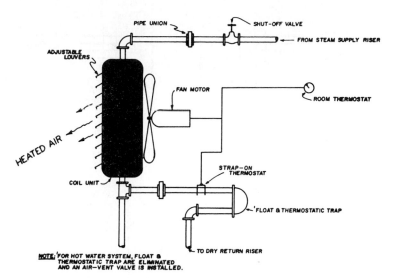

FIGURE 5-2 Typical steam unit heater installation diagram. (*Source: R. O. Parmley,* HVAC Field Manual, *McGraw-Hill, New York, 1988. Used with permission of publisher.*)

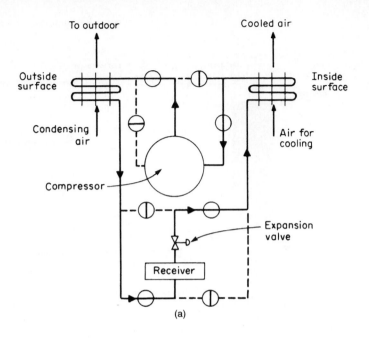

(a)

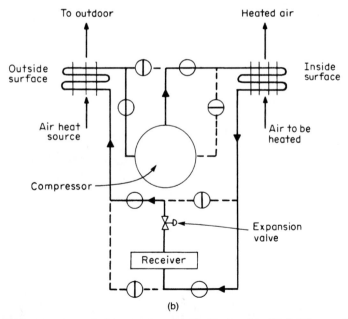

(b)

FIGURE 5-3 Air-to-air heat pump. (*a*) Cooling cycle; (*b*) heating cycle. (*Source: A. M. Khashab,* Heating, Ventilating and Air Conditioning Systems Estimating Manual, *McGraw-Hill, New York, 1977. Used with permission.*)

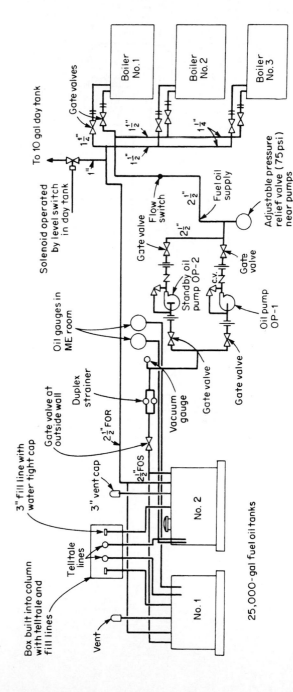

FIGURE 5-4 Boiler fuel-oil diagram. (*Source: A. M. Khashab, Heating, Ventilating and Air Conditioning Systems Estimating Manual, McGraw-Hill, New York, 1977. Used with permission.*)

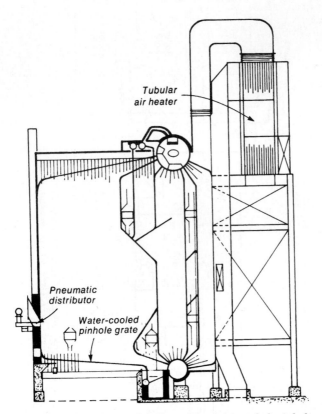

FIGURE 5-5 Field-erected boiler with water-cooled pinhole grate. (*Source: T. C. Elliott et al.,* Standard Handbook of Powerplant Engineering, *McGraw-Hill, New York, 1989. Used with permission of publisher.*)

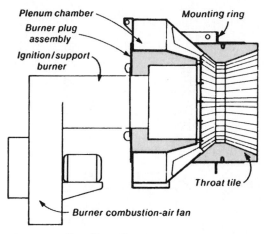

FIGURE 5-6 Standing pilot ensures reignition if fuel flow is interrupted. (*Source: T. C. Elliott et al., Standard Handbook of Powerplant Engineering, McGraw-Hill, New York, 1989. Used with permission of publisher.*)

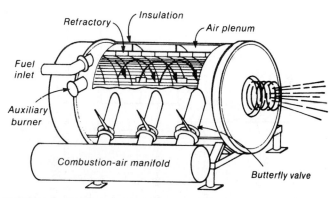

FIGURE 5-7 Cyclonic burner. (*Source: T. C. Elliott et al., Standard Handbook of Powerplant Engineering, McGraw-Hill, New York, 1989. Used with permission of publisher.*)

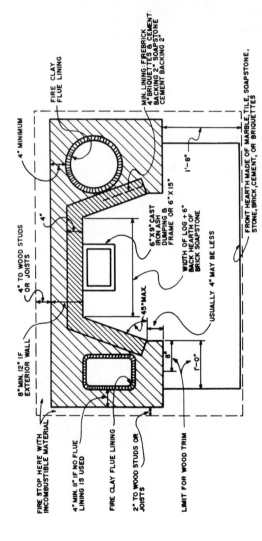

FIGURE 5-8 Plan view of a typical fireplace. (*Source: R. O. Parmley, HVAC Field Manual, McGraw-Hill, New York, 1988. Used with permission of publisher.*)

Labels in figure:

FIRE CLAY FLUE LINING

4" MINIMUM

4" TO WOOD STUDS OR JOISTS

MIN. LINING: FIREBRICK 4" BRIQUETTES & CEMENT BACKING 2" SOAPSTONE CEMENT BACKING 2"

4"

8" MIN. 12" IF EXTERIOR WALL

6"X9" CAST IRON ASH DUMPING & FRAME OR 6"X15"

45° MAX.

WIDTH OF LOG + 6" BACK HEARTH OF BRICK SOAPSTONE

USUALLY 4" MAY BE LESS

1'-8"

FRONT HEARTH MADE OF MARBLE, TILE, SOAPSTONE, STONE, BRICK, CEMENT, OR BRIQUETTES

8"

1'-0"

FIRE STOP HERE WITH INCOMBUSTIBLE MATERIAL

4" MIN. 8" IF NO FLUE LINING IS USED

FIRE CLAY FLUE LINING

2" TO WOOD STUDS OR JOISTS

LIMIT FOR WOOD TRIM

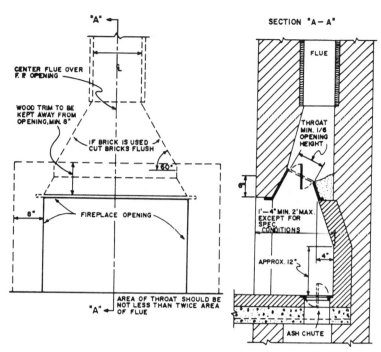

FIGURE 5-9 Front elevation and cross section of a typical fireplace. (*Source: R. O. Parmley,* HVAC Field Manual, *McGraw-Hill, New York, 1988. Used with permission of publisher.*)

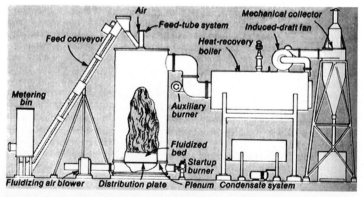

FIGURE 5-10 Fluidized-bed combustor burns hogged wood waste. (*Source: T. C. Elliott et al.,* Standard Handbook of Powerplant Engineering, *McGraw-Hill, New York, 1989. Used with permission of publisher.*)

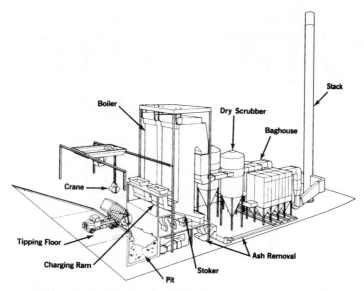

FIGURE 5-11 Typical mass burning refuse-to-energy system. (*Source: T. C. Elliott et al.,* Standard Handbook of Powerplant Engineering, *McGraw-Hill, New York, 1989. Used with permission of publisher.*)

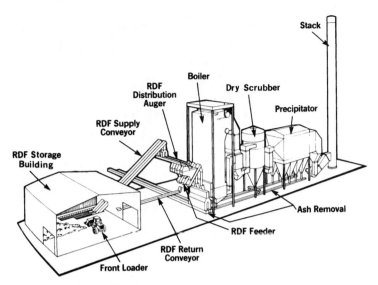

FIGURE 5-12 Typical RDF refuse-to-energy system. (*Source: T. C. Elliott et al.*, Standard Handbook of Powerplant Engineering, *McGraw-Hill, New York, 1989. Used with permission of publisher.*)

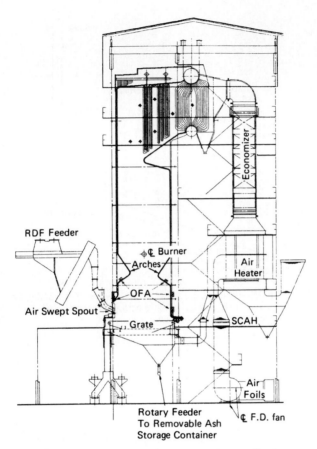

FIGURE 5-13 RDF boiler with arches to boost effectiveness of overfire air system. (*Source: T. C. Elliott et al.,* Standard Handbook of Powerplant Engineering, *McGraw-Hill, New York, 1989. Used with permission of publisher.*)

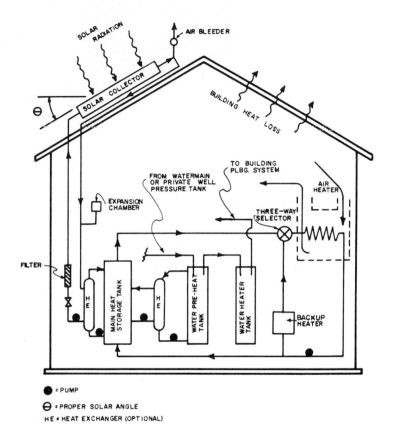

FIGURE 5-14 Schematic diagram of a typical residential solar heat system (water-cooled collector). (*Source: R. O. Parmley,* HVAC Field Manual, *McGraw-Hill, New York, 1988. Used with permission of publisher.*)

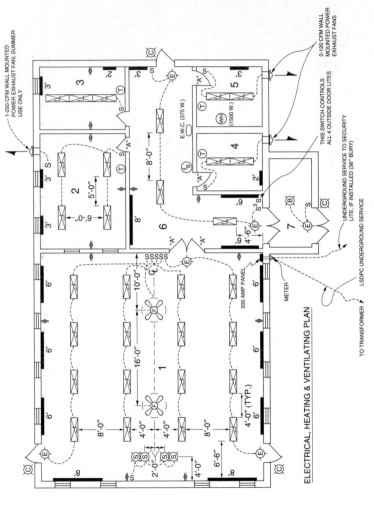

ELECTRICAL, HEATING & VENTILATING PLAN

FIGURE 5-15 Typical layout of an electrically heated building. (*Source: R. O. Parmley, HVAC Field Manual, McGraw-Hill, New York, 1988. Used with permission of publisher.*)

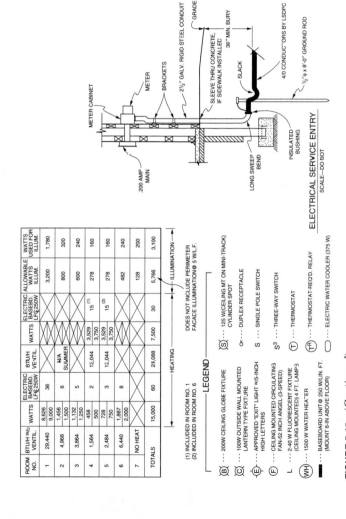

ROOM NO.	BTU/H w/o VENTIL.	WATTS	ELECTRIC BASEBD. LF@250W	BTU/H VE-VTIL.	WATTS	ELECTRIC BASEBD. LF@250W	ALLOWABLE WATTS ILLUM.	WATTS USED FOR ILLUM.
1	29,440	8,626	36	N/A SUMMER			3,200	1,780
		9,000						
2	4,968	1,456	6				800	320
		1,500						
3	3,864	1,132	5				600	240
		1,250						
4	1,564	458	2	12,044	3,529	15 (1)	278	160
		500			3,750			
5	2,484	728	3	12,044	3,529	15 (2)	278	160
		750			3,750			
6	6,440	1,887	8				482	240
		2,000						
7	NO HEAT						128	200
TOTALS		15,000	60	24,088	7,500	30	5,766	3,100

|← HEATING →|← ILLUMINATION →|

(1) INCLUDED IN ROOM NO. 1
(2) INCLUDED IN ROOM NO. 6

DOES NOT INCLUDE PERIMETER
FACADE ILLUMINATION@ 5 W.L.F.

LEGEND

B --- 200W CEILING GLOBE FIXTURE

C --- 100W OUTSIDE WALL MOUNTED
LANTERN TYPE FIXTURE

E --- APPROVED "EXIT" LIGHT W/5-INCH
HIGH LETTERS

F --- CEILING MOUNTED CIRCULATING
FAN-52 INCH ANGEL (3-SPEED)

L --- 2-40 W FLUORESCENT FIXTURE
(CEILING MOUNTED) 4 FT. LAMPS

WH --- 1500 W WATER HEATER

■ --- BASEBOARD UNIT@ 250 W/LIN. FT.
(MOUNT 6-IN ABOVE FLOOR)

S --- 125 W/CEILING MT ON MINI-TRACK)
CYLINDER SPOT

--- DUPLEX RECEPTACLE

S --- SINGLE POLE SWITCH

S³ --- THREE-WAY SWITCH

T --- THERMOSTAT

Tᴿ --- THERMOSTAT-REQ'D. RELAY

--- ELECTRIC WATER COOLER (375 W)

FIGURE 5-15 (*Continued*)

METER CABINET

METER

BRACKETS

2½" GALV. RIGID STEEL CONDUIT

GRADE

SLEEVE THRU CONCRETE,
IF SIDEWALK INSTALLED

36" MIN. BURY

SLACK

4/0 CONDUC"ORS BY LSDPC

200 AMP MAIN

LONG SWEEP BEND

INSULATED BUSHING

½"⌀ x 8'-0" GROUND ROD

ELECTRICAL SERVICE ENTRY

SCALE—DO flOT

NOTES

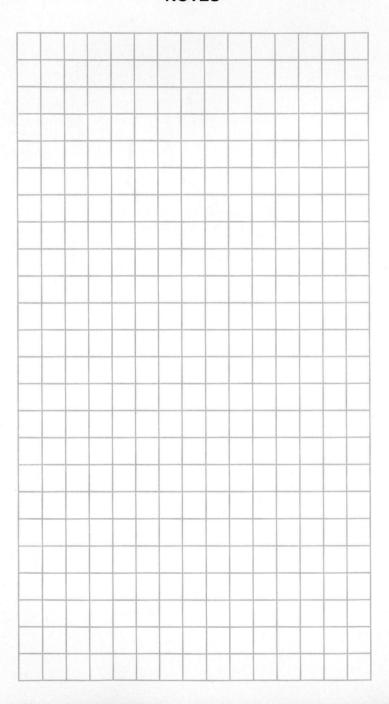

Ventilating Systems

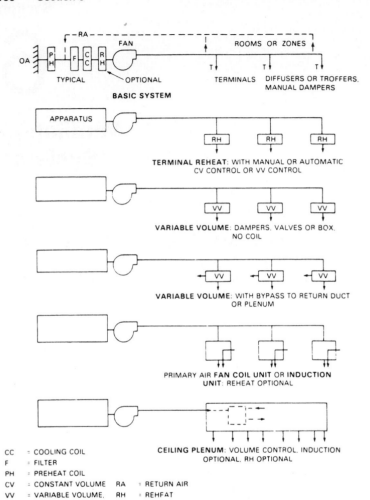

CC = COOLING COIL
F   = FILTER
PH = PREHEAT COIL
CV = CONSTANT VOLUME   RA = RETURN AIR
VV = VARIABLE VOLUME.   RH = REHEAT

FIGURE 6-1   Single-path air systems. (*From* HVAC Duct Construction Standards—Metal and Flexible, *1st ed., © 1985 by the Sheet-Metal and Air Conditioning Contractor's National Association. Used with permission of the copyright holder.*)

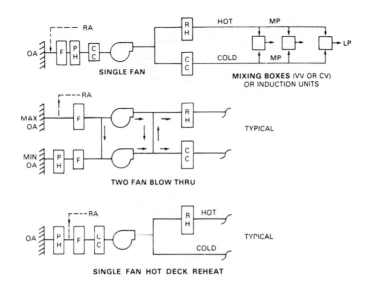

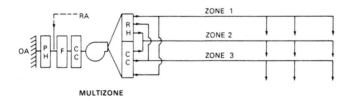

F	= FILTER		VV	= VARIABLE VOLUME	
CC	= COOLING COIL		CV	= CONSTANT VOLUME	
PH	= PREHEAT COIL		LP	= LOW PRESSURE DUCT	
OA	= OUTSIDE AIR		MP	= MEDIUM PRESSURE OR	
RH	= REHEAT COIL			HIGH PRESSURE DUCT	
			RA	= RETURN AIR	

FIGURE 6-2 Dual-path air systems. (*From* HVAC Duct Construction Standards—Metal and Flexible, *1st ed., © 1985 by the Sheet-Metal and Air Conditioning Contractor's National Association. Used with permission of the copyright holder.*)

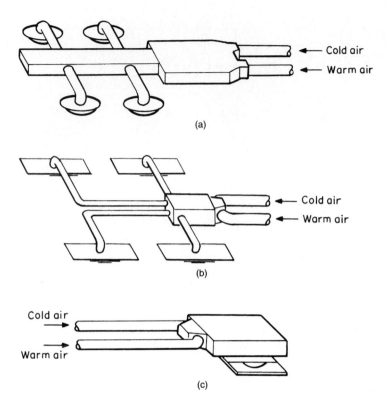

(a)

(b)

Cold air

Warm air

(c)

FIGURE 6-3 Dual-duct mixing box. (*a*) High-pressure type (discharges air to a low-velocity duct which supplies a series of round diffusers by flexible ducts); (*b*) mechanical constant-volume type (supplies a series of square diffusers by flexible ducts connected to the box); (*c*) mechanical constant-volume type (supplies a diffuser directly connected to the box). (*Source: A. M. Khashab,* Heating, Ventilating and Air Conditioning Systems Estimating Manual, *McGraw-Hill, New York, 1977. Used with permission.*)

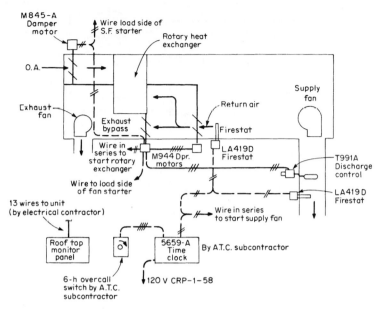

FIGURE 6-4 Rooftop heat-recovery-unit control. (*Source: A. M. Khashab, Heating, Ventilating and Air Conditioning Systems Estimating Manual, McGraw-Hill, New York, 1977. Used with permission.*)

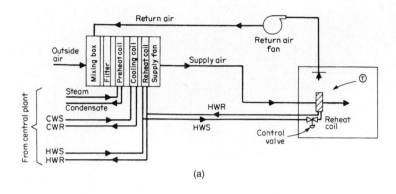

(a)

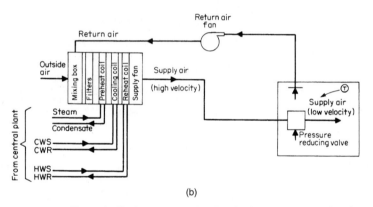

(b)

FIGURE 6-5 Typical all-air systems. (*a*) Single-duct system with reheat; (*b*) single-duct high-velocity system; (*c*) single-duct variable-volume system; (*d*) multizone system. (*Source: A. M. Khashab,* Heating, Ventilating and Air Conditioning Systems Estimating Manual, *McGraw-Hill, New York, 1977. Used with permission.*)

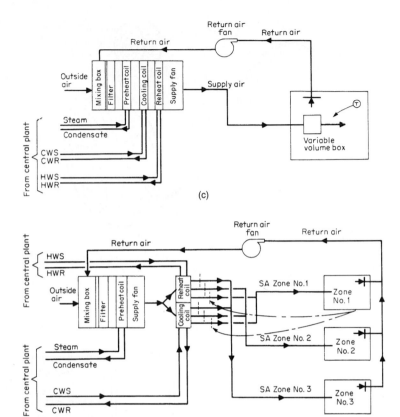

(c)

(d)

FIGURE 6-5 (*Continued*)

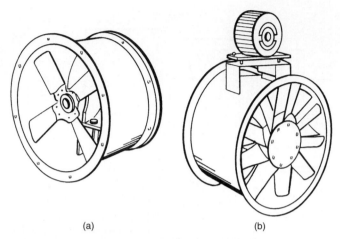

(a)   (b)

FIGURE 6-6   Axial fans. (*a*) Tube-axial (direct driven); (*b*) vane-axial. (*Source: A. M. Khashab,* Heating, Ventilating and Air Conditioning Systems Estimating Manual, *McGraw-Hill, New York, 1977. Used with permission.*)

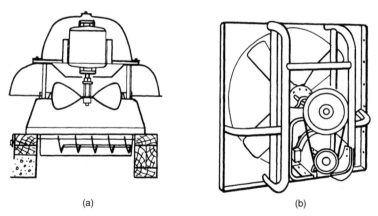

(a)   (b)

FIGURE 6-7   Propeller fans. (*a*) Roof type; (*b*) V-belt driven. (*Source: A. M. Khashab,* Heating, Ventilating and Air Conditioning Systems Estimating Manual, *McGraw-Hill, New York, 1977. Used with permission.*)

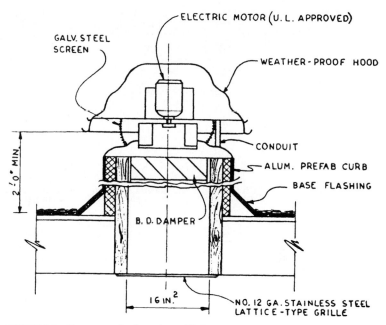

FIGURE 6-8   Power roof exhaust ventilator.

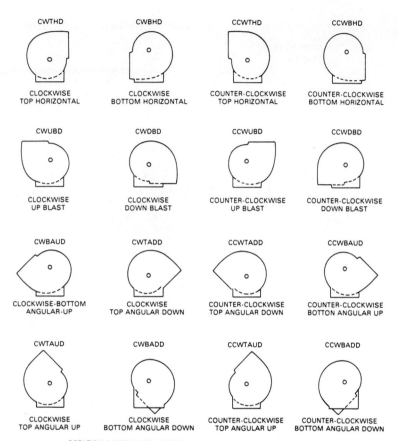

ROTATION & DISCHARGE IS DETERMINED FROM THE DRIVE OF A FAN
WHEN FAN IS RESTING ON THE FLOOR. THE DRIVE SIDE IS CONSIDERED
TO BE THE SIDE OPPOSITE THE INLET ON A SINGLE INLET FAN.

FIGURE 6-9 Fan rotation and discharge positions. (*From* HVAC Duct Construction Standards—Metal and Flexible, *1st ed., © 1985 by the Sheet-Metal and Air Conditioning Contractor's National Association. Used with permission of the copyright holder.*)

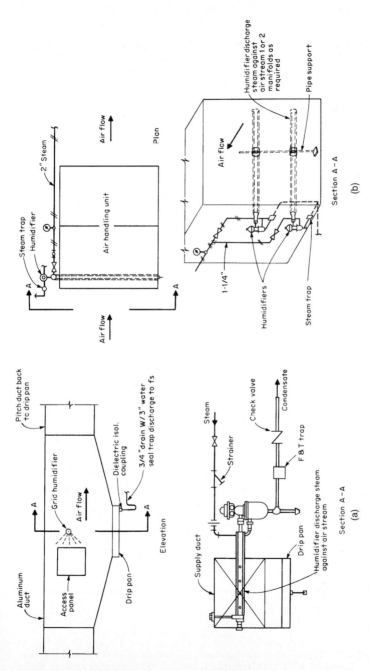

FIGURE 6-10 Steam humidifiers. (a) Duct humidifier; (b) air-handling-unit humidifier. (*Source: A. M. Khashab, Heating, Ventilating and Air Conditioning Systems Estimating Manual, McGraw-Hill, New York, 1977. Used with permission.*)

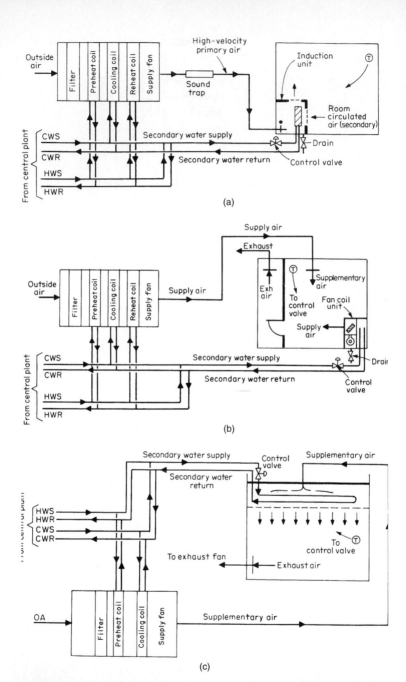

FIGURE 6-11 Typical air–water systems. (a) Induction-unit system; (b) fan-coil-unit system with supplementary air; (c) radiant-panel system with supplementary air. (*Source: A. M. Khashab,* Heating, Ventilating and Air Conditioning Systems Estimating Manual, *McGraw-Hill, New York, 1977. Used with permission.*)

NOTES

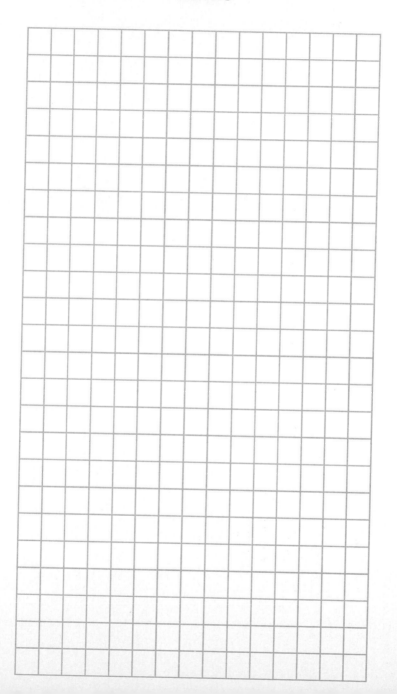

NOTES

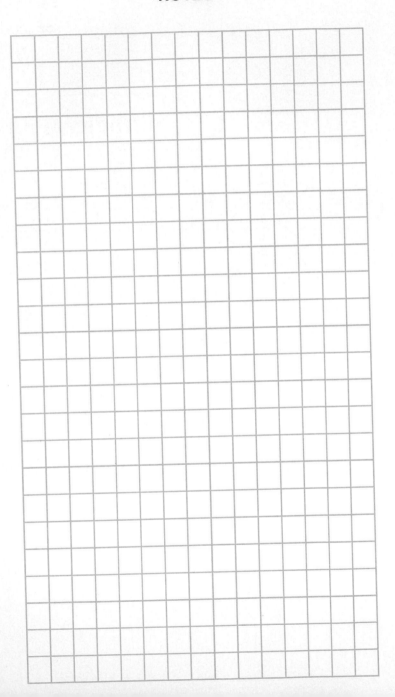

Air Conditioning

TABLE 7-1   Air Conditioning Formulas and Data

Formula	Legend	Formula	Legend
$T_1 = \dfrac{T_{cl}}{12,000}$	T_1 = load expressed in tons of refrigeration T_{cl} = total cooling load, expressed in Btu/h 1 ton of refrigeration = amount of cooling that can be done by 1 ton of ice melting in a 24-h period	q_{ef} = watts × 3.42 × 1.25	q_{ef} = gain from fluorescent lighting (25% increase for ballast), expressed in Btu's
$Q = \dfrac{q_s}{1.08(T_a - T_d)}$	Q = outside air entering conditioned room (make-up air), expressed in cfm (ft³/min) q_s = sensible load due to outside ventilation air, expressed in Btu/h T_a = indoor temperature (dry bulb)	$q_{fan} = R \times Q_s$	q_{fan} = gain from fan motor, expressed in Btu's R = 0.05 for small systems 0.35 for large systems Q_s = total sensible load, expressed in Btu/h
		q_{os} = number of persons × H_s	q_{os} = sensible heat load from occupants, expressed in Btu/h H_s = sensible heat per person—see ASHRAE tables
	T_d = dry-bulb temperature of air discharged from air-handling unit	q_{ol} = number of persons × H_1	q_{ol} = latent heat load from occupants, expressed in Btu/h H_1 = latent heat per person—see ASHRAE tables
$q_s = 1.08Q\,(T_o - T_i)$	T_o = design dry-bulb temperature of outside air T_i = design dry-bulb temperature of conditioned air	$Q = \dfrac{qs}{1.08\,\Delta T}$	ΔT = temperature difference between room and air leaving coil (usually 18°F)
$q_1 = 0.67Q\,(G_o - G_i)$	q_1 = latent load due to outside ventilation air, expressed in Btu/h G_o = moisture content of outside air, expressed in g/lb of air G_i = moisture content of inside air, expressed in g/lb of air	$V = \dfrac{Q}{Ac}$	V = air velocity, ft/min A_c = coil face area (ft²)
q_e = watts × 3.42	q_e = gain from electric light fixtures, expression in Btu's	$Q_g = \dfrac{24 \times \text{tons of refrigeration}}{\Delta T_1}$	Q_g = amount of chilled water in gal/min to be circulated ΔT_1 = temperature rise (°F) of water on passing through cooling coil (usually 8°F or 10°F)

SOURCE: R. O. Parmley, *HVAC Field Manual*, McGraw-Hill, New York, 1988. Used with permission of publisher.

TABLE 7-2 Outside Design Temperatures and Latitudes for Various Cities in the United States

STATE	CITY	DESIGN TEMP. DB	DESIGN TEMP. WB	NORTH LATITUDE, DEGREES
ALABAMA	ANNISTON	95	75	33
	BIRMINGHAM	95	78	33
	MOBILE	95	80	31
ALASKA	JUNEAU	65	52	58
ARIZONA	PHOENIX	105	76	33
	TUCSON	105	72	32
ARKANSAS	LITTLE ROCK	95	78	35
CALIFORNIA	FRESNO	105	74	37
	LOS ANGELES	90	70	34
	SACRAMENTO	100	72	38
	SAN FRANCISCO	85	65	38
COLORADO	DENVER	95	64	40
	PUEBLO	95	65	38
CONNECTICUT	HARTFORD	93	75	42
DELAWARE	WILMINGTON	95	78	40
DISTRICT OF COLUMBIA	WASHINGTON	95	78	39
FLORIDA	JACKSONVILLE	95	78	30
	MIAMI	91	79	26
GEORGIA	ATLANTA	95	76	34
	SAVANNAH	95	78	32
HAWAII	HONOLULU	83	73	21
IDAHO	BOISE	95	65	44
	POCATELLO	95	65	43
ILLINOIS	CHICAGO	95	75	43
	PEORIA	96	76	43
	SPRINGFIELD	98	77	42
INDIANA	EVANSVILLE	95	78	38
	FORT WAYNE	95	75	41
	INDIANAPOLIS	95	76	40
IOWA	DES MOINES	95	78	42
	DUBUQUE	95	78	42
	SIOUX CITY	95	76	42
KANSAS	KANSAS CITY	100	76	39
	WICHITA	100	75	38
KENTUCKY	ASHLAND	95	76	38
	LOUISVILLE	95	78	38

TABLE 7-2 Outside Design Temperatures and Latitudes for Various Cities in the United States (Continued)

		95	80	30
LOUISIANA	NEW ORLEANS	95	80	30
	SHREVEPORT	100	78	32
MAINE	PORTLAND	90	73	44
MARYLAND	BALTIMORE	95	78	39
MASSACHUSETTS	BOSTON	92	75	42
	HOLYOKE	93	75	42
MICHIGAN	DETROIT	95	75	42
	GRAND RAPIDS	93	73	43
MINNESOTA	DULUTH	95	75	47
	MINNEAPOLIS	95	75	45
MISSISSIPPI	JACKSON	100	76	32
MISSOURI	KANSAS CITY	100	75	39
	SPRINGFIELD	95	78	37
	ST. LOUIS	90	66	39
MONTANA	BILLINGS	95	67	46
	HELENA	95	78	46
NEBRASKA	LINCOLN	95	78	41
	OMAHA	95	65	41
NEVADA	RENO	90	73	39
NEW HAMPSHIRE	CONCORD	95	78	43
NEW JERSEY	TRENTON	95	70	40
NEW MEXICO	ALBUQUERQUE	95	65	35
	SANTA FE	93	73	36
NEW YORK	ALBANY	93	75	43
	BUFFALO	95	75	43
	NEW YORK	93	78	41
NORTH CAROLINA	ASHEVILLE	95	73	36
	GREENSBORO	95	75	36
NORTH DAKOTA	BISMARCK	95	77	47
OHIO	CINCINNATI	101	77	39
	CLEVELAND	101	68	42
OKLAHOMA	OKLAHOMA CITY	95	78	35
	TULSA	95	75	36
OREGON	PORTLAND	93	75	45
PENNSYLVANIA	PHILADELPHIA	95	78	40
	PITTSBURGH	95	70	40
RHODE ISLAND	PROVIDENCE	95	76	42
SOUTH CAROLINA	CHARLESTON	95	78	33
	GREENVILLE			35
SOUTH DAKOTA	RAPID CITY			44
TENNESSEE	CHATTANOOGA			35
	MEMPHIS			35

TEXAS	DALLAS	100	78	33
	EL PASO	100	69	32
	GALVESTON	95	80	29
	HOUSTON	95	80	30
	SAN ANTONIO	100	78	29
UTAH	SALT LAKE CITY	95	64	41
VIRGINIA	NORFOLK	95	78	37
	RICHMOND	95	78	38
	ROANOKE	95	78	37
WASHINGTON	SEATTLE	85	65	48
	SPOKANE	95	65	48
WEST VIRGINIA	CHARLESTON	95	75	42
WISCONSIN	EAU CLAIRE	95	75	45
	MADISON	95	75	43
	MILWAUKEE	95	75	43
WYOMING	CHEYENNE	95	65	41

SOURCE: The Trane Company, LaCrosse, WI. Reproduced by permission.

TABLE 7-3   Design Temperature Differences

ITEM NO.	ITEM	TEMPERATURE DIFFERENCE* F
1	WALLS, EXTERIOR	17
2	GLASS IN EXTERIOR WALLS	17
3	GLASS IN PARTITIONS	10
4	STORE SHOW WINDOWS HAVING A LARGE LIGHTING LOAD	30
5	PARTITIONS	10
6	PARTITIONS, OR GLASS IN PARTITIONS, ADJACENT TO LAUNDRIES, KITCHENS, OR BOILER ROOMS	25
7	FLOORS ABOVE UNCONDITIONED ROOMS	10
8	FLOORS ON GROUND	0
9	FLOORS ABOVE BASEMENTS	0
10	FLOORS ABOVE ROOMS OR BASEMENTS USED AS LAUNDRIES, KITCHENS, OR BOILER ROOMS	35
11	FLOORS ABOVE VENTED SPACES	17
12	FLOORS ABOVE UNVENTED SPACES	0
13	CEILINGS WITH UNCONDITIONED ROOMS ABOVE	10
14	CEILINGS WITH ROOMS ABOVE USED AS LAUNDRIES, KITCHENS, ETC.	20
15	CEILING WITH ROOF DIRECTLY ABOVE (NO ATTIC)	17
16	CEILING WITH TOTALLY ENCLOSED ATTIC ABOVE	17
17	CEILING WITH CROSS-VENTILATED ATTIC ABOVE	17

*These temperature differences are based on the assumption that the air conditioning system is being designed to maintain an inside temperature 17 F lower than the outdoor temperature. For air conditioning systems designed to maintain a greater temperature difference than 17 F between the inside and outside, add to the values in the above table, the difference between the assumed design temperature difference and 17 F.

SOURCE: The Trane Company, LaCrosse, WI. Reproduced by permission.

TABLE 7-4 Equivalent Full-Load Hours of Operation per Year

	ATLANTA	BALTI-MORE	BOSTON	CHICAGO	DALLAS	DENVER	DETROIT	LOS ANGELES	MIAMI	MILWAU-KEE	MINNE-APOLIS	NEW ORLEANS	OKLA CITY	PHILA-DELPHIA	PHNX MIX	PORTLAND ORE.	SAN FRAN	SAINT LOUIS	WASH B.C.	NEW YORK
RESTAURANTS	1750	1620	1050	1250	2240	1050	1250	1150	2020	1050	1050	2020	2240	1480	2240	1050	450	2020	1520	1430
DRUG STORES	1700	1580	1030	1220	2170	1030	1220	1120	1950	1030	1030	1950	2170	1440	2170	1030	400	1950	1580	1400
CAFETERIAS	1370	1270	825	990	1750	825	990	910	1580	825	825	1580	1750	1160	1750	825	350	1580	1270	1120
JEWELRY STORES	1020	950	620	750	1300	620	750	700	1170	620	620	1170	1300	875	1300	620	250	1170	990	850
BARBER SHOPS	1020	950	620	750	1300	620	750	700	1170	620	620	1170	1300	875	1300	620	250	1170	950	850
NIGHT CLUBS	1010	940	610	730	1280	810	730	675	1150	610	610	1150	1280	860	1280	610	240	1150	940	840
THEATERS	650-1000	600-1000	400-650	500-800	850-1400	400-650	500-800	475-750	800-1300	400-650	400-650	800-1300	850-1400	550-920	850-1400	400-650	200-400	800-1300	600-1000	550-800
DRESS SHOPS	840	870	565	675	1200	565	675	630	1080	565	565	1080	1200	800	1200	565	225	1080	870	780
LARGE OFFICES	915	850	550	660	1180	550	660	610	1060	550	550	1060	1180	775	1180	550	200	1060	830	750
DEPARTMENT STORES	850	790	515	650	1100	515	650	600	1000	515	515	1000	1100	725	1100	515	175	1000	790	700
SPECIALTY SHOPS (5 & 10)	840	780	510	640	1080	510	640	590	975	510	510	975	1080	710	1080	510	175	975	730	700
RESIDENCES	810	750	480	600	1050	480	600	550	950	480	480	950	1050	690	1050	480	170	950	730	670
SHOE STORES	650	600	400	500	850	400	500	475	775	400	400	775	850	575	850	400	170	778	750	550
BEAUTY SHOPS	625	580	380	450	800	380	450	425	750	380	380	750	800	540	800	380	150	750	400	525
SMALL OFFICES	540	500	425	450	700	425	450	410	650	425	425	650	700	480	700	425	150	730	580	475
RECREATION SPACES	520	480	450	400	675	450	400	380	650	450	450	650	675	470	675	450	125	650	500	450
FUNERAL PARLORS	480	425	380	375	600	380	375	375	575	380	380	575	600	410	600	380	100	575	480	400

SOURCE: The Trane Company, LaCrosse, WI. Reproduced by permission.

TABLE 7-5   Overall Heat-Transfer Coefficients for Glass
and Glass Blocks

DESCRIPTION	OUTDOOR EXPOSURE
SINGLE-GLASS WINDOWS	1.06
DOUBLE-GLASS WINDOWS	0.64
TRIPLE-GLASS WINDOWS	0.34
GLASS BLOCK	0.56
GLASS BLOCK	0.48

SOURCE: The Trane Company, LaCrosse, WI. Reproduced by permission.

TABLE 7-6   Cooling-Load Low Check Figures

Building Classification	Cooling-Load Low Figure			
	Occupancy, ft²/person	Lights, W/ft²	Refrigeration, ft²/ton	Air quantities, cfm/ft²
Apartment (high-rise)	325	1.0	450	0.7
Auditoriums, churches, theaters	15	1.0	400	1.0
Educational facilities (schools, colleges, universities)	30	2.0	240	0.9
Manufacturing				
Light	200	9.0*	200	1.6
Heavy	300	15.0*	100	2.5
Hospitals				
Patient rooms	75	1.0	275	0.4†
Public areas	100	1.0	175	1.0
Hotels, motels, dormitories	200	1.0	350	1.0
Libraries and museums	80	1.0	340	1.0
Office buildings				
General	130	4.0	360	0.25† exterior / 0.80 interior
Private	150	2.0		0.25†
Residential				
Large	600	1.0	600	0.7
Medium	600	0.7	700	0.6
Restaurants				
Large	17	1.5	135	1.4
Medium			150	1.5
Shopping centers				
Department stores				
Basement	30	2.0	340	0.7
Main floor	45	3.5	350	0.9
Upper floors	75	2.0	400	0.8
Malls	100	1.0	365	1.1

*Includes other loads.
†Based on induction system.

SOURCE: A. M. Khashab, *Heating, Ventilating and Air Conditioning Systems Estimating Manual,* McGraw-Hill, New York, 1977. Used with permission.

TABLE 7-7   Cooling-Load Average Check Figures

Building Classification	Occupancy, ft²/person	Lights, W/ft²	Refrigeration, ft²/ton	Air quantities, cfm/ft²
	Cooling-Load Average Figure			
Apartment (high-rise)	175	2.0	400	1.0
Auditoriums, churches, theaters	11	2.0	250	2.0
Educational facilities (schools, colleges, universities)	25	4.0	185	1.4
Manufacturing				
Light	150	10.0*	150	2.5
Heavy	250	45.0*	80	4.0
Hospitals				
Patient rooms	50	1.5	220	0.5†
Public areas	80	1.5	140	1.1
Hotels, motels, dormitories	150	2.0	300	1.3
Libraries and museums	60	1.5	280	1.2
Office buildings				
General	110	6.0*	280	0.5† exterior / 1.1 interior
Private	125	5.8		0.5†
Residential				
Large	400	2.0	500	1.0
Medium	360	1.5	550	0.9
Restaurants				
Large	15	1.7	100	1.75
Medium			120	1.50
Shopping centers				
Dept. stores				
Basement	25	3.0	285	1.0
Main floor	25	6.0*	245	1.4
Upper floors	55	2.5	340	1.0
Malls	75	1.5	230	1.8

*Includes other loads.
†Based on induction system.
SOURCE: A. M. Khashab, *Heating, Ventilating and Air Conditioning Systems Estimating Manual,* McGraw-Hill, New York, 1977. Used with permission.

TABLE 7-8 Cooling-Load High Check Figures

| Building Classification | Cooling-Load High Figure | | | |
	Occupancy, ft²/person	Lights, W/ft²	Refrigeration, ft²/ton	Air quantities, cfm/ft²
Apartment (high-rise)	100	4.0	350	1.5
Auditoriums, churches, theaters	6	3.0	90	3.0
Educational facilities (schools, colleges, universities)	20	6.0	150	2.0
Manufacturing				
Light	100	12.0*	100	3.8
Heavy	200	60.0*	60	6.5
Hospitals				
Patient rooms	25	2.0	165	0.7†
Public areas	50	2.0	110	1.3
Hotels, motels, dormitories	100	3.0	220	1.5
Libraries and museums	40	3.0	200	1.4
Office buildings				
General	80	9.0*	190	0.8† exterior / 1.8 interior
	100	8.0		0.8†
Residential				
Large	200	4.0	380	1.5
Medium	200	3.0	400	1.3
Restaurants				
Large	13	2.0	80	2.4
Medium			100	2.0
Shopping centers				
Dept. stores				
Basement	20	4.0	225	1.2
Main floor	16	9.0*	150	2.0
Upper floors	40	3.5*	280	1.2
Malls	50	2.0	160	2.5

*Includes other loads.
†Based on induction system.

SOURCE: A. M. Khashab, *Heating, Ventilating and Air Conditioning Systems Estimating Manual,* McGraw-Hill, New York, 1977. Used with permission.

TABLE 7-9   Atmospheric Pressure and Barometer Readings at Different Altitudes

Altitude above Sea Level, Ft	Atmospheric Pressure, Psi	Barometer Reading, In. Hg	Altitude above Sea Level, Ft	Atmospheric Pressure, Psi	Barometer Reading, In. Hg
0	14.69	29.92	7,500	11.12	22.65
500	14.42	29.38	8,000	10.91	22.22
1,000	14.16	28.86	8,500	10.70	21.80
1,500	13.91	28.33	9,000	10.50	21.38
2,000	13.66	27.82	9,500	10.30	20.98
2,500	13.41	27.31	10,000	10.10	20.58
3,000	13.16	26.81	10,500	9.90	20.18
3,500	12.92	26.32	11,000	9.71	19.75
4,000	12.68	25.84	11,500	9.52	19.40
4,500	12.45	25.36	12,000	9.34	19.03
5,000	12.22	24.89	12,500	9.15	18.65
5,500	11.99	24.43	13,000	8.97	18.29
6,000	11.77	23.98	13,500	8.80	17.93
6,500	11.55	23.53	14,000	8.62	17.57
7,000	11.33	23.09	14,500	8.45	17.22
			15,000	8.28	16.88

SOURCE: *Compressed Air and Gas Handbook,* 5th ed., 1988. Used with permission from CAGI, 1300 Sumner Ave., Cleveland, Ohio 44115.

TABLE 7-10 Moisture Content of Saturated Air at Various Temperatures

°F	Grains per cu. ft.	°F	Grains per cu. ft.	°F	Grains per cu. ft.	°F	Grains per cu. ft.
−60	.01470	−15	.218	+30	1.935	+75	9.356
−59	.01573	−14	.234	+31	2.023	+76	9.749
−58	.01677	−13	.243	+32	2.113	+77	9.962
−57	.01795	−12	.257	+33	2.194	+78	10.38
56	.01914	−11	.270	+34	2.279	+79	10.601
−55	.02047	−10	.285	+35	2.366	+80	11.04
−54	.02184	−9	.300	+36	2.457	+81	11.27
−53	.02320	−8	.316	+37	2.550	+82	11.75
−52	.02485	−7	.332	+38	2.646	+83	11.98
−51	.02650	−6	.350	+39	2.746	+84	12.49
−50	.02826	−5	.370	+40	2.849	+85	12.73
−49	.03004	−4	.389	+41	2.955	+86	13.27
−48	.03207	−3	.411	+42	3.064	+87	13.53
−47	.03412	−2	.434	+43	3.177	+88	14.08
−46	.03622	−1	.457	+44	3.294	+89	14.36
−45	.03865	0	.481	+45	3.414	+90	14.94
−44	.04111	+1	.505	+46	3.539	+91	15.23
−43	.04375	+2	.529	+47	3.667	+92	15.84
−42	.04650	+3	.554	+48	3.800	+93	16.15
−41	.04947	+4	.582	+49	3.936	+94	16.79
−40	.05249	+5	.640	+50	4.076	+95	17.12
−39	.05583	+6	.639	+51	4.222	+96	17.80
−38	.05922	+7	.671	+52	4.372	+97	18.44
−37	.06292	+8	.704	+53	4.526	+98	18.85
−36	.06677	+9	.739	+54	4.685	+99	19.24
−35	.07085	+10	.776	+55	4.849	+100	19.95
−34	.07517	+11	.816	+56	5.016	+101	20.33
−33	.07962	+12	.856	+57	5.191	+102	21.11
−32	.08447	+13	.898	+58	5.370	+103	21.54
−31	.08942	+14	.941	+59	5.555	+104	22.32
−30	.09449	+15	.986	+60	5.745	+105	22.75
−29	.09982	+16	1.032	+61	5.941	+106	23.60
−28	.10616	+17	1.080	+62	6.142	+107	24.26
−27	.11258	+18	1.128	+63	6.349	+108	24.93
−26	.11914	+19	1.181	+64	6.563	+109	25.41
−25	.12611	+20	1.235	+65	6.782	+110	26.34
−24	.13334	+21	1.294	+66	7.069	+111	27.07
−23	.14113	+22	1.355	+67	7.241	+112	27.81
−22	.14901	+23	1.418	+68	7.480	+113	28.57
−21	.15739	+24	1.483	+69	7.726	+114	29.34
−20	.166	+25	1.551	+70	7.980	+115	30.14
−19	.174	+26	1.623	+71	8.240	+116	30.95
−18	.184	+27	1.697	+72	8.508	+117	31.79
−17	.196	+28	1.773	+73	8.782	+118	32.63
−16	.207	+29	1.853	+74	9.153	+119	33.51

7000 grains moisture = 1 pound.

SOURCE: *Compressed Air and Gas Handbook,* 5th ed., 1988. Used with permission from CAGI, 1300 Sumner Ave., Cleveland, Ohio 44115.

TABLE 7-11 Specific Gravity of Moist Air at Standard Sea-Level Pressure

Temperature, deg F	Relative Humidity, Per Cent									
	10	20	30	40	50	60	70	80	90	100
0	.9999	.9999	.9999	.9998	.9998	.9997	.9997	.9996	.9996	.9995
10	.9999	.9998	.9998	.9997	.9996	.9995	.9994	.9994	.9993	.9992
20	.9999	.9997	.9996	.9995	.9994	.9992	.9991	.9990	.9988	.9987
30	.9998	.9996	.9994	.9992	.9990	.9987	.9985	.9983	.9981	.9979
40	.9997	.9994	.9991	.9987	.9984	.9981	.9978	.9975	.9972	.9969
50	.9995	.9991	.9986	.9982	.9977	.9972	.9968	.9963	.9959	.9954
60	.9993	.9987	.9980	.9974	.9967	.9960	.9954	.9947	.9941	.9934
70	.9991	.9981	.9972	.9963	.9953	.9944	.9935	.9925	.9916	.9907
80	.9987	.9974	.9961	.9948	.9935	.9922	.9909	.9896	.9883	.9870
90	.9982	.9964	.9946	.9928	.9910	.9892	.9875	.9857	.9839	.9821
100	.9976	.9951	.9927	.9903	.9878	.9854	.9830	.9805	.9781	.9756
110	.9967	.9935	.9902	.9869	.9837	.9804	.9771	.9738	.9706	.9673
120	.9957	.9913	.9870	.9827	.9783	.9740	.9697	.9653	.9610	.9567
130	.9943	.9886	.9830	.9773	.9716	.9659	.9602	.9545	.9488	.9432

SOURCE: *Compressed Air and Gas Handbook*, 5th ed., 1988. Used with permission from CAGI, 1300 Sumner Ave., Cleveland, Ohio 44115.

TABLE 7-12   Weight of Water Vapor in One Cubic Foot of Air
at Various Temperatures and Percentages of Saturation

Temperature deg F	Per cent of Saturation									
	10	20	30	40	50	60	70	80	90	100
	Weight, Grains									
−10	0.028	0.057	0.086	0.114	0.142	0.171	0.200	0.228	0.256	0.285
0	0.048	0.096	0.144	0.192	0.240	0.289	0.337	0.385	0.433	0.481
10	0.078	0.155	0.233	0.210	0.388	0.466	0.543	0.621	0.698	0.776
20	0.124	0.247	0.370	0.494	0.618	0.741	0.864	0.988	1.112	1.235
30	0.194	0.387	0.580	0.774	0.968	1.161	1.354	1.548	1.742	1.935
32	0.211	0.422	0.634	0.845	1.056	1.268	1.479	1.690	1.902	2.113
35	0.237	0.473	0.710	0.947	1.183	1.420	1.656	1.893	2.129	2.366
40	0.285	0.570	0.855	1.140	1.424	1.709	1.994	2.279	2.564	2.749
45	0.341	0.683	1.024	1.366	1.707	2.048	2.390	2.731	3.073	3.414
50	0.408	0.815	1.223	1.630	2.038	2.446	2.853	3.261	3.668	4.076
55	0.485	0.970	1.455	1.940	2.424	2.909	3.394	3.879	4.364	4.849
60	0.574	1.149	1.724	2.298	2.872	3.447	4.022	4.596	5.170	5.745
62	0.614	1.228	1.843	2.457	3.071	3.685	4.299	4.914	5.528	6.142
64	0.656	1.313	1.969	2.625	3.282	3.938	4.594	5.250	5.907	6.563
66	0.701	1.402	2.103	2.804	3.504	4.205	4.906	5.607	6.208	7.009
68	0.748	1.496	2.244	2.992	3.740	4.488	5.236	5.974	7.732	7.480
70	0.798	1.596	2.394	3.192	3.990	5.788	5.586	6.384	7.182	7.980
72	0.851	1.702	2.552	3.403	4.254	5.105	5.956	6.806	7.657	8.508
74	0.907	1.813	2.720	3.626	4.553	5.440	6.346	7.523	8.159	9.066
76	0.966	1.931	2.896	3.862	4.828	5.793	6.758	7.724	8.690	9.655
78	1.028	2.055	3.083	4.111	5.138	6.166	7.194	8.222	9.249	10.277
80	1.093	2.187	3.280	4.374	5.467	6.560	7.654	8.747	9.841	10.934
82	1.163	2.325	3.488	4.650	5.813	6.976	8.138	9.301	10.463	11.626
84	1.236	2.471	3.707	4.942	6.178	7.414	8.649	9.885	11.120	12.356
86	1.313	2.625	3.938	5.251	6.564	7.877	9.189	10.502	11.814	13.127
88	1.394	2.787	4.181	5.575	6.968	8.362	9.576	11.150	12.543	13.937
90	1.479	2.958	4.437	5.916	7.395	8.874	10.353	11.832	13.311	14.790
92	1.569	3.138	4.707	7.276	7.844	9.413	10.982	12.551	13.120	15.689
94	1.663	3.327	4.990	6.654	8.317	9.980	11.644	13.307	14.971	16.634
96	1.763	3.525	5.288	7.050	8.813	10.576	12.338	14.101	15.863	17.626
98	1.867	3.734	5.601	7.468	9.336	11.203	13.070	14.937	16.804	18.671
100	1.977	3.953	5.930	7.906	9.883	11.860	13.836	15.813	17.789	19.766

SOURCE: *Compressed Air and Gas Handbook,* 5th ed., 1988. Used with permission from CAGI, 1300 Sumner Ave., Cleveland, Ohio 44115.

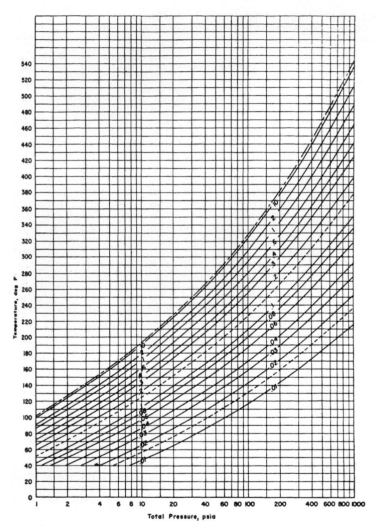

FIGURE 7-1   Pounds of vapor per pound of dry air at saturation. (*Source: Compressed Air and Gas Handbook, 5th ed., 1988. Used with permission from CAGI, 1300 Sumner Ave., Cleveland, Ohio 44115.*)

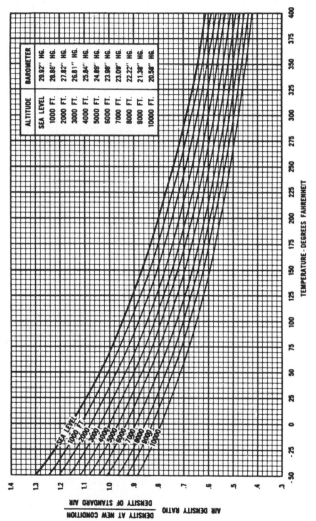

FIGURE 7-2 Graph for determining air density at temperatures and altitudes other than standard. (*Reproduced by permission of The Trane Company, LaCrosse, WI.*)

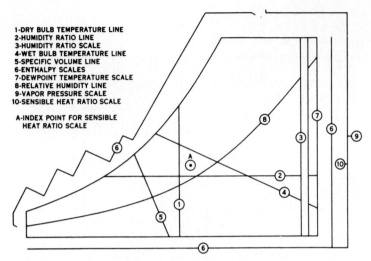

1-DRY BULB TEMPERATURE LINE
2-HUMIDITY RATIO LINE
3-HUMIDITY RATIO SCALE
4-WET BULB TEMPERATURE LINE
5-SPECIFIC VOLUME LINE
6-ENTHALPY SCALES
7-DEWPOINT TEMPERATURE SCALE
8-RELATIVE HUMIDITY LINE
9-VAPOR PRESSURE SCALE
10-SENSIBLE HEAT RATIO SCALE

A-INDEX POINT FOR SENSIBLE
  HEAT RATIO SCALE

FIGURE 7-3   Lines and scales on the Trane psychrometic chart. (*Reproduced by permission of The Trane Company, LaCrosse, WI.*)

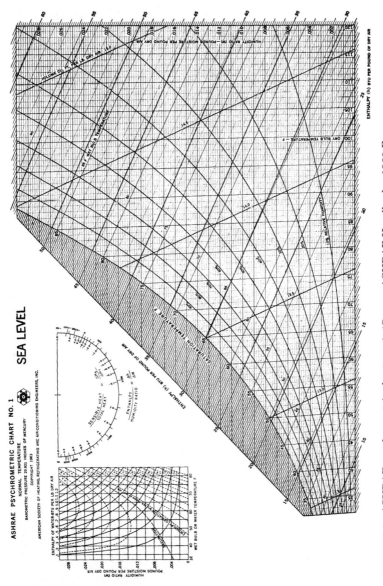

FIGURE 7-4 ASHRAE psychrometric chart no. 1. (*Source:* ASHRAE Handbook—1989 Fundamentals. *Reprinted by permission.*)

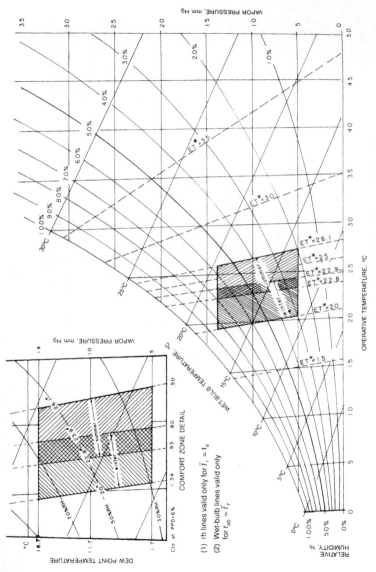

FIGURE 7-5 Standard effective temperature and the ASHRAE comfort zones. (*Source: ASHRAE Handbook—1989 Fundamentals. Reprinted by permission.*)

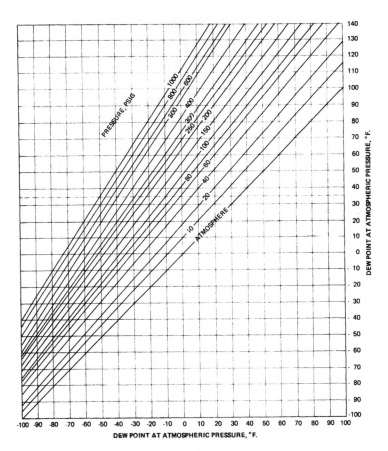

Dew Point Conversion:

To obtain the dew point temperature expected if the gas were expanded to a lower pressure proceed as follows: —

1. Using "dew point at pressure," locate this temperature on scale at right hand side of chart.

2. Read horizontally to intersection of curve corresponding to the operating pressure at which the gas was dried.

3. From that point read vertically downward to curve corresponding to the expanded lower pressure.

4. From that point read horizontally to scale on right hand side of chart to obtain dew point temperature at the expanded lower pressure.

5. If dew point temperatures of atmospheric pressure are desired, after step 2, above read vertically downward to scale at bottom of chart which gives "Dew Point at Atmospheric Pressure."

FIGURE 7-6 Dew point conversion. (*Source:* Compressed Air and Gas Handbook, *5th ed., 1988. Used with permission from CAGI, 1300 Sumner Ave., Cleveland, Ohio 44115.*)

TABLE 7-13   Data for Brine Calculations Needed by Refrigeration Engineers

Salinometer reading at 64°F	Percentage of calcium chloride by weight	Freezing point, °F	Specific gravity at 64°F
28	6.60	25.8	1.050
34	8.49	23.6	1.065
40	9.43	22.1	1.073
48	11.32	19.1	1.089
52	12.26	17.7	1.097
60	14.15	13.8	1.114
64	15.09	11.9	1.122
68	16.03	10.0	1.131
76	17.92	5.4	1.149
80	18.86	3.1	1.158
84	19.80	−0.8	1.167
88	20.75	−4.4	1.176
92	21.69	−8.0	1.186
96	22.63	−11.6	1.196
100	23.58	−15.2	1.205
104	24.52	−19.6	1.215
108	25.46	−24.4	1.225
112	26.40	−29.3	1.236

SOURCE: S. M. Elonka and Q. W. Minich, *Standard Refrigeration and Air Conditioning Questions and Answers,* McGraw-Hill, New York, 1973. Used with permission of publisher.

TABLE 7-14 Capacity Requirements for Ammonia Compressors

Pressure and corresponding saturated temperature of suction gas		Discharge pressure, psig															
		135		145		155		165		175		185		195		205	
		Corresponding saturated temperature, °F															
		78.7		82.5		86.2		89.7		93.0		96.2		99.3		102.3	
Pressure, psig	Temperature, °F	Cfm/Ton	Bhp/Ton	Cfm/Ton	Bhp/Ton	Cfm/Ton	Bhp/Ton	Cfm/Ton	Bhp/Ton	Cfm/Ton	Bhp/Ton	Cfm/Ton	Bhp/Ton	Cfm/Ton	Bhp/Ton	Cfm/Ton	Bhp/Ton
0	−28.0	12.34	2.51														
5	−17.2	8.42	2.02	8.64	2.12	8.89	2.22	9.16	2.32	9.40	2.41	9.66	2.50				
10	−8.4	6.36	1.694	6.53	1.782	6.70	1.869	6.84	1.955	7.01	2.04	7.16	2.12	7.35	2.20	7.53	2.28
15	−1.0	5.09	1.449	5.21	1.531	5.34	1.611	5.45	1.690	5.56	1.765	5.67	1.839	5.79	1.911	5.93	1.980
20	5.5	4.28	1.270	4.39	1.341	4.46	1.412	4.55	1.482	4.64	1.552	4.72	1.621	4.80	1.688	4.90	1.755
25	11.3	3.69	1.127	3.75	1.193	3.84	1.259	3.91	1.324	3.98	1.387	4.06	1.449	4.12	1.509	4.20	1.568
30	16.6	3.24	1.011	3.31	1.073	3.38	1.133	3.44	1.193	3.50	1.251	3.56	1.307	3.63	1.363	3.69	1.419
35	21.4	2.92	0.910	2.96	0.971	3.00	1.029	3.06	1.085	3.11	1.140	3.16	1.193	3.22	1.244	3.27	1.293
40	25.8	2.62	0.825	2.66	0.883	2.71	0.938	2.75	0.992	2.81	1.044	2.85	1.094	2.90	1.144	2.94	1.193
45	30.0	2.39	0.755	2.43	0.807	2.47	0.859	2.50	0.910	2.54	0.961	2.58	1.099	2.62	1.057	2.65	1.103
50	33.8	2.18	0.689	2.21	0.739	2.25	0.789	2.28	0.838	2.33	0.886	2.36	0.934	2.39	0.980	2.42	1.026

Note: Based on two-cylinder compressors with 6-in. bore and 6-in. stroke.

Values above the heavy line are for ratios of compression above 8.0, and for these conditions it is recommended that multistage compression be considered.

$\dfrac{Cfm}{Ton}$ = average values of swept volume, in cubic feet per minute, required per ton refrigeration based on a suction superheat of 10°F and on liquid ammonia entering the refrigerant control valve at the saturated temperature corresponding to the discharge pressure.

$\dfrac{Bhp}{Ton}$ = average values of actual compressor brake horsepow v per ton refrigeration.

SOURCE: S. M. Elonka and Q. W. Minich, *Standard Refrigeration and Air Conditioning Questions and Answers*, McGraw-Hill, New York, 1973. Used with permission of publisher.

TABLE 7-15 Thermodynamic Properties of Moist Air, Standard Atmospheric Pressure, 14.696 psi (29.921 inHg)

Temp. t, °F	Humidity Ratio lb_w/lb_da W_s	Volume ft³/lb dry air v_a	v_{as}	v_s	Enthalpy Btu/lb dry air h_a	h_{as}	h_s	Entropy Btu/(lb dry air) °F s_a	s_{as}	s_s	Condensed Water Enthalpy Btu/lb h_w	Entropy Btu/lb·°F s_w	Vapor Press. in. Hg p_s	Temp. °F
-80	0.0000049	9.553	0.000	9.553	-19.221	0.005	-19.215	-0.04594	0.00001	-0.04592	-193.45	-0.4067	0.000236	-80
-79	0.0000053	9.579	0.000	9.579	-18.980	0.005	-18.975	-0.04531	0.00002	-0.04529	-193.06	-0.4056	0.000255	-79
-78	0.0000057	9.604	0.000	9.604	-18.740	0.006	-18.734	-0.04468	0.00002	-0.04466	-192.66	-0.4046	0.000275	-78
-77	0.0000062	9.629	0.000	9.629	-18.500	0.007	-18.493	-0.04405	0.00002	-0.04403	-192.27	-0.4036	0.000296	-77
-76	0.0000067	9.655	0.000	9.655	-18.259	0.007	-18.252	-0.04342	0.00002	-0.04340	-191.87	-0.4025	0.000319	-76
-75	0.0000072	9.680	0.000	9.680	-18.019	0.007	-18.011	-0.04279	0.00002	-0.04277	-191.47	-0.4015	0.000344	-75
-74	0.0000078	9.705	0.000	9.705	-17.778	0.008	-17.770	-0.04217	0.00002	-0.04215	-191.07	-0.4005	0.000371	-74
-73	0.0000084	9.731	0.000	9.731	-17.538	0.009	-17.529	-0.04155	0.00002	-0.04152	-190.68	-0.3994	0.000400	-73
-72	0.0000090	9.756	0.000	9.756	-17.298	0.010	-17.288	-0.04093	0.00003	-0.04090	-190.27	-0.3984	0.000430	-72
-71	0.0000097	9.781	0.000	9.781	-17.057	0.010	-17.047	-0.04031	0.00003	-0.04028	-189.87	-0.3974	0.000463	-71
-70	0.0000104	9.807	0.000	9.807	-16.806	0.011	-16.817	-0.03969	0.00003	-0.03966	-189.47	-0.3963	0.000498	-70
-69	0.0000112	9.832	0.000	9.832	-16.577	0.012	-16.565	-0.03907	0.00003	-0.03904	-189.07	-0.3953	0.000536	-69
-68	0.0000120	9.857	0.000	9.858	-16.336	0.013	-16.324	-0.03846	0.00003	-0.03843	-188.66	-0.3943	0.000576	-68
-67	0.0000129	9.883	0.000	9.883	-16.096	0.013	-16.083	-0.03785	0.00004	-0.03781	-188.26	-0.3932	0.000619	-67
-66	0.0000139	9.908	0.000	9.908	-15.856	0.015	-15.841	-0.03724	0.00004	-0.03720	-187.85	-0.3922	0.000665	-66
-65	0.0000149	9.933	0.000	9.934	-15.616	0.015	-15.600	-0.03663	0.00004	-0.03659	-187.44	-0.3912	0.000714	-65
-64	0.0000160	9.959	0.000	9.959	-15.375	0.017	-15.359	-0.03602	0.00005	-0.03597	-187.04	-0.3901	0.000766	-64
-63	0.0000172	9.984	0.000	9.984	-15.117	0.018	-15.135	-0.03541	0.00005	-0.03536	-186.63	-0.3891	0.000822	-63
-62	0.0000184	10.009	0.000	10.010	-14.895	0.019	-14.876	-0.03481	0.00005	-0.03476	-186.22	-0.3881	0.000882	-62
-61	0.0000198	10.035	0.000	10.035	-14.654	0.021	-14.634	-0.03420	0.00006	-0.03415	-185.81	-0.3870	0.000945	-61
-60	0.0000212	10.060	0.000	10.060	-14.414	0.022	-14.392	-0.03360	0.00006	-0.03354	-185.39	-0.3860	0.001013	-60
-59	0.0000227	10.085	0.000	10.086	-14.174	0.024	-14.150	-0.03300	0.00006	-0.03294	-184.98	-0.3850	0.001086	-59
-58	0.0000243	10.111	0.000	10.111	-13.933	0.025	-13.908	-0.03240	0.00007	-0.03233	-184.57	-0.3839	0.001163	-58
-57	0.0000260	10.136	0.000	10.137	-13.693	0.027	-13.666	-0.03180	0.00007	-0.03173	-184.15	-0.3829	0.001246	-57

-56	0.0000279	10.161	0.000	10.162	-13.453	0.029	-13.424	-0.03121	0.00008	-0.03113	-183.74	-0.3819	0.001333	-56
-55	0.0000298	10.187	0.000	10.187	-13.213	0.031	-13.182	-0.03061	0.00008	-0.03053	-183.32	-0.3808	0.001427	-55
-54	0.0000319	10.212	0.001	10.213	-12.972	0.033	-12.939	-0.03002	0.00009	-0.02993	-182.90	-0.3798	0.001526	-54
-53	0.0000341	10.237	0.001	10.238	-12.732	0.035	-12.697	-0.02943	0.00009	-0.02934	-182.48	-0.3788	0.001632	-53
-52	0.0000365	10.263	0.001	10.263	-12.492	0.038	-12.454	-0.02884	0.00010	-0.02874	-182.06	-0.3778	0.001745	-52
-51	0.0000390	10.288	0.001	10.289	-12.251	0.041	-12.211	-0.02825	0.00011	-0.02814	-181.64	-0.3767	0.001865	-51
-50	0.0000416	10.313	0.001	10.314	-12.011	0.043	-11.968	-0.02766	0.00011	-0.02755	-181.22	-0.3757	0.001992	-50
-49	0.0000445	10.339	0.001	10.340	-11.771	0.046	-11.725	-0.02708	0.00012	-0.02696	-180.80	-0.3747	0.002128	-49
-48	0.0000475	10.364	0.001	10.365	-11.531	0.050	-11.481	-0.02649	0.00013	-0.02636	-180.37	-0.3736	0.002272	-48
-47	0.0000507	10.389	0.001	10.390	-11.290	0.053	-11.237	-0.02591	0.00014	-0.02577	-179.95	-0.3726	0.002425	-47
-46	0.0000541	10.415	0.001	10.416	-11.050	0.056	-10.994	-0.02533	0.00015	-0.02518	-179.52	-0.3716	0.002587	-46
-45	0.0000577	10.440	0.001	10.441	-10.810	0.060	-10.750	-0.02475	0.00016	-0.02459	-179.10	-0.3705	0.002760*	-45
-44	0.0000615	10.465	0.001	10.465	-10.570	0.064	-10.505	-0.02417	0.00017	-0.02400	-178.67	-0.3695	0.002943	-44
-43	0.0000656	10.491	0.001	10.492	-10.329	0.068	-10.261	-0.02359	0.00018	-0.02342	-178.24	-0.3685	0.003137	-43
-42	0.0000699	10.516	0.001	10.517	-10.089	0.073	-10.016	-0.02302	0.00019	-0.02283	-177.81	-0.3675	0.003343	-42
-41	0.0000744	10.541	0.001	10.543	-9.849	0.078	-9.771	-0.02244	0.00020	-0.02224	-177.38	-0.3664	0.003562	-41
-40	0.0000793	10.567	0.001	10.568	-9.609	0.083	-9.526	-0.02187	0.00021	-0.02166	-176.95	-0.3654	0.003793	-40
-39	0.0000844	10.592	0.001	10.593	-9.368	0.088	-9.280	-0.02130	0.00022	-0.02107	-176.52	-0.3644	0.004039	-39
-38	0.0000898	10.617	0.002	10.619	-9.128	0.094	-9.034	-0.02073	0.00024	-0.02049	-176.08	-0.3633	0.004299	-38
-37	0.0000956	10.643	0.002	10.644	-8.888	0.100	-8.788	-0.02016	0.00025	-0.01991	-175.65	-0.3623	0.004575	-37
-36	0.0001017	10.668	0.002	10.670	-8.648	0.106	-8.541	-0.01959	0.00027	-0.01932	-175.21	-0.3613	0.004866	-36
-35	0.0001081	10.693	0.002	10.695	-8.407	0.113	-8.294	-0.01902	0.00028	-0.01874	-174.78	-0.3603	0.005175	-35
-34	0.0001150	10.719	0.002	10.721	-8.167	0.120	-8.047	-0.01846	0.00030	-0.01816	-0.3592	-174.34	0.005502	-34
-33	0.0001222	10.744	0.002	10.746	-7.927	0.128	-7.799	-0.01790	0.00032	-0.01758	-173.90	-0.3582	0.005848	-33

TABLE 7-15 Thermodynamic Properties of Moist Air, Standard Atmospheric Pressure, 14.696 psi (29.921 inHg) *(Continued)*

Temp. t, °F	Humidity Ratio lb_w/lb_da [a] W_s	Volume ft³/lb dry air			Enthalpy Btu/lb dry air			Entropy Btu/(lb dry air) °F			Condensed Water			Temp. °F
		v_a	v_{as}	v_s	h_a	h_{as}	h_s	s_a	s_{as}	s_s	Enthalpy Btu/lb h_w	Entropy Btu/lb·°F s_w	Vapor Press. in. Hg p_s	
−32	0.0001298	10.769	0.002	10.772	−7.687	0.136	−7.551	−0.01733	0.00034	−0.01699	−173.46	−0.3572	0.006214	−32
−31	0.0001379	10.795	0.002	10.797	−7.447	0.145	−7.302	−0.01677	0.00036	−0.01641	−173.02	−0.3561	0.006601	−31
−30	0.0001465	10.820	0.003	10.822	−7.206	0.154	−7.053	−0.01621	0.00038	−0.01583	−172.58	−0.3551	0.007009	−30
−29	0.0001555	10.845	0.003	10.848	−6.966	0.163	−6.803	−0.01565	0.00040	−0.01525	−172.14	−0.3541	0.007442	−29
−28	0.0001650	10.871	0.003	10.873	−6.726	0.173	−6.553	−0.01510	0.00043	−0.01467	−171.70	−0.3531	0.007898	−28
−27	0.0001751	10.896	0.003	10.899	−6.486	0.184	−6.302	−0.01454	0.00045	−0.01409	−171.25	−0.3520	0.008381	−27
−26	0.0001858	10.921	0.003	10.924	−6.245	0.195	−6.051	−0.01399	0.00048	−0.01351	−170.81	−0.3510	0.008890	−26
−25	0.0001970	10.947	0.003	10.950	−6.005	0.207	−5.798	−0.01343	0.00051	−0.01293	−170.36	−0.3500	0.009428	−25
−24	0.0002088	10.972	0.004	10.976	−5.765	0.220	−5.545	−0.01288	0.00054	−0.01235	−169.92	−0.3489	0.009995	−24
−23	0.0002214	10.997	0.004	11.001	−5.525	0.233	−5.292	−0.01233	0.00057	−0.01176	−169.47	−0.3479	0.010594	−23
−22	0.0002346	11.022	0.004	11.027	−5.284	0.247	−5.038	−0.01178	0.00060	−0.01118	−169.02	−0.3469	0.011226	−22
−21	0.0002485	11.048	0.004	11.052	−5.044	0.261	−4.783	−0.01123	0.00063	−0.01060	−168.57	−0.3459	0.011893	−21
−20	0.0002632	11.073	0.005	11.078	−4.804	0.277	−4.527	−0.01069	0.00067	−0.01002	−168.12	−0.3448	0.012595	−20
−19	0.0002786	11.098	0.005	11.103	−4.564	0.293	−4.271	−0.01014	0.00071	−0.00943	−167.67	−0.3438	0.013336	−19
−18	0.0002950	11.124	0.005	11.129	−4.324	0.311	−4.013	−0.00960	0.00075	−0.00885	−167.21	−0.3428	0.014117	−18
−17	0.0003121	11.149	0.006	11.155	−4.084	0.329	−3.754	−0.00905	0.00079	−0.00826	−166.76	−0.3418	0.014939	−17
−16	0.0003303	11.174	0.006	11.180	−3.843	0.348	−3.495	−0.00851	0.00083	−0.00768	−166.30	−0.3407	0.015806	−16
−15	0.0003493	11.200	0.006	11.206	−3.603	0.368	−3.235	−0.00797	0.00088	−0.00709	−165.85	−0.3397	0.016718	−15
−14	0.0003694	11.225	0.007	11.232	−3.363	0.390	−2.973	−0.00743	0.00093	−0.00650	−165.39	−0.3387	0.017679	−14
−13	0.0003905	11.250	0.007	11.257	−3.123	0.412	−2.710	−0.00689	0.00098	−0.00591	−164.93	−0.3377	0.018690	−13
−12	0.0004128	11.276	0.007	11.283	−2.882	0.436	−2.447	−0.00635	0.00103	−0.00532	−164.47	−0.3366	0.019754	−12
−11	0.0004362	11.301	0.008	11.309	−2.642	0.460	−2.182	−0.00582	0.00109	−0.00473	−164.01	−0.3356	0.020873	−11
−10	0.0004608	11.326	0.008	11.335	−2.402	0.487	−1.915	−0.00528	0.00115	−0.00414	−163.55	−0.3346	0.022050	−10
−9	0.0004867	11.351	0.009	11.360	−2.162	0.514	−1.647	−0.00475	0.00121	−0.00354	−163.09	−0.3335	0.023289	−9

-8	0.0005139	11.377	0.009	11.386	-1.922	0.543	-1.378	-0.00422	0.00127	-0.00294	-162.63	-0.3325	0.024591	-8
-7	0.0005425	11.402	0.010	11.412	-1.681	0.574	-1.108	-0.00369	0.00134	-0.00234	-162.17	-0.3315	0.025959	-7
-6	0.0005726	11.427	0.010	11.438	-1.441	0.606	-0.835	-0.00316	0.00141	-0.00174	-161.70	-0.3305	0.027397	-6
-5	0.0006041	11.453	0.011	11.464	-1.201	0.640	-0.561	-0.00263	0.00145	-0.00114	-161.23	-0.3294	0.028907	-5
-4	0.0006373	11.478	0.012	11.490	-0.961	0.675	-0.286	-0.00210	0.00157	-0.00053	-160.77	-0.3284	0.030494	-4
-3	0.0006722	11.503	0.012	11.516	-0.721	0.712	-0.008	-0.00157	0.00165	0.00008	-160.30	-0.3274	0.032160	-3
-2	0.0007088	11.529	0.013	11.542	-0.480	0.751	0.271	-0.00105	0.00174	0.00069	-159.83	-0.3264	0.033909	-2
-1	0.0007472	11.554	0.014	11.568	-0.240	0.792	0.552	-0.00052	0.00183	0.00130	-159.36	-0.3253	0.035744	-1
0	0.0007875	11.579	0.015	11.594	0.0	0.835	0.835	0.00000	0.00152	0.00192	-158.89	-0.3243	0.037671	0
1	0.0008298	11.604	0.015	11.620	0.240	0.880	1.121	0.00052	0.00202	0.00254	-158.42	-0.3233	0.039694	1
2	0.0008742	11.630	0.016	11.646	0.480	0.928	1.408	0.00104	0.00212	0.00317	-157.95	-0.3223	0.041814	2
3	0.0009207	11.655	0.017	11.672	0.721	0.978	1.699	0.00156	0.00223	0.00380	-157.47	-0.3212	0.044037	3
4	0.0009695	11.680	0.018	11.699	0.961	1.030	1.991	0.00208	0.00235	0.00443	-157.00	-0.3202	0.046370	4
5	0.0010207	11.706	0.019	11.725	1.201	1.085	2.286	0.00260	0.00247	0.00506	-156.52	-0.3192	0.048814	5
6	0.0010743	11.731	0.020	11.751	1.441	1.143	2.584	0.00311	0.00259	0.00570	-156.05	-0.3182	0.051375	6
7	0.0011306	11.756	0.021	11.778	1.681	1.203	2.884	0.00363	0.00635	0.00272	-155.57	-0.3171	0.054060	7
8	0.0011895	11.782	0.022	11.804	1.922	1.266	3.188	0.00414	0.00286	0.00700	-155.09	-0.3161	0.056872	8
9	0.0012512	11.807	0.224	11.831	2.162	1.332	3.494	0.00466	0.00300	0.00766	-154.61	-0.3151	0.059819	9
10	0.0013158	11.832	0.025	11.857	2.402	1.402	3.804	0.00517	0.00315	0.00832	-154.13	-0.3141	0.062901	10
11	0.0013835	11.857	0.026	11.884	2.642	1.474	4.117	0.00568	0.00330	0.00898	-153.65	-0.3130	0.066131	11
12	0.0014544	11.883	0.028	11.910	2.882	1.550	4.433	0.00619	0.00347	0.00966	-153.17	-0.3120	0.069511	12
13	0.0015286	11.908	0.029	11.937	3.123	1.630	4.753	0.00670	0.00364	0.01033	-152.68	-0.3110	0.073049	13
14	0.0016062	11.933	0.031	11.964	3.363	1.714	5.077	0.00721	0.00381	0.01102	-152.20	-0.3100	0.076751	14
15	0.0016874	11.959	0.032	11.991	3.603	1.801	5.404	0.00771	0.00400	0.01171	-151.71	-0.3089	0.080623	15

TABLE 7-15 Thermodynamic Properties of Moist Air, Standard Atmospheric Pressure, 14.696 psi (29.921 inHg) (Continued)

Temp. t, °F	Humidity Ratio lb_w/lb_{da} W_s	Volume ft³/lb dry air			Enthalpy Btu/lb dry air			Entropy Btu/(lb dry air) °F			Condensed Water		Vapor Press. in. Hg p_s	Temp. °F
		v_a	v_{as}	v_s	h_a	h_{as}	h_s	s_a	s_{as}	s_s	Enthalpy Btu/lb h_w	Entropy Btu/lb·°F s_w		
16	0.0017724	11.984	0.034	12.018	3.843	1.892	5.736	0.00822	0.00419	0.01241	−151.22	−0.3079	0.084673	16
17	0.0018613	12.009	0.036	12.045	4.084	1.988	6.072	0.00872	0.00439	0.01312	−150.74	−0.3069	0.088907	17
18	0.0019543	12.035	0.038	12.072	4.324	2.088	6.412	0.00923	0.00460	0.01383	−150.25	−0.3059	0.093334	18
19	0.0020515	12.060	0.040	12.099	4.564	2.193	6.757	0.00973	0.00482	0.01455	−149.76	−0.3049	0.097962	19
20	0.0021531	12.085	0.042	12.127	4.804	2.303	7.107	0.01023	0.00505	0.01528	−149.27	−0.3038	0.102798	20
21	0.0022592	12.110	0.044	12.154	5.044	2.417	7.462	0.01073	0.00529	0.01602	−148.78	−0.3028	0.107849	21
22	0.0023703	12.136	0.046	12.182	5.285	2.537	7.822	0.01123	0.00554	0.01677	−148.28	−0.3018	0.113130	22
23	0.0024863	12.161	0.048	12.209	5.525	2.662	8.187	0.01173	0.00580	0.01753	−147.79	−0.3008	0.118645	23
24	0.0026073	12.186	0.051	12.237	5.765	2.793	8.558	0.01223	0.00607	0.01830	−147.30	−0.2997	0.124396	24
25	0.0027339	12.212	0.054	12.265	6.005	2.930	8.935	0.01272	0.00636	0.01908	−146.80	−0.2987	0.130413	25
26	0.0028660	12.237	0.056	12.293	6.246	3.073	9.318	0.01322	0.00665	0.01987	−146.30	−0.2977	0.136684	26
27	0.0030039	12.262	0.059	12.321	6.486	3.222	9.708	0.01371	0.00696	0.02067	−145.81	−0.2967	0.143233	27
28	0.0031480	12.287	0.062	12.349	6.726	3.378	10.104	0.01420	0.00728	0.02148	−145.31	−0.2956	0.150066	28
29	0.0032984	12.313	0.065	12.378	6.966	3.541	10.507	0.01470	0.00761	0.02231	−144.81	−0.2946	0.157198	29
30	0.0034552	12.338	0.068	12.406	7.206	3.711	10.917	0.01519	0.00796	0.02315	−144.31	−0.2936	0.164631	30
31	0.0036190	12.363	0.072	12.435	7.447	3.888	11.335	0.01568	0.00832	0.02400	−143.80	−0.2926	0.172390	31
32	0.0037895	12.389	0.075	12.464	7.687	4.073	11.760	0.01617	0.00870	0.02487	−143.30	−0.2915	0.180479	32
32*	0.003790	12.389	0.075	12.464	7.687	4.073	11.760	0.01617	0.00870	0.02487	0.02	0.0000	0.18050	32
33	0.003947	12.414	0.079	12.492	7.927	4.243	12.170	0.01665	0.00905	0.02570	1.03	0.0020	0.18791	33
34	0.004109	12.439	0.082	12.521	8.167	4.420	12.587	0.01714	0.00940	0.02655	2.04	0.0041	0.19559	34
35	0.004277	12.464	0.085	12.550	8.408	4.603	13.010	0.01763	0.00977	0.02740	3.05	0.0061	0.20356	35
36	0.004452	12.490	0.089	12.579	8.648	4.793	13.441	0.01811	0.01016	0.02827	4.05	0.0081	0.21181	36
37	0.004633	12.515	0.093	12.608	8.888	4.990	13.878	0.01860	0.01055	0.02915	5.06	0.0102	0.22035	37
38	0.004820	12.540	0.097	12.637	9.128	5.194	14.322	0.01908	0.01096	0.03004	6.06	0.0122	0.22920	38
39	0.005014	12.566	0.101	12.667	9.369	5.405	14.773	0.01956	0.01139	0.03095	7.07	0.0142	0.23835	39

40	0.005216	12.591	0.105	12.696	9.609	5.624	15.233	0.02004	0.01183	0.03187	8.07	0.0162	0.24784	40
41	0.005424	12.616	0.110	12.726	9.849	5.851	15.700	0.02052	0.01228	0.03281	9.08	0.0182	0.25765	41
42	0.005640	12.641	0.114	12.755	10.089	6.086	16.175	0.02100	0.0 275	0.03375	10.08	0.0202	0.26781	42
43	0.005863	12.667	0.119	12.785	10.330	6.330	16.660	0.02148	0.0 324	0.03472	11.09	0.0222	0.27831	43
44	0.006094	12.692	0.124	12.815	10.570	6.582	17.152	0.02196	0.01374	0.03570	12.09	0.0242	0.28918	44
45	0.006334	12.717	0.129	12.846	10.810	6.843	17.653	0.02244	0.01426	0.03669	13.09	0.0262	0.30042	45
46	0.006581	12.743	0.134	12.877	11.050	7.114	18.164	0.02291	0.01479	0.03770	14.10	0.0282	0.31206	46
47	0.006838	12.768	0.140	12.908	11.291	7.394	18.685	0.02339	0.01534	0.03873	15.10	0.0302	0.32408	47
48	0.007103	12.793	0.146	12.939	11.531	7.684	19.215	0.02386	0.01592	0.03978	16.10	0.0321	0.33651	48
49	0.007378	12.818	0.152	12.970	11.771	7.984	19.756	0.02433	0.C1651	0.04084	17.10	0.0341	0.34937	49
50	0.007661	12.844	0.158	13.001	12.012	8.295	20.306	0.02480	0.C1712	0.04192	18.11	0.0361	0.36264	50
51	0.007955	12.869	0.164	13.033	12.252	8.616	20.868	0.02528	0.C1775	0.04302	19.11	0.0381	0.37636	51
52	0.008259	12.894	0.171	13.065	12.492	8.949	21.441	0.02575	0.01840	0.04415	20.11	0.0400	0.39054	52
53	0.008573	12.920	0.178	13.097	12.732	9.293	22.025	0.02622	0.01907	0.04529	21.11	0.0420	0.40518	53
54	0.008897	12.945	0.185	13.129	12.973	9.648	22.621	0.02668	0.01976	0.04645	22.11	0.0439	0.42030	54
55	0.009233	12.970	0.192	13.152	13.213	10.016	23.229	0.02715	0.02048	0.04763	23.11	0.0459	0.43592	55
56	0.009580	12.995	0.200	13.195	13.453	10.397	23.850	0.02762	0.32122	0.04884	24.11	0.0478	0.45205	56
57	0.009938	13.021	0.207	13.228	13.694	10.790	24.484	0.02808	0.32198	0.05006	25.11	0.0497	0.46870	57
58	0.010309	13.046	0.216	13.262	13.934	11.197	25.131	0.02855	0.02277	0.05132	26.11	0.0517	0.48589	58
59	0.010692	13.071	0.224	13.295	14.174	11.618	25.792	0.02901	0.02358	0.05259	27.11	0.0536	0.50363	59
60	0.011087	13.096	0.233	13.329	14.415	12.052	26.467	0.02947	0.02442	0.05389	28.11	0.0555	0.52193	60
61	0.011496	13.122	0.242	13.564	14.655	12.502	27.157	0.02994	0.02528	0.05522	29.12	0.0575	0.54082	61
62	0.011919	13.147	0.251	13.398	14.895	12.966	27.862	0.03040	0.02617	0.05657	30.11	0.0594	0.56032	62
63	0.012355	13.172	0.261	13.433	15.135	13.446	28.582	0.03086	0.02709	0.05795	31.11	0.0613	0.58041	63

TABLE 7-15 Thermodynamic Properties of Moist Air, Standard Atmospheric Pressure, 14.696 psi (29.921 inHg) *(Continued)*

Temp. t, °F	Humidity Ratio lb$_w$/lb$_{da}$[a] W_s	Volume ft³/lb dry air			Enthalpy Btu/lb dry air			Entropy Btu/(lb dry air) °F			Condensed Water			
											Enthalpy Btu/lb h_w	Entropy Btu/lb·°F s_w	Vapor Press. in. Hg p_s	Temp. °F
		v_a	v_{as}	v_s	h_a	h_{as}	h_s	s_a	s_{as}	s_s				
64	0.012805	13.198	0.271	13.468	15.376	13.942	29.318	0.03132	0.02804	0.05936	32.11	0.0632	0.60113	64
65	0.013270	13.223	0.281	13.504	15.616	14.454	30.071	0.03178	0.02902	0.06080	33.11	0.0651	0.62252	65
66	0.013750	13.248	0.292	13.540	15.856	14.983	30.840	0.03223	0.03003	0.06226	34.11	0.0670	0.64454	66
67	0.014246	13.273	0.303	13.577	16.097	15.530	31.626	0.03269	0.03107	0.06376	35.11	0.0689	0.66725	67
68	0.014758	13.299	0.315	13.613	16.337	16.094	32.431	0.03315	0.03214	0.06529	36.11	0.0708	0.69065	68
69	0.015286	13.324	0.326	13.650	16.577	16.677	33.254	0.03360	0.03325	0.06685	37.11	0.0727	0.71479	69
70	0.015832	13.349	0.339	13.688	16.818	17.279	34.097	0.03406	0.03438	0.06844	38.11	0.0746	0.73966	70
71	0.016395	13.375	0.351	13.726	17.058	17.901	34.959	0.03451	0.03556	0.07007	39.11	0.0765	0.76528	71
72	0.016976	13.400	0.365	13.764	17.299	18.543	35.841	0.03496	0.03677	0.07173	40.11	0.0783	0.79167	72
73	0.017575	13.425	0.378	13.803	17.539	19.204	36.743	0.03541	0.03801	0.07343	41.11	0.0802	0.81882	73
74	0.018194	13.450	0.392	13.843	17.779	19.889	37.668	0.03586	0.03930	0.07516	42.11	0.0821	0.84684	74
75	0.018833	13.476	0.407	13.882	18.020	20.595	38.615	0.03631	0.04062	0.07694	43.11	0.0840	0.87567	75
76	0.019491	13.501	0.422	13.923	18.260	21.323	39.583	0.03676	0.04199	0.07875	44.10	0.0858	0.90533	76
77	0.020170	13.526	0.437	13.963	18.500	22.075	40.576	0.03721	0.04339	0.08060	45.10	0.0877	0.93589	77
78	0.020871	13.551	0.453	14.005	18.741	22.851	41.592	0.03766	0.04484	0.08250	46.10	0.0896	0.96733	78
79	0.021594	13.577	0.470	14.046	18.981	23.652	42.633	0.03811	0.04633	0.08444	47.10	0.0914	0.99970	79
80	0.022340	13.602	0.487	14.089	19.222	24.479	43.701	0.03855	0.04787	0.08642	48.10	0.0933	1.03302	80
81	0.023109	13.627	0.505	14.132	19.462	25.332	44.794	0.03900	0.04945	0.08844	49.10	0.0951	1.06728	81
82	0.023902	13.653	0.523	14.175	19.702	26.211	45.913	0.03944	0.05108	0.09052	50.10	0.0970	1.10252	82
83	0.024720	13.678	0.542	14.220	19.943	27.120	47.062	0.03988	0.05276	0.09264	51.09	0.0988	1.13882	83
84	0.025563	13.703	0.561	14.264	20.183	28.055	48.238	0.04033	0.05448	0.09481	52.09	0.1006	1.17608	84
85	0.026433	13.728	0.581	14.310	20.424	29.021	49.445	0.04077	0.05626	0.09703	53.09	0.1025	1.21445	85
86	0.027329	13.754	0.602	14.356	20.664	30.017	50.681	0.04121	0.05809	0.09930	54.09	0.1043	1.25388	86
87	0.028254	13.779	0.624	14.403	20.905	31.045	51.949	0.04165	0.05998	0.10163	55.09	0.1061	1.29443	87

88	0.029208	13.804	0.646	14.450	21.145	32.105	53.250	0.04209	0.06192	0.10401	56.09	0.1080	1.33613	88
89	0.030189	13.829	0.669	14.498	21.385	33.197	54.582	0.04253	0.06392	0.10645	57.09	0.1098	1.37893	89
90	0.031203	13.855	0.692	14.547	21.626	34.325	55.951	0.04297	0.05598	0.10895	58.08	0.1116	1.42298	90
91	0.032247	13.880	0.717	14.597	21.866	35.489	57.355	0.06810	0.04340	0.11150	59.08	0.1134	1.46824	91
92	0.033323	13.905	0.742	14.647	22.107	36.687	58.794	0.04384	0.07028	0.11412	60.08	0.1152	1.51471	92
93	0.034433	13.930	0.768	14.699	22.347	37.924	60.271	0.04427	0.07253	0.11680	61.08	0.1170	1.56248	93
94	0.035577	13.956	0.795	14.751	22.588	39.199	61.787	0.04471	0.07484	0.11955	62.08	0.1188	1.61154	94
95	0.036757	13.981	0.823	14.804	22.828	40.515	63.343	0.04514	0.07722	0.12237	63.08	0.1206	1.66196	95
96	0.037972	14.006	0.852	14.858	23.069	41.871	64.940	0.04558	0.07968	0.12525	64.07	0.1224	1.71372	96
97	0.039225	14.032	0.881	14.913	23.309	43.269	66.578	0.04601	0.08220	0.12821	65.07	0.1242	1.76685	97
98	0.040516	14.057	0.912	14.969	23.550	44.711	68.260	0.04644	0.38480	0.13124	66.07	0.1260	1.82141	98
99	0.041848	14.082	0.944	15.026	23.790	46.198	69.988	0.04687	0.08747	0.13434	67.07	0.1278	1.87745	99
100	0.043219	14.107	0.976	15.084	24.031	47.730	71.761	0.04730	0.09022	0.13752	68.07	0.1296	1.93492	100
101	0.044634	14.133	1.010	15.143	24.271	49.312	73.583	0.04773	0.09306	0.14079	69.07	0.1314	1.99396	101
102	0.046090	14.158	1.045	15.203	24.512	50.940	75.452	0.04816	0.09597	0.14413	70.06	0.1332	2.05447	102
103	0.047592	14.183	1.081	15.264	24.752	52.621	77.373	0.04859	0.09897	0.14756	71.06	0.1349	2.11661	103
104	0.049140	14.208	1.118	15.326	24.993	54.354	79.346	0.04901	0.10206	0.15108	72.06	0.1367	2.18037	104
105	0.050737	14.234	1.156	15.390	25.233	56.142	81.375	0.04944	0.10525	0.15469	73.06	0.1385	2.24581	105
106	0.052383	14.259	1.196	15.455	25.474	57.986	83.460	0.04987	0.10852	0.15839	74.06	0.1402	2.31297	106
107	0.054077	14.284	1.236	15.521	25.714	59.884	85.599	0.05029	0.11189	0.16218	75.06	0.1420	2.38173	107
108	0.055826	14.309	1.279	15.588	25.955	61.844	87.799	0.05071	0.11537	0.16608	76.05	0.1438	2.45232	108
109	0.057628	14.335	1.322	15.657	26.195	63.866	90.061	0.05114	0.11894	0.17008	77.05	0.1455	2.52473	109
110	0.059486	14.360	1.367	15.727	26.436	65.950	92.386	0.05156	0.12262	0.17418	78.05	0.1473	2.59891	110
111	0.061401	14.385	1.414	15.799	26.677	68.099	94.776	0.05198	0.12641	0.17839	79.05	0.1490	2.67500	111

TABLE 7-15 Thermodynamic Properties of Moist Air, Standard Atmospheric Pressure, 14.696 psi (29.921 inHg) *(Continued)*

Temp. t, °F	Humidity Ratio lb$_w$/lb$_{da}$[a] W_s	Volume ft³/lb dry air			Enthalpy Btu/lb dry air			Entropy Btu/(lb dry air) °F			Condensed Water			Temp. °F
		v_a	v_{as}	v_s	h_a	h_{as}	h_s	s_a	s_{as}	s_s	Enthalpy Btu/lb h_w	Entropy Btu/lb·°F s_w	Vapor Press. in. Hg p_s	
112	0.063378	14.411	1.462	15.872	26.917	70.319	97.237	0.05240	0.13032	0.18272	80.05	0.1508	2.75310	112
113	0.065411	14.436	1.511	15.947	27.158	72.603	99.760	0.05282	0.13433	0.18716	81.05	0.1525	2.83291	113
114	0.067512	14.461	1.562	16.023	27.398	74.964	102.362	0.05324	0.13847	0.19172	82.04	0.1543	2.91491	114
115	0.069676	14.486	1.615	16.101	27.639	77.396	105.035	0.05366	0.14274	0.19640	83.04	0.1560	2.99883	115
116	0.071908	14.512	1.670	16.181	27.879	79.906	107.786	0.05408	0.14713	0.20121	84.04	0.1577	3.08488	116
117	0.074211	14.537	1.726	16.263	28.120	82.497	110.617	0.05450	0.15165	0.20615	85.04	0.1595	3.17305	117
118	0.076586	14.562	1.784	16.346	28.361	85.169	113.530	0.05492	0.15631	0.21122	86.04	0.1612	3.26335	118
119	0.079036	14.587	1.844	16.432	28.601	87.927	116.528	0.05533	0.16111	0.21644	87.04	0.1629	3.35586	119
120	0.081560	14.613	1.906	16.519	28.842	90.770	119.612	0.05575	0.16605	0.22180	88.04	0.1647	3.45052	120
121	0.084169	14.638	1.971	16.609	29.083	93.709	122.792	0.05616	0.17115	0.22731	89.04	0.1664	3.54764	121
122	0.086860	14.663	2.037	16.700	29.323	96.742	126.065	0.05658	0.17640	0.23298	90.03	0.1681	3.64704	122
123	0.089633	14.688	2.106	16.794	29.564	99.868	129.432	0.05699	0.18181	0.23880	91.03	0.1698	3.74871	123
124	0.092500	14.714	2.176	16.890	29.805	103.102	132.907	0.05740	0.18739	0.24480	92.03	0.1715	3.85298	124
125	0.095456	14.739	2.250	16.989	30.045	106.437	136.482	0.05781	0.19314	0.25096	93.03	0.1732	3.95961	125
126	0.098504	14.764	2.325	17.090	30.286	109.877	140.163	0.05823	0.19907	0.25729	94.03	0.1749	4.06863	126
127	0.101657	14.789	2.404	17.193	30.527	113.438	143.965	0.05864	0.20519	0.26382	95.03	0.1766	4.18046	127
128	0.104910	14.815	2.485	17.299	30.767	117.111	147.878	0.05905	0.21149	0.27054	96.03	0.1783	4.29477	128
129	0.108270	14.840	2.569	17.409	31.008	120.908	151.916	0.05946	0.21800	0.27745	97.03	0.1800	4.41181	129
130	0.111738	14.865	2.655	17.520	31.249	124.828	156.076	0.05986	0.22470	0.28457	98.03	0.1817	4.53148	130
131	0.115322	14.891	2.745	17.635	31.489	128.880	160.370	0.06027	0.23162	0.29190	99.02	0.1834	4.65397	131
132	0.119023	14.916	2.837	17.753	31.730	133.066	164.796	0.06068	0.23876	0.29944	100.02	0.1851	4.77919	132
133	0.122855	14.941	2.934	17.875	31.971	137.403	169.374	0.06109	0.24615	0.30723	101.02	0.1868	4.90755	133
134	0.126804	14.966	3.033	17.999	32.212	141.873	174.084	0.06149	0.25375	0.31524	102.02	0.1885	5.03844	134
135	0.130895	14.992	3.136	18.127	32.452	146.504	178.957	0.06190	0.26161	0.32351	103.02	0.1902	5.17258	135

136	0.135124	15.017	3.242	18.259	32.693	151.294	183.987	0.06230	0.26973	0.33203	104.02	0.1919	5.30973	136
137	0.139494	15.042	3.352	18.394	32.934	156.245	189.179	0.06271	0.27811	0.34082	105.02	0.1935	5.44985	137
138	0.144019	15.067	3.467	18.534	33.175	161.374	194.548	0.06311	0.28678	0.34989	106.02	0.1952	5.59324	138
139	0.148696	15.093	3.585	18.678	33.415	166.677	200.092	0.06351	0.28678	0.35924	107.02	0.1969	5.73970	139
140	0.153538	15.118	3.708	18.825	33.656	172.168	205.824	0.06391	0.30498	0.36890	108.02	0.1985	5.88945	140
141	0.158552	15.143	3.835	18.978	33.897	177.857	211.754	0.06431	0.31456	0.37887	109.02	0.2002	6.04256	141
142	0.163748	15.168	3.967	19.135	34.138	183.754	217.892	0.06471	0.32446	0.38918	110.02	0.2019	6.19918	142
143	0.169122	15.194	4.103	19.297	34.379	189.855	224.233	0.06511	0.33470	0.39981	111.02	0.2035	6.35898	143
144	0.174694	15.219	4.245	19.464	34.620	196.183	230.802	0.06551	0.34530	0.41081	112.02	0.2052	6.52241	144
145	0.180467	15.244	4.392	19.637	34.860	202.740	237.600	0.06591	0.35626	0.42218	113.02	0.2068	6.68932	145
146	0.186460	15.269	4.545	19.815	35.101	209.550	244.651	0.06631	0.36764	0.43395	114.02	0.2085	6.86009	146
147	0.192668	15.295	4.704	19.999	35.342	216.607	251.949	0.06671	0.37941	0.44611	115.02	0.2101	7.03435	147
148	0.199110	15.320	4.869	20.189	35.583	223.932	259.514	0.06710	0.39160	0.45871	116.02	0.2118	7.21239	148
149	0.205792	15.345	5.040	20.385	35.824	231.533	267.356	0.06750	0.40424	0.47174	117.02	0.2134	7.39413	149
150	0.212730	15.370	5.218	20.589	36.064	239.426	275.490	0.06790	0.41735	0.48524	118.02	0.2151	7.57977	150
151	0.219945	15.396	5.404	20.799	36.305	247.638	283.943	0.06829	0.43096	0.49925	119.02	0.2167	7.76958	151
152	0.227429	15.421	5.596	21.017	36.546	256.158	292.705	0.06868	0.44507	0.51375	120.02	0.2184	7.96306	152
153	0.235218	15.446	5.757	21.243	36.787	265.028	301.816	0.06908	0.45973	0.52881	121.02	0.2200	8.16087	153
154	0.243309	15.471	6.065	21.477	37.028	274.245	311.273	0.06947	0.47494	0.54441	122.02	0.2216	8.36256	154
155	0.251738	15.497	6.223	21.720	37.269	283.849	321.118	0.06986	0.49077	0.56064	123.02	0.2233	8.56871	155
156	0.260512	15.522	6.450	21.972	37.510	293.849	331.359	0.07025	0.50723	0.57749	124.02	0.2249	8.77915	156
157	0.269644	15.547	6.686	22.233	37.751	304.261	342.012	0.07065	0.52434	0.59499	125.02	0.2265	8.99378	157
158	0.279166	15.572	6.933	22.505	37.992	315.120	353.112	0.07104	0.54217	0.61320	126.02	0.2281	9.21297	158
159	0.289101	15.598	7.190	22.788	38.233	326.452	364.685	0.07143	0.56074	0.63216	127.02	0.2297	9.43677	159

TABLE 7-15　Thermodynamic Properties of Moist Air, Standard Atmospheric Pressure, 14.696 psi (29.921 inHg) (Continued)

Temp. t, °F	Humidity Ratio lb_w/lb_da^a W_s	Volume ft³/lb dry air			Enthalpy Btu/lb dry air			Entropy Btu/(lb dry air) °F			Condensed Water			
											Enthalpy Btu/lb	Entropy Btu/lb·°F	Vapor Press. in.Hg	Temp. °F
		v_a	v_{as}	v_s	h_a	h_{as}	h_s	s_a	s_{as}	s_s	h_w	s_w	p_s	
160	0.29945	15.623	7.459	23.082	38.474	338.263	376.737	0.07181	0.58007	0.65188	128.02	0.2314	9.6648	160
161	0.31027	15.648	7.740	23.388	38.715	350.610	389.325	0.07220	0.60025	0.67245	129.02	0.2330	9.8978	161
162	0.32156	15.673	8.034	23.707	38.956	363.501	402.457	0.07259	0.62128	0.69388	130.03	0.2346	10.1353	162
163	0.33336	15.699	8.341	24.040	39.197	376.979	416.175	0.07298	0.64325	0.71623	131.03	0.2362	10.3776	163
164	0.34572	15.724	8.664	24.388	39.438	391.095	430.533	0.07337	0.66622	0.73959	132.03	0.2378	10.6250	164
165	0.35865	15.749	9.001	24.750	39.679	405.865	445.544	0.07375	0.69022	0.76397	133.03	0.2394	10.8771	165
166	0.37220	15.774	9.355	25.129	39.920	421.352	461.271	0.07414	0.71535	0.78949	134.03	0.2410	11.1343	166
167	0.38639	15.800	9.726	25.526	40.161	437.578	477.739	0.07452	0.74165	0.81617	135.03	0.2426	11.3965	167
168	0.40131	15.825	10.117	25.942	40.402	454.630	495.032	0.07491	0.76925	0.84415	136.03	0.2442	11.6641	168
169	0.41698	15.850	10.527	26.377	40.643	472.554	513.197	0.07529	0.79821	0.87350	137.04	0.2458	11.9370	169
170	0.43343	15.875	10.959	26.834	40.884	491.372	532.256	0.07567	0.82858	0.90425	138.04	0.2474	12.2149	170
171	0.45079	15.901	11.414	27.315	41.125	511.231	552.356	0.07606	0.86058	0.93664	139.04	0.2490	12.4988	171
172	0.46905	15.926	11.894	27.820	41.366	532.138	573.504	0.07644	0.89423	0.97067	140.04	0.2506	12.7880	172
173	0.48829	15.951	12.400	28.352	41.607	554.160	595.767	0.07682	0.92962	1.00644	141.04	0.2521	13.0823	173
174	0.50867	15.976	12.937	28.913	41.848	577.489	619.337	0.07720	0.96707	1.04427	142.04	0.2537	13.3831	174
175	0.53019	16.002	13.504	29.505	42.089	602.139	644.229	0.07758	1.00657	1.08416	143.05	0.2553	13.6894	175
176	0.55294	16.027	14.103	30.130	42.331	628.197	670.528	0.07796	1.04828	1.12624	144.05	0.2569	14.0010	176
177	0.57710	16.052	14.741	30.793	42.572	655.876	698.448	0.07834	1.09253	1.17087	145.05	0.2585	14.3191	177
178	0.60274	16.078	15.418	31.496	42.813	685.260	728.073	0.07872	1.13943	1.21815	146.05	0.2600	14.6430	178
179	0.63002	16.103	16.139	32.242	43.054	716.524	759.579	0.07910	1.18927	1.26837	147.06	0.2616	14.9731	179
180	0.65911	16.128	16.909	33.037	43.295	749.871	793.166	0.07947	1.24236	1.32183	148.06	0.2632	15.3097	180
181	0.69012	16.153	17.730	33.883	43.536	785.426	828.962	0.07985	1.29888	1.37873	149.06	0.2647	15.6522	181
182	0.72331	16.178	18.609	34.787	43.778	823.487	867.265	0.08023	1.35932	1.43954	150.06	0.2663	16.0014	182
183	0.75885	16.204	19.551	35.755	44.019	864.259	908.278	0.08060	1.42396	1.50457	151.07	0.2679	16.3569	183

184	0.79703	16.229	20.564	36.793	44.260	908.061	952.321	0.08098	1.49332	1.57430	152.07	0.2694	16.7190	184
185	0.83817	16.254	21.656	37.910	44.501	955.261	999.763	0.08135	1.56797	1.64932	153.07	0.2710	17.0880	185
186	0.88251	16.280	22.834	39.113	44.742	1006.149	1050.892	0.08172	1.64834	1.73006	154.08	0.2725	17.4634	186
187	0.93057	16.305	24.111	40.416	44.984	1061.314	1106.298	0.08210	1.73534	1.81744	155.08	0.2741	17.8462	187
188	0.98272	16.330	25.498	41.828	45.225	1121.174	1166.399	0.08247	1.82963	1.91210	156.08	0.2756	18.2357	188
189	1.03951	16.355	27.010	43.365	45.466	1186.382	1231.848	0.08284	1.93221	2.01505	157.09	0.2772	18.6323	189
190	1.10154	16.381	28.661	45.042	45.707	1257.614	1303.321	0.08321	2.04412	2.12733	158.09	0.2787	19.0358	190
191	1.16965	16.406	30.476	46.882	45.949	1335.834	1381.783	0.08359	2.16684	2.25043	159.09	0.2803	19.4468	191
192	1.24471	16.431	32.477	48.908	46.190	1422.047	1468.238	0.08396	2.30193	2.38589	160.10	0.2818	19.8652	192
193	1.32788	16.456	34.695	51.151	46.431	1517.581	1564.013	0.08433	2.45144	2.53576	161.10	0.2834	20.2913	193
194	1.42029	16.481	37.161	53.642	46.673	1623.758	1670.430	0.08470	2.61738	2.70208	162.11	0.2849	20.7244	194
195	1.52396	16.507	39.928	56.435	46.914	1742.879	1789.793	0.08506	2.80332	2.88838	163.11	0.2864	21.1661	195
196	1.64070	16.532	43.046	59.578	47.155	1877.032	1924.188	0.08543	3.01244	3.09787	164.12	0.2880	21.6152	196
197	1.77299	16.557	46.580	63.137	47.397	2029.069	2076.466	0.08580	3.24914	3.33494	165.12	0.2895	22.0714	197
198	1.92472	16.583	50.636	67.218	47.638	2203.464	2251.102	0.08617	3.52030	3.60647	166.13	0.2910	22.5367	198
199	2.09975	16.608	55.316	71.923	47.879	2404.668	2452.547	0.08653	3.83275	3.91929	167.13	0.2926	23.0092	199
200	2.30454	16.633	60.793	77.426	48.121	2640.084	2688.205	0.08690	4.19787	4.28477	168.13	0.2941	23.4906	200

SOURCE: *ASHRAE Handbook—1989 Fundamentals*. Reprinted by permission.

TABLE 7-16 Thermodynamic Properties of Water at Saturation

Temp °F	Absolute Pressure, p psi	Absolute Pressure in. Hg	Specific Volume, ft³/lb Sat. Solid v_i	Evap. v_{ig}	Sat. Vapor v_g	Enthalpy, Btu/lb Sat. Solid h_i	Evap. h_{ig}	Sat. Vapor h_g	Entropy, Btu/lb·°F Sat. Solid s_i	Evap. s_{ig}	Sat. Vapor s_g	Temp °F
−80	0.000116	0.000236	0.01732	1953234	1953234	−193.50	1219.19	1025.69	−0.4067	3.2112	2.8045	−80
−79	0.000125	0.000254	0.01732	1814052	1814052	−193.11	1219.24	1026.13	−0.4056	3.2029	2.7972	−79
−78	0.000135	0.000275	0.01732	1685445	1685445	−192.71	1219.28	1026.57	−0.4046	3.1946	2.7900	−78
−77	0.000145	0.000296	0.01732	1566663	1566663	−192.31	1219.33	1027.02	−0.4036	3.1864	2.7828	−77
−76	0.000157	0.000319	0.01732	1456752	1456752	−191.92	1219.38	1027.46	−0.4025	3.1782	2.7757	−76
−75	0.000169	0.000344	0.01733	1355059	1355059	−191.52	1219.42	1027.90	−0.4015	3.1701	2.7685	−75
−74	0.000182	0.000371	0.01733	1260977	1260977	−191.12	1219.47	1028.34	−0.4005	3.1619	2.7615	−74
−73	0.000196	0.000399	0.01733	1173848	1173848	−190.72	1219.51	1028.79	−0.3994	3.1539	2.7544	−73
−72	0.000211	0.000430	0.01733	1093149	1093149	−190.32	1219.55	1029.23	−0.3984	3.1459	2.7475	−72
−71	0.000227	0.000463	0.01733	1018381	1018381	−189.92	1219.59	1029.67	−0.3974	3.1379	2.7405	−71
−70	0.000245	0.000498	0.01733	949067	949067	−189.52	1219.63	1030.11	−0.3963	3.1299	2.7336	−70
−69	0.000263	0.000536	0.01733	884803	884803	−189.11	1219.67	1030.55	−0.3953	3.1220	2.7267	−69
−68	0.000283	0.000576	0.01733	825187	825187	−188.71	1219.71	1031.00	−0.3943	3.1141	2.7199	−68
−67	0.000304	0.000619	0.01734	769864	769864	−188.30	1219.74	1031.44	−0.3932	3.1063	2.7131	−67
−66	0.000326	0.000664	0.01734	718508	718508	−187.90	1219.78	1031.88	−0.3922	3.0985	2.7063	−66
−65	0.000350	0.000714	0.01734	670800	670800	−187.49	1219.82	1032.32	−0.3912	3.0907	2.6996	−65
−64	0.000376	0.000766	0.01734	626503	626503	−187.08	1219.85	1032.77	−0.3901	3.0830	2.6929	−64
−63	0.000404	0.000822	0.01734	585316	585316	−186.67	1219.88	1033.21	−0.3891	3.0753	2.6862	−63
−62	0.000433	0.000882	0.01734	547041	547041	−186.26	1219.91	1033.65	−0.3881	3.0677	2.6796	−62
−61	0.000464	0.000945	0.01734	511446	511446	−185.85	1219.95	1034.09	−0.3870	3.0601	2.6730	−61
−60	0.000498	0.001013	0.01734	478317	478317	−185.44	1219.98	1034.54	−0.3860	3.0525	2.6665	−60
−59	0.000533	0.001086	0.01735	447495	447495	−185.03	1220.01	1034.98	−0.3850	3.0449	2.6600	−59
−58	0.000571	0.001163	0.01735	418803	418803	−184.61	1220.03	1035.42	−0.3839	3.0374	2.6535	−58
−57	0.000612	0.001246	0.01735	392068	392068	−184.20	1220.06	1035.86	−0.3829	3.0299	2.6470	−57

−56	0.000655	0.001333	C.01735	367172	367172	−183.78	1220.09	1036.30	−0.3819	3.0225	2.6406	−56
−55	0.000701	0.001427	C.01735	343970	343970	−183.37	1220.11	1036.75	−0.3808	3.0151	2.6342	−55
−54	0.000750	0.001526	0.01735	322336	322336	−182.95	1220.14	1037.19	−0.3798	3.0077	2.6279	−54
−53	0.000802	0.001632	0.01735	302157	302157	−182.53	1220.16	1037.63	−0.3788	3.0004	2.6216	−53
−52	0.000857	0.001745	0.01735	283335	283335	−182.11	1220.18	1038.07	−0.3778	2.9931	2.6153	−52
−51	0.000916	0.001865	0.01736	265773	265773	−181.69	1220.21	1038.52	−0.3767	2.9858	2.6091	−51
−50	0.000979	0.001992	0.01736	249381	249381	−181.27	1220.23	1038.96	−0.3757	2.9786	2.6029	−50
−49	0.001045	0.002128	0.01736	234067	234067	−180.85	1220.25	1039.40	−0.3747	2.9714	2.5967	−49
−48	0.001116	0.002272	0.01736	219766	219766	−180.42	1220.26	1039.84	−0.3736	2.9642	2.5906	−48
−47	0.001191	0.002425	0.01736	206398	206398	−180.00	1220.28	1040.28	−0.3726	2.9570	2.5844	−47
−46	0.001271	0.002587	0.01736	193909	193909	−179.57	1220.30	1040.73	−0.3716	2.9499	2.5784	−46
−45	0.001355	0.002760	0.01736	182231	182231	−179.14	1220.31	1041.17	−0.3705	2.9429	2.5723	−45
−44	0.001445	0.002943	0.01736	171304	171304	−178.72	1220.33	1041.61	−0.3695	2.9353	2.5663	−44
−43	0.001541	0.003137	0.01737	161084	161084	−178.29	1220.34	1042.05	−0.3685	2.9288	2.5603	−43
−42	0.001642	0.003343	0.01737	151518	151518	−177.86	1220.36	1042.50	−0.3675	2.9218	2.5544	−42
−41	0.001749	0.003562	0.01737	142566	142566	−177.43	1220.37	1042.94	−0.3664	2.9149	2.5485	−41
−40	0.001863	0.003793	0.01737	134176	134176	−177.00	1220.38	1043.38	−0.3654	2.9080	2.5426	−40
−39	0.001984	0.004039	0.01737	126322	126322	−176.57	1220.39	1043.82	−0.3644	2.9011	2.5367	−39
−38	0.002111	0.004299	0.01737	118959	118959	−176.13	1220.40	1044.27	−0.3633	2.8942	2.5309	−38
−37	0.002247	0.004574	0.01737	112058	112058	−175.70	1220.40	1044.71	−0.3623	2.8874	2.5251	−37
−36	0.002390	0.004866	0.01738	105592	105592	−175.26	1220.41	1045.15	−0.3613	2.8806	2.5193	−36
−35	0.002542	0.005175	0.01738	99522	99522	−174.83	1220.42	1045.59	−0.3603	2.8738	2.5136	−35
−34	0.002702	0.005502	0.01738	93828	93828	−174.39	1220.42	046.03	−0.3592	2.8671	2.5078	−34
−33	0.002872	0.005848	0.01738	88489	88489	−173.95	1220.43	046.48	−0.3582	2.8604	2.5022	−33

TABLE 7-16   Thermodynamic Properties of Water at Saturation (Continued)

Temp °F	Absolute Pressure, p psi	in. Hg	Specific Volume, ft³/lb Sat. Solid v_i	Evap. v_{ig}	Sat. Vapor v_g	Enthalpy, Btu/lb Sat. Solid h_i	Evap. h_{ig}	Sat. Vapor h_g	Entropy, Btu/lb·°F Sat. Solid s_i	Evap. s_{ig}	Sat. Vapor s_g	Temp °F
−32	0.003052	0.006213	0.01738	83474	83474	−173.51	1220.43	1046.92	−0.3572	2.8537	2.4965	−32
−31	0.003242	0.006600	0.01738	78763	78763	−173.07	1220.43	1047.36	−0.3561	2.8470	2.4909	−31
−30	0.003443	0.007009	0.01738	74341	74341	−172.63	1220.43	1047.80	−0.3551	2.8404	2.4853	−30
−29	0.003655	0.007441	0.01738	70187	70187	−172.19	1220.43	1048.25	−0.3541	2.8338	2.4797	−29
−28	0.003879	0.007898	0.01739	66282	66282	−171.74	1220.43	1048.69	−0.3531	2.8272	2.4742	−28
−27	0.004116	0.008380	0.01739	62613	62613	−171.30	1220.43	1049.13	−0.3520	2.8207	2.4687	−27
−26	0.004366	0.008890	0.01739	59161	59161	−170.86	1220.43	1049.57	−0.3510	2.8142	2.4632	−26
−25	0.004630	0.009428	0.01739	55915	55915	−170.41	1220.42	1050.01	−0.3500	2.8077	2.4577	−25
−24	0.004909	0.009995	0.01739	52861	52861	−169.96	1220.42	1050.46	−0.3489	2.8013	2.4523	−24
−23	0.005203	0.010594	0.01739	49986	49986	−169.51	1220.41	1050.90	−0.3479	2.7948	2.4469	−23
−22	0.005514	0.011226	0.01739	47281	47281	−169.07	1220.41	1051.34	−0.3469	2.7884	2.4415	−22
−21	0.005841	0.011892	0.01740	44733	44733	−168.62	1220.40	1051.78	−0.3459	2.7820	2.4362	−21
−20	0.006186	0.012595	0.01740	42333	42333	−168.16	1220.39	1052.22	−0.3448	2.7757	2.4309	−20
−19	0.006550	0.013336	0.01740	40073	40073	−167.71	1220.38	1052.67	−0.3438	2.7694	2.4256	−19
−18	0.006933	0.014117	0.01740	37943	37943	−167.26	1220.37	1053.11	−0.3428	2.7631	2.4203	−18
−17	0.007337	0.014939	0.01740	35934	35934	−166.81	1220.36	1053.55	−0.3418	2.7568	2.4151	−17
−16	0.007763	0.015806	0.01740	34041	34041	−166.35	1220.34	1053.99	−0.3407	2.7506	2.4098	−16
−15	0.008211	0.016718	0.01740	32256	32256	−165.90	1220.33	1054.43	−0.3397	2.7444	2.4046	−15
−14	0.008683	0.017678	0.01741	30572	30572	−165.44	1220.31	1054.87	−0.3387	2.7382	2.3995	−14
−13	0.009179	0.018689	0.01741	28983	28983	−164.98	1220.30	1055.32	−0.3377	2.7320	2.3943	−13
−12	0.009702	0.019753	0.01741	27483	27483	−164.52	1220.28	1055.76	−0.3366	2.7259	2.3892	−12
−11	0.010242	0.020873	0.01741	26067	26067	−164.06	1220.26	1056.20	−0.3356	2.7197	2.3841	−11
−10	0.010830	0.022050	0.01741	24730	24730	−163.60	1220.24	1056.64	−0.3346	2.7136	2.3791	−10
−9	0.011438	0.023288	0.01741	23467	23467	−163.14	1220.22	1057.08	−0.3335	2.7076	2.3740	−9

-8	0.012077	0.024590	0.01741	22274	22274	-162.68	1220.20	1057.53	-0.3325	2.7015	2.3690	-8
-7	0.012749	0.025958	0.01742	21147	21147	-162.21	1220.18	1057.97	-0.3315	2.6955	2.3640	-7
-6	0.013456	0.027396	0.01742	20081	20081	-162.75	1220.16	1058.41	-0.3305	2.6895	2.3591	-6
-5	0.014197	0.028906	0.01742	19074	19074	-161.28	1220.13	1058.85	-0.3294	2.6836	2.3541	-5
-4	0.014977	0.030493	0.01742	18121	18121	-160.82	1220.11	1059.29	-0.3284	2.6776	2.3492	-4
-3	0.015795	0.032159	0.01742	17220	17220	-160.35	1220.08	1059.73	-0.3274	2.6717	2.3443	-3
-2	0.016654	0.033908	0.01742	16367	16367	-159.88	1220.05	1060.17	-0.3264	2.6653	2.3394	-2
-1	0.017556	0.035744	0.01742	15561	15561	-159.41	1220.02	1060.62	-0.3253	2.6599	2.3346	-1
0	0.018502	0.037671	0.01743	14797	14797	-158.94	1220.00	1061.06	-0.3243	2.6541	2.3298	0
1	0.019495	0.039693	0.01743	14073	14073	-158.47	1219.96	1061.50	-0.3233	2.6482	2.3249	1
2	0.020537	0.041813	0.01743	13388	13388	-157.99	1219.93	1061.94	-0.3223	2.6424	2.3202	2
3	0.021629	0.044037	0.01743	12740	12740	-157.52	1219.90	1062.38	-0.3212	2.6367	2.3154	3
4	0.022774	0.046369	0.01743	12125	12125	-157.05	1219.87	1062.82	-0.3202	2.6309	2.3107	4
5	0.023975	0.048813	0.01743	11543	11543	-156.57	1219.83	1063.26	-0.3192	2.6252	2.3060	5
6	0.025233	0.051375	0.01743	10991	10991	-156.09	1219.80	1063.70	-0.3182	2.6194	2.3013	6
7	0.026552	0.054059	0.01744	10468	10468	-155.62	1219.76	1064.14	-0.3171	2.6138	2.2966	7
8	0.027933	0.056872	0.01744	9971	9971	-155.14	1219.72	1064.58	-0.3161	2.6081	2.2920	8
9	0.029379	0.059817	0.01744	9500	9500	-154.66	1219.68	1065.03	-0.3151	2.6024	2.2873	9
10	0.030894	0.062901	0.01744	9054	9054	-154.18	1219.64	1065.47	-0.3141	2.5968	2.2827	10
11	0.032480	0.066131	0.01744	8630	8630	-153.70	1219.60	1065.91	-0.3130	2.5912	2.2782	11
12	0.034140	0.069511	0.01744	8228	8228	-153.21	1219.56	1066.35	-0.3120	2.5856	2.2736	12
13	0.035878	0.073047	0.01745	7846	7846	-152.73	1219.52	1066.79	-0.3110	2.5801	2.2691	13
14	0.037696	0.076748	0.01745	7483	7483	-152.24	1219.47	1067.23	-0.3100	2.5745	2.2645	14
15	0.039597	0.080621	0.01745	7139	7139	-151.76	1219.43	1067.67	-0.3089	2.5690	2.2600	15

TABLE 7-16 Thermodynamic Properties of Water at Saturation *(Continued)*

Temp °F	Absolute Pressure, p psi	in. Hg	Specific Volume, ft³/lb Sat. Solid v_i	Evap. v_{ig}	Sat. Vapor v_g	Enthalpy, Btu/lb Sat. Solid h_i	Evap. h_{ig}	Sat. Vapor h_g	Entropy, Btu/lb·°F Sat. Solid s_i	Evap. s_{ig}	Sat. Vapor s_g	Temp °F
16	0.041586	0.084671	0.01745	6811	6811	−151.27	1219.38	1068.11	−0.3079	2.5635	2.2556	16
17	0.043666	0.088905	0.01745	6501	6501	−150.78	1219.33	1068.55	−0.3069	2.5580	2.2511	17
18	0.045841	0.093332	0.01745	6205	6205	−150.30	1219.28	1068.99	−0.3059	2.5526	2.2467	18
19	0.048113	0.097960	0.01745	5924	5924	−149.81	1219.23	1069.43	−0.3049	2.5471	2.2423	19
20	0.050489	0.102796	0.01746	5657	5657	−149.32	1219.18	1069.87	−0.3038	2.5417	2.2379	20
21	0.052970	0.107849	0.01746	5404	5404	−148.82	1219.13	1070.31	−0.3028	2.5363	2.2335	21
22	0.055563	0.113128	0.01746	5162	5162	−148.33	1219.08	1070.75	−0.3018	2.5309	2.2292	22
23	0.058271	0.118641	0.01746	4932	4932	−147.84	1219.02	1071.19	−0.3008	2.5256	2.2248	23
24	0.061099	0.124398	0.01746	4714	4714	−147.34	1218.97	1071.63	−0.2997	2.5203	2.2205	24
25	0.064051	0.130419	0.01746	4506	4506	−146.85	1218.91	1072.07	−0.2987	2.5149	2.2162	25
26	0.067133	0.136684	0.01747	4308	4308	−146.35	1218.85	1072.50	−0.2977	2.5096	2.2119	26
27	0.070349	0.143233	0.01747	4119	4119	−145.85	1218.80	1072.94	−0.2967	2.5044	2.2077	27
28	0.073706	0.150066	0.01747	3940	3940	−145.35	1218.74	1073.38	−0.2956	2.4991	2.2035	28
29	0.077207	0.157195	0.01747	3769	3769	−144.85	1218.68	1073.82	−0.2946	2.4939	2.1992	29
30	0.080860	0.164632	0.01747	3606	3606	−144.35	1218.61	1074.26	−0.2936	2.4886	2.1951	30
31	0.084669	0.172387	0.01747	3450	3450	−143.85	1218.55	1074.70	−0.2926	2.4834	2.1909	31
32	0.088640	0.180474	0.01747	3302	3302	−143.35	1218.49	1075.14	−0.2915	2.4783	2.1867	32
32*	0.08865	0.18049	0.01602	3302.07	3302.09	−0.02	1075.15	1075.14	0.0000	2.1867	2.1867	32
33	0.09229	0.18791	0.01602	3178.15	3178.16	0.99	1074.59	1075.58	0.0020	2.1811	2.1832	33
34	0.09607	0.19559	0.01602	3059.47	3059.49	2.00	1074.02	1076.01	0.0041	2.1756	2.1796	34
35	0.09998	0.20355	0.01602	2945.66	2945.68	3.00	1073.45	1076.45	0.0061	2.1700	2.1761	35
36	0.10403	0.21180	0.01602	2836.60	2836.61	4.01	1072.88	1076.89	0.0081	2.1645	2.1726	36
37	0.10822	0.22035	0.01602	2732.13	2732.15	5.02	1072.32	1077.33	0.0102	2.1590	2.1692	37
38	0.11257	0.22919	0.01602	2631.88	2631.89	6.02	1071.75	1077.77	0.0122	2.1535	2.1657	38
39	0.11707	0.23835	0.01602	2535.86	2535.88	7.03	1071.18	1078.21	0.0142	2.1481	2.1623	39

40	0.12172	0.24783	0.01602	2443.67	2443.69	8.03	1070.62	1078.65	0.0162	2.1426	2.1589	40
41	0.12654	0.25765	0.01602	2355.22	2355.24	9.04	1070.05	1079.09	0.0182	2.1372	2.1554	41
42	0.13153	0.26780	0.01602	2270.42	2270.43	10.04	1069.48	1079.52	0.0202	2.1318	2.1521	42
43	0.13669	0.27831	0.01602	2189.02	2189.04	11.04	1068.92	1079.96	0.0222	2.1265	2.1487	43
44	0.14203	0.28918	0.01602	2110.92	2110.94	12.05	1068.35	1080.40	0.0242	2.1211	2.1454	44
45	0.14755	0.30042	0.01602	2035.91	2035.92	13.05	1067.79	1080.84	0.0262	2.1158	2.1420	45
46	0.15326	0.31205	0.01602	1963.85	1963.87	14.05	1067.22	1081.28	0.0282	2.1105	2.1387	46
47	0.15917	0.32407	0.01602	1894.71	1894.73	15.06	1066.66	1081.71	0.0302	2.1052	2.1354	47
48	0.16527	0.33650	0.01602	1828.28	1828.30	16.06	1066.09	1082.15	0.0321	2.1000	2.1321	48
49	0.17158	0.34935	0.01602	1764.44	1764.46	17.06	1065.53	1082.59	0.0341	2.0947	2.1288	49
50	0.17811	0.36263	0.01602	1703.18	1703.20	18.06	1064.96	1083.03	0.0361	2.0895	2.1256	50
51	0.18484	0.37635	0.01602	1644.25	1644.26	19.06	1064.40	1083.46	0.0381	2.0843	2.1224	51
52	0.19181	0.39054	0.01603	1587.64	1587.65	20.07	1063.83	1083.90	0.0400	2.0791	2.1191	52
53	0.19900	0.40518	0.01603	1533.22	1533.24	21.07	1063.27	1084.34	0.0420	2.0740	2.1159	53
54	0.20643	0.42031	0.01603	1480.89	1480.91	22.07	1062.71	1084.77	0.0439	2.0689	2.1128	54
55	0.21410	0.43592	0.01603	1430.61	1430.62	23.07	1062.14	1085.21	0.0459	2.0637	2.1096	55
56	0.22202	0.45204	0.01603	1382.19	1382.21	24.07	1061.58	1085.65	0.0478	2.0586	2.1064	56
57	0.23020	0.46869	0.01603	1335.65	1335.67	25.07	1061.01	1086.08	0.0497	2.0536	2.1033	57
58	0.23864	0.48588	0.01603	1290.85	1290.87	26.07	1060.45	1086.52	0.0517	2.0485	2.1002	58
59	0.24735	0.50362	0.01603	1247.76	1247.78	27.07	1059.89	1086.96	0.0536	2.0435	2.0971	59
60	0.25635	0.52192	0.01604	1206.30	1206.32	28.07	1059.32	1087.39	0.0555	2.0385	2.0940	60
61	0.26562	0.54081	0.01604	1166.38	1166.40	29.07	1058.76	1087.83	0.0575	2.0334	2.0909	61
62	0.27519	0.56029	0.01604	1127.93	1127.94	30.07	1058.19	1088.27	0.0594	2.0285	2.0878	62
63	0.28506	0.58039	0.01604	1090.94	1090.95	31.07	1057.63	1088.70	0.0613	2.0235	2.0848	63

TABLE 7-16   Thermodynamic Properties of Water at Saturation (Continued)

Temp °F	Absolute Pressure, p psi	in. Hg	Specific Volume, ft³/lb Sat. Solid v_i	Evap. v_{ig}	Sat. Vapor v_g	Enthalpy, Btu/lb Sat. Solid h_i	Evap. h_{ig}	Sat. Vapor h_g	Entropy, Btu/lb·°F Sat. Solid s_i	Evap. s_{ig}	Sat. Vapor s_g	Temp °F
64	0.29524	0.60112	0.01604	1055.32	1055.33	32.07	1057.07	1089.14	0.0632	2.0186	2.0818	64
65	0.30574	0.62249	0.01604	1020.98	1021.00	33.07	1056.50	1089.57	0.0651	2.0136	2.0787	65
66	0.31656	0.64452	0.01604	987.95	987.97	34.07	1055.94	1090.01	0.0670	2.0087	2.0758	66
67	0.32772	0.66724	0.01605	956.11	956.12	35.07	1055.37	1090.44	0.0689	2.0039	2.0728	67
68	0.33921	0.69065	0.01605	925.44	925.45	36.07	1054.81	1090.88	0.0708	1.9990	2.0698	68
69	0.35107	0.71478	0.01605	895.86	895.87	37.07	1054.24	1091.31	0.0727	1.9941	2.0668	69
70	0.36328	0.73964	0.01605	867.34	867.36	38.07	1053.68	1091.75	0.0746	1.9893	2.0639	70
71	0.37586	0.76526	0.01605	839.87	839.88	39.07	1053.11	1092.18	0.0765	1.9845	2.0610	71
72	0.38882	0.79164	0.01606	813.37	813.39	40.07	1052.55	1092.61	0.0783	1.9797	2.0580	72
73	0.40217	0.81883	0.01606	787.85	787.87	41.07	1051.98	1093.05	0.0802	1.9749	2.0552	73
74	0.41592	0.84682	0.01606	763.19	763.21	42.06	1051.42	1093.48	0.0821	1.9702	2.0523	74
75	0.43008	0.87564	0.01606	739.42	739.44	43.06	1050.85	1093.92	0.0840	1.9654	2.0494	75
76	0.44465	0.90532	0.01606	716.51	716.53	44.06	1050.29	1094.35	0.0858	1.9607	2.0465	76
77	0.45966	0.93587	0.01607	694.38	694.40	45.06	1049.72	1094.78	0.0877	1.9560	2.0437	77
78	0.47510	0.96732	0.01607	673.05	673.06	46.06	1049.16	1095.22	0.0896	1.9513	2.0409	78
79	0.49100	0.99968	0.01607	652.44	652.46	47.06	1048.59	1095.65	0.0914	1.9466	2.0380	79
80	0.50736	1.03298	0.01607	632.54	632.56	48.06	1048.03	1096.08	0.0933	1.9420	2.0352	80
81	0.52419	1.06725	0.01608	613.35	613.37	49.06	1047.46	1096.51	0.0951	1.9373	2.0324	81
82	0.54150	1.10250	0.01608	594.82	594.84	50.05	1046.89	1096.95	0.0970	1.9327	2.0297	82
83	0.55931	1.13877	0.01608	576.90	576.92	51.05	1046.33	1097.38	0.0988	1.9281	2.0269	83
84	0.57763	1.17606	0.01608	559.63	559.65	52.05	1045.76	1097.81	0.1006	1.9235	2.0242	84
85	0.59647	1.21442	0.01609	542.93	542.94	53.05	1045.19	1098.24	0.1025	1.9189	2.0214	85
86	0.61584	1.25385	0.01609	526.80	526.81	54.05	1044.63	1098.67	0.1043	1.9144	2.0187	86
87	0.63575	1.29440	0.01609	511.21	511.22	55.05	1044.06	1099.11	0.1061	1.9098	2.0160	87

88	0.65622	1.33608	0.01609	496.14	496.15	56.05	1043.49	1099.54	0.1080	1.9053	2.0133	88
89	0.67726	1.37892	0.01610	481.60	481.61	57.04	1042.92	1099.97	0.1098	1.9008	2.0106	89
90	0.69889	1.42295	0.01610	467.52	467.53	58.04	1042.36	1100.40	0.1116	1.8963	2.0079	90
91	0.72111	1.46820	0.01610	453.91	453.93	59.04	1041.79	1100.83	0.1134	1.8918	2.0053	91
92	0.74394	1.51468	0.01611	440.76	440.78	60.04	1041.22	1101.26	0.1152	1.8874	2.0026	92
93	0.76740	1.56244	0.01611	428.04	428.06	61.04	1040.65	1101.69	0.1170	1.8829	2.0000	93
94	0.79150	1.61151	0.01611	415.74	415.76	62.04	1040.08	1102.12	0.1188	1.8785	1.9973	94
95	0.81625	1.66189	0.01612	403.84	403.86	63.03	1039.51	1102.55	0.1206	1.8741	1.9947	95
96	0.84166	1.71364	0.01612	392.33	392.34	64.03	1038.95	1102.98	0.1224	1.8697	1.9921	96
97	0.86776	1.76678	0.01612	381.20	381.21	65.03	1038.38	1103.41	0.1242	1.8653	1.9895	97
98	0.89456	1.82134	0.01612	370.42	370.44	66.03	1037.81	1103.84	0.1260	1.8610	1.9870	98
99	0.92207	1.87736	0.01613	359.99	360.01	67.03	1037.24	1104.26	0.1278	1.8566	1.9844	99
100	0.95031	1.93485	0.01613	349.91	349.92	68.03	1036.67	1104.69	0.1296	1.8523	1.9819	100
101	0.97930	1.99387	0.01613	340.14	340.15	69.02	1036.10	1105.12	0.1314	1.8479	1.9793	101
102	1.00904	2.05443	0.01614	330.69	330.71	70.02	1035.53	1105.55	0.1332	1.8436	1.9768	102
103	1.03956	2.11667	0.01614	321.53	321.55	71.02	1034.95	1105.98	0.1349	1.8393	1.9743	103
104	1.07088	2.18034	0.01614	312.67	312.69	72.02	1034.38	1106.40	0.1367	1.8331	1.9718	104
105	1.10301	2.24575	0.01615	304.08	304.10	73.02	1033.81	1106.83	0.1385	1.8308	1.9693	105
106	1.13597	2.31285	0.01615	295.76	295.77	74.02	1033.24	1107.26	0.1402	1.8256	1.9668	106
107	1.16977	2.38168	0.01616	287.71	287.73	75.01	1032.67	1107.68	0.1420	1.8223	1.9643	107
108	1.20444	2.45226	0.01616	279.91	279.92	76.01	1032.10	1108.11	0.1438	1.8181	1.9619	108
109	1.23999	2.52464	0.01616	272.34	272.36	77.01	1031.52	1108.54	0.1455	1.8139	1.9594	109
110	1.27644	2.59885	0.01617	265.02	265.03	78.01	1030.95	1108.96	0.1473	1.8097	1.9570	110
111	1.31381	2.67494	0.01617	257.91	257.93	79.01	1030.38	1109.39	0.1490	1.8055	1.9546	111

TABLE 7-16 Thermodynamic Properties of Water at Saturation (Continued)

Temp °F	Absolute Pressure, p psi	in. Hg	Specific Volume, ft³/lb Sat. Solid v_i	Evap. v_{ig}	Sat. Vapor v_g	Enthalpy, Btu/lb Sat. Solid h_i	Evap. h_{ig}	Sat. Vapor h_g	Entropy, Btu/lb·°F Sat. Solid s_i	Evap. s_{ig}	Sat. Vapor s_g	Temp °F
112	1.35212	2.75293	0.01617	251.02	251.04	80.01	1029.80	1109.81	0.1508	1.8014	1.9521	112
113	1.39138	2.83288	0.01618	244.36	244.38	81.01	1029.23	1110.24	0.1525	1.7972	1.9497	113
114	1.43162	2.91481	0.01618	237.89	237.90	82.00	1028.66	1110.66	0.1543	1.7931	1.9474	114
115	1.47286	2.99878	0.01619	231.62	231.63	83.00	1028.08	1111.09	0.1560	1.7890	1.9450	115
116	1.51512	3.08481	0.01619	225.53	225.55	84.00	1027.51	1111.51	0.1577	1.7849	1.9426	116
117	1.55842	3.17296	0.01619	219.63	219.65	85.00	1026.93	1111.93	0.1595	1.7808	1.9402	117
118	1.60277	3.26327	0.01620	213.91	213.93	86.00	1026.36	1112.36	0.1612	1.7767	1.9379	118
119	1.64820	3.35577	0.01620	208.36	208.37	87.00	1025.78	1112.78	0.1629	1.7726	1.9356	119
120	1.69474	3.45052	0.01620	202.98	202.99	88.00	1025.20	1113.20	0.1647	1.7686	1.9332	120
121	1.74240	3.54755	0.01621	197.75	197.76	89.00	1024.63	1113.62	0.1664	1.7645	1.9309	121
122	1.79117	3.64691	0.01621	192.68	192.69	90.00	1024.05	1114.05	0.1681	1.7605	1.9286	122
123	1.84117	3.74863	0.01622	187.76	187.78	90.99	1023.47	1114.47	0.1698	1.7565	1.9263	123
124	1.89233	3.85282	0.01622	182.98	182.99	91.99	1022.90	1114.89	0.1715	1.7525	1.9240	124
125	1.94470	3.95945	0.01623	178.34	178.36	92.99	1022.32	1115.31	0.1732	1.7485	1.9217	125
126	1.99831	4.06860	0.01623	173.85	173.86	93.99	1021.74	1115.73	0.1749	1.7445	1.9195	126
127	2.05318	4.18032	0.01623	169.47	169.49	94.99	1021.16	1116.15	0.1766	1.7406	1.9172	127
128	2.10934	4.29465	0.01624	165.23	165.25	95.99	1020.58	1116.57	0.1783	1.7366	1.9150	128
129	2.16680	4.41165	0.01624	161.11	161.12	96.99	1020.00	1116.99	0.1800	1.7327	1.9127	129
130	2.22560	4.53136	0.01625	157.11	157.12	97.99	1019.42	1117.41	0.1817	1.7288	1.9105	130
131	2.28576	4.65384	0.01625	153.22	153.23	98.99	1018.84	1117.83	0.1834	1.7249	1.9083	131
132	2.34730	4.77914	0.01626	149.44	149.46	99.99	1018.26	1118.25	0.1851	1.7210	1.9061	132
133	2.41025	4.90730	0.01626	145.77	145.78	100.99	1017.68	1118.67	0.1868	1.7171	1.9039	133
134	2.47463	5.03839	0.01627	142.21	142.23	101.99	1017.10	1119.08	0.1885	1.7132	1.9017	134
135	2.54048	5.17246	0.01627	138.74	138.76	102.99	1016.52	1119.50	0.1902	1.7093	1.8995	135

136	2.60782	5.30956	0.01627	135.37	135.39	103.98	1015.93	1119.92	0.1919	1.7055	1.8974	136
137	2.67667	5.44975	0.01628	132.10	132.12	104.98	1015.35	1120.34	0.1935	1.7017	1.8952	137
138	2.74707	5.59308	0.01628	128.92	128.94	105.98	1014.77	1120.75	0.1952	1.6978	1.8930	138
139	2.81903	5.73961	0.01629	125.83	125.85	106.98	1014.18	1121.17	0.1969	1.6940	1.8909	139
140	2.89260	5.88939	0.01629	122.82	122.84	107.98	1013.60	1121.58	0.1985	1.6902	1.8888	140
141	2.96780	6.04250	0.01630	119.90	119.92	108.98	1013.01	1122.00	0.2002	1.6864	1.8867	141
142	3.04465	6.19897	0.01630	117.05	117.07	109.98	1012.43	1122.41	0.2019	1.6827	1.8845	142
143	3.12320	6.35888	0.01631	114.29	114.31	110.98	1011.84	1122.83	0.2035	1.6789	1.8824	143
144	3.20345	6.52229	0.01631	111.60	111.62	111.98	1011.26	1123.24	0.2052	1.6752	1.8803	144
145	3.28546	6.68926	0.01632	108.99	109.00	112.98	1010.67	1123.66	0.2068	1.6714	1.8783	145
146	3.36924	6.85984	0.01632	106.44	106.45	113.98	1010.09	1124.07	0.2085	1.6677	1.8762	146
147	3.45483	7.03410	0.01633	103.96	103.98	114.98	1009.50	1124.48	0.2101	1.6640	1.8741	147
148	3.54226	7.21211	0.01633	101.55	101.57	115.98	1008.91	1124.89	0.2118	1.6603	1.8721	148
149	3.63156	7.39393	0.01634	99.21	99.22	116.98	1008.32	1125.31	0.2134	1.6566	1.8700	149
150	3.72277	7.57962	0.01634	96.93	96.94	117.98	1007.73	1125.72	0.2151	1.6529	1.8680	150
151	3.81591	7.76925	0.01635	94.70	94.72	118.99	1007.14	1126.13	0.2167	1.6492	1.8659	151
152	3.91101	7.96289	0.01635	92.54	92.56	119.99	1006.55	1126.54	0.2184	1.6455	1.8639	152
153	4.00812	8.16061	0.01636	90.44	90.46	120.99	1005.96	1126.95	0.2200	1.6419	1.8619	153
154	4.10727	8.36247	0.01636	88.39	88.41	121.99	1005.37	1127.36	0.2216	1.6383	1.8599	154
155	4.20848	8.56854	0.01637	86.40	86.41	122.99	1004.78	1127.77	0.2233	1.6346	1.8579	155
156	4.31180	8.77890	0.01637	84.45	84.47	123.99	1004.19	1128.18	0.2249	1.6310	1.8559	156
157	4.41725	8.99360	0.01638	82.56	82.58	124.99	1003.60	1128.59	0.2265	1.6274	1.8539	157
158	4.52488	9.21274	0.01638	80.72	80.73	125.99	1003.00	1128.99	0.2281	1.5238	1.8519	158
159	4.63472	9.43637	0.01639	78.92	78.94	126.99	1002.41	1129.40	0.2297	1.5202	1.8500	159

TABLE 7-16 Thermodynamic Properties of Water at Saturation (Continued)

Temp °F	Absolute Pressure, p psi	in. Hg	Specific Volume, ft³/lb			Enthalpy, Btu/lb			Entropy, Btu/lb·°F			Temp °F
			Sat. Solid v_i	Evap. v_{ig}	Sat. Vapor v_g	Sat. Solid h_i	Evap. h_{ig}	Sat. Vapor h_g	Sat. Solid s_i	Evap. s_{ig}	Sat. Vapor s_g	
160	4.7468	9.6646	0.01639	77.175	77.192	127.99	1001.82	1129.81	0.2314	1.6167	1.8480	160
161	4.8612	9.8974	0.01640	75.471	75.488	128.99	1001.22	1130.22	0.2330	1.6131	1.8461	161
162	4.9778	10.1350	0.01640	73.812	73.829	130.00	1000.63	1130.62	0.2346	1.6095	1.8441	162
163	5.0969	10.3774	0.01641	72.196	72.213	131.00	1000.03	1131.03	0.2362	1.6060	1.8422	163
164	5.2183	10.6246	0.01642	70.619	70.636	132.00	999.43	1131.43	0.2378	1.6025	1.8403	164
165	5.3422	10.8768	0.01642	69.084	69.101	133.00	998.84	1131.84	0.2394	1.5989	1.8383	165
166	5.4685	11.1340	0.01643	67.587	67.604	134.00	998.24	1132.24	0.2410	1.5954	1.8364	166
167	5.5974	11.3963	0.01643	66.130	66.146	135.00	997.64	1132.64	0.2426	1.5919	1.8345	167
168	5.7287	11.6638	0.01644	64.707	64.723	136.01	997.04	1133.05	0.2442	1.5884	1.8326	168
169	5.8627	11.9366	0.01644	63.320	63.336	137.01	996.44	1133.45	0.2458	1.5850	1.8308	169
170	5.9993	12.2148	0.01645	61.969	61.986	138.01	995.84	1133.85	0.2474	1.5815	1.8289	170
171	6.1386	12.4983	0.01646	60.649	60.666	139.01	995.24	1134.25	0.2490	1.5780	1.8270	171
172	6.2806	12.7874	0.01646	59.363	59.380	140.01	994.64	1134.66	0.2506	1.5746	1.8251	172
173	6.4253	13.0821	0.01647	58.112	58.128	141.02	994.04	1135.06	0.2521	1.5711	1.8233	173
174	6.5729	13.3825	0.01647	56.887	56.904	142.02	993.44	1135.46	0.2537	1.5677	1.8214	174
175	6.7232	13.6886	0.01648	55.694	55.711	143.02	992.83	1135.86	0.2553	1.5643	1.8196	175
176	6.8765	14.0006	0.01648	54.532	54.549	144.02	992.23	1136.26	0.2569	1.5609	1.8178	176
177	7.0327	14.3186	0.01649	53.397	53.414	145.03	991.63	1136.65	0.2585	1.5575	1.8159	177
178	7.1918	14.6426	0.01650	52.290	52.307	146.03	991.02	1137.05	0.2600	1.5541	1.8141	178
179	7.3539	14.9727	0.01650	51.210	51.226	147.03	990.42	1137.45	0.2616	1.5507	1.8123	179
180	7.5191	15.3091	0.01651	50.155	50.171	148.04	989.81	1137.85	0.2632	1.5473	1.8105	180
181	7.6874	15.6518	0.01651	49.126	49.143	149.04	989.20	1138.24	0.2647	1.5440	1.8087	181
182	7.8589	16.0008	0.01652	48.122	48.138	150.04	988.60	1138.64	0.2663	1.5406	1.8069	182
183	8.0335	16.3564	0.01653	47.142	47.158	151.05	987.99	1139.03	0.2679	1.5373	1.8051	183

184	8.2114	16.7185	0.01653	46.185	46.202	152.05	987.38	1139.43	0.2694	1.5339	1.8034	184
185	8.3926	17.0874	0.01654	45.251	45.267	153.05	986.77	1139.82	0.2710	1.5306	1.8016	185
186	8.5770	17.4630	0.01654	44.339	44.356	154.06	986.16	1140.22	0.2725	1.5273	1.7998	186
187	8.7649	17.8455	0.01655	43.448	43.465	155.06	985.55	1140.61	0.2741	1.5240	1.7981	187
188	8.9562	18.2350	0.01656	42.579	42.595	156.07	984.94	1141.00	0.2756	1.5207	1.7963	188
189	9.1510	18.6316	0.01656	41.730	41.746	157.07	984.32	1141.39	0.2772	1.5174	1.7946	189
190	9.3493	19.0353	0.01657	40.901	40.918	158.07	983.71	1141.78	0.2787	1.5141	1.7929	190
191	9.5512	19.4464	0.01658	40.092	40.108	159.08	983.10	1142.18	0.2803	1.5109	1.7911	191
192	9.7567	19.8648	0.01658	39.301	39.317	160.08	982.48	1142.57	0.2818	1.5076	1.7894	192
193	9.9659	20.2907	0.01659	38.528	38.544	161.09	981.87	1142.95	0.2834	1.5043	1.7877	193
194	10.1788	20.7242	0.01659	37.774	37.790	162.09	981.25	1143.34	0.2849	1.5011	1.7860	194
195	10.3955	21.1653	0.01660	37.035	37.052	163.10	980.63	1143.73	0.2864	1.4979	1.7843	195
196	10.6160	21.6143	0.01661	36.314	36.331	164.10	980.02	1144.12	0.2880	1.4946	1.7826	196
197	10.8404	22.0712	0.01661	35.611	35.628	165.11	979.40	1144.51	0.2895	1.4914	1.7809	197
198	11.0687	22.5361	0.01662	34.923	34.940	166.11	978.78	1144.89	0.2910	1.4882	1.7792	198
199	11.3010	23.0091	0.01663	34.251	34.268	167.12	978.16	1145.28	0.2926	1.4850	1.7776	199
200	11.5374	23.4904	0.01663	33.594	33.610	168.13	977.54	1145.66	0.2941	1.4818	1.7759	200
201	11.7779	23.9800	0.01664	32.951	32.968	169.13	976.92	1146.05	0.2956	1.4786	1.7742	201
202	12.0225	24.4780	0.01665	32.324	32.340	170.14	976.29	1146.43	0.2971	1.4755	1.7726	202
203	12.2713	24.9847	0.01665	31.710	31.726	171.14	975.67	1146.81	0.2986	1.4723	1.7709	203
204	12.5244	25.5000	0.01666	31.110	31.127	172.15	975.05	1147.20	0.3002	1.4591	1.7693	204
205	12.7819	26.0241	0.01667	30.523	30.540	173.16	974.42	1147.58	0.3017	1.4560	1.7677	205
206	13.0436	26.5571	0.01667	29.949	29.965	174.16	973.80	1147.96	0.3032	1.4628	1.7660	206
207	13.3099	27.0991	0.01668	29.388	29.404	175.17	973.17	1148.34	0.3047	1.4597	1.7644	207

TABLE 7-16 Thermodynamic Properties of Water at Saturation (Continued)

Temp °F	Absolute Pressure, p psi	Absolute Pressure, p in. Hg	Specific Volume, ft³/lb Sat. Solid v_i	Specific Volume, ft³/lb Evap. v_{ig}	Specific Volume, ft³/lb Sat. Vapor v_g	Enthalpy, Btu/lb Sat. Solid h_i	Enthalpy, Btu/lb Evap. h_{ig}	Enthalpy, Btu/lb Sat. Vapor h_g	Entropy, Btu/lb·°F Sat. Solid s_i	Entropy, Btu/lb·°F Evap. s_{ig}	Entropy, Btu/lb·°F Sat. Vapor s_g	Temp °F
208	13.5806	27.6503	0.01669	28.839	28.856	176.18	972.54	1148.72	0.3062	1.4566	1.7628	208
209	13.8558	28.2108	0.01669	28.303	28.319	177.18	971.92	1149.10	0.3077	1.4535	1.7612	209
210	14.1357	28.7806	0.01670	27.778	27.795	178.18	971.29	1149.48	0.3092	1.4503	1.7596	210
212	14.7096	29.9489	0.01671	26.763	26.780	180.20	970.03	1150.23	0.3122	1.4442	1.7564	212
214	15.3025	31.1563	0.01673	25.790	25.807	182.22	968.76	1150.98	0.3152	1.4380	1.7532	214
216	15.9152	32.4036	0.01674	24.861	24.878	184.24	967.50	1151.73	0.3182	1.4319	1.7501	216
218	16.5479	33.6919	0.01676	23.970	23.987	186.25	966.23	1152.48	0.3212	1.4258	1.7469	218
220	17.2013	35.0218	0.01677	23.118	23.134	188.27	964.95	1153.22	0.3241	1.4197	1.7438	220
222	17.8759	36.3956	0.01679	22.299	22.316	190.29	963.67	1153.96	0.3271	1.4136	1.7407	222
224	18.5721	37.8131	0.01680	21.516	21.533	192.31	962.39	1154.70	0.3301	1.4076	1.7377	224
226	19.2905	39.2758	0.01682	20.765	20.782	194.33	961.11	1155.43	0.3330	1.4016	1.7347	226
228	20.0316	40.7848	0.01683	20.045	20.062	196.35	959.82	1156.16	0.3359	1.3957	1.7316	228
230	20.7961	42.3412	0.01684	19.355	19.372	198.37	958.52	1156.89	0.3389	1.3898	1.7287	230
232	21.5843	43.9461	0.01686	18.692	18.709	200.39	957.22	1157.62	0.3418	1.3839	1.7257	232
234	22.3970	45.6006	0.01688	18.056	18.073	202.41	955.92	1158.34	0.3447	1.3780	1.7227	234
236	23.2345	47.3060	0.01689	17.446	17.463	204.44	954.62	1159.06	0.3476	1.3722	1.7198	236
238	24.0977	49.0633	0.01691	16.860	16.877	206.46	953.31	1159.77	0.3505	1.3664	1.7169	238
240	24.9869	50.8738	0.01692	16.298	16.314	208.49	952.00	1160.48	0.3534	1.3606	1.7140	240
242	25.9028	52.7386	0.01694	15.757	15.774	210.51	950.68	1161.19	0.3563	1.3548	1.7111	242
244	26.8461	54.6591	0.01695	15.238	15.255	212.54	949.35	1161.90	0.3592	1.3491	1.7083	244
246	27.8172	56.6364	0.01697	14.739	14.756	214.57	948.03	1162.60	0.3621	1.3434	1.7055	246
248	28.8169	58.6717	0.01698	14.259	14.276	216.60	946.70	1163.29	0.3649	1.3377	1.7026	248
250	29.8457	60.7664	0.01700	13.798	13.815	218.63	945.36	1163.99	0.3678	1.3321	1.6998	250
252	30.9043	62.9218	0.01702	13.355	13.372	220.66	944.02	1164.68	0.3706	1.3264	1.6971	252

254	31.9934	65.1391	0.01703	12.928	12.945	222.69	942.68	1165.37	0.3735	1.3208	1.6943	254
256	33.1135	67.4197	0.01705	1.6916	12.517	12.535	224.72	941.33	1166.05	0.3763	1.3152	256
258	34.2653	69.7649	0.01707	12.123	12.140	226.76	939.97	1166.73	0.3792	1.3097	1.6889	258
260	35.4496	72.1760	0.01708	11.742	11.759	228.79	938.61	1167.40	0.3820	1.3042	1.6862	260
262	36.6669	74.6545	0.01710	11.376	11.393	230.83	937.25	1168.08	0.3848	1.2987	1.6835	262
264	37.9180	77.2017	0.01712	11.024	11.041	232.87	935.88	1168.74	0.3876	1.2932	1.6808	264
266	39.2035	79.8190	0.01714	10.684	10.701	234.90	934.50	1169.41	0.3904	1.2877	1.6781	266
268	40.5241	82.5078	0.01715	10.357	10.374	236.94	933.12	1170.07	0.3932	1.2823	1.6755	268
270	41.8806	85.2697	0.01717	10.042	10.059	238.98	931.74	1173.72	0.3960	1.2769	1.6729	270
272	43.2736	88.1059	0.01719	9.737	9.755	241.03	930.35	1171.38	0.3988	1.2715	1.6703	272
274	44.7040	91.0181	0.01721	9.445	9.462	243.07	928.95	1172.02	0.4016	1.2661	1.6677	274
276	46.1723	94.0076	0.01722	9.162	9.179	245.11	927.55	1172.67	0.4044	1.2608	1.6651	276
278	47.6794	97.0761	0.01724	8.890	8.907	247.16	926.15	1173.31	0.4071	1.2554	1.6626	278
280	49.2260	100.2250	0.01726	8.627	8.644	249.20	924.74	1173.94	0.4099	1.2501	1.6600	280
282	50.8128	103.4558	0.01728	8.373	8.390	251.25	923.32	1174.57	0.4127	1.2448	1.6575	282
284	52.4406	106.7701	0.01730	8.128	8.146	253.30	921.90	1175.20	0.4154	1.2396	1.6550	284
286	54.1103	110.1695	0.01731	7.892	7.910	255.35	920.47	1175.82	0.4182	1.2345	1.6525	286
288	55.8225	113.6556	0.01733	7.664	7.681	257.40	919.03	1176.44	0.4209	1.2291	1.6500	288
290	57.5780	117.2299	0.01735	7.444	7.461	259.45	917.59	1177.05	0.4236	1.2239	1.6476	290
292	59.3777	120.8941	0.01737	7.231	7.248	261.51	916.15	1177.66	0.4264	1.2187	1.6451	292
294	61.2224	124.6498	0.01739	7.026	7.043	263.56	914.69	1178.26	0.4291	1.2136	1.6427	294
296	63.1128	128.4987	0.01741	6.827	6.844	265.62	913.24	1178.86	0.4318	1.2084	1.6402	296
298	65.0498	132.4425	0.01743	6.635	6.652	267.68	911.77	1179.45	0.4345	1.2033	1.6378	298
300	67.0341	136.4827	0.01745	6.450	6.467	269.74	910.30	1130.04	0.4372	1.1982	1.6354	300

SOURCE: *ASHRAE Handbook—1989 Fundamentals.* Reprinted by permission.

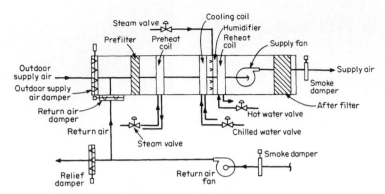

FIGURE 7-7 Central heating–ventilating–air-conditioning unit. (*Source: A. M. Khashab,* Heating, Ventilating and Air Conditioning Systems Estimating Manual, *McGraw-Hill, New York, 1977. Used with permission.*)

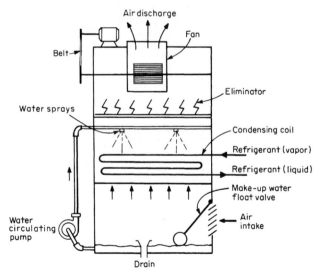

FIGURE 7-8 Evaporative condenser (draw-through type). (*Source: A. M. Khashab,* Heating, Ventilating and Air Conditioning Systems Estimating Manual, *McGraw-Hill, New York, 1977. Used with permission.*)

TABLE 7-17   Quick Checklist for Air-Conditioning Refrigeration Systems

Item	Range
Condensing-water inlet temp	60–95°F
Condensing-water outlet temp	85–115°F
Water-temp rise in condenser	15–25°F
Freon-12 head pressure	100–150 psig
Freon-12 suction pressure	25–50 psig
Temp air enters cooling coil	75–90°F
Temp drop through cooling coil	12–25°F
Temp air leaves cooling coil	50–70°F
Air circulated per ton refrigeration	300–600 cfm
Room temp for comfort cooling	70–86°F
Condensing water	3–5 gpm per ton
Temp chilled-water enters coolers	50–65°F
Temp chilled-water leaves coolers	40–50°F

SOURCE: S. M. Elonka and Q. W. Minich, *Standard Refrigeration and Air Conditioning Questions and Answers,* McGraw-Hill, New York, 1973. Used with permission of publisher.

TABLE 7-18   Troubleshooter's Guide to Refrigeration Problems

TROUBLESHOOTER'S GUIDE TO REFRIGERATION PROBLEMS

Trouble: Compressor Fails to Start

Probable cause	Symptoms	Recommended action
1. Power failure	1. Electric circuit test shows no current on line side of motor starter	1. Check for blown line fuse or broken lead
2. Disconnect switch open	2. Electric circuit test shows no current on line side of motor starter	2. Determine why switch was opened. If everything OK, close switch

TABLE 7-18   Troubleshooter's Guide to Refrigeration Problems (*Continued*)

Trouble: Compressor Fails to Start (*contd*)

Probable cause	Symptoms	Recommended action
3. Fuse blown	3. Electric circuit test shows current on line side but not on motor side of fuse	3. Replace fuse. Check load on motor
4. Low voltage	4. Electric circuit tester glows but not at full brilliance	4. Check with voltmeter, then call power company
5. Burned-out motor	5. Full voltage at motor terminals but motor will not run	5. Repair or replace
6. Inoperative motor starter	6. Test for burned-out holding coil or broken contacts	6. Repair or replace
7. Open control circuit 　a. dual pressure control 　b. oil failure control 　c. motor starter thermal overloads 　d. thermostat not set for cooling 　e. open circuit from "interlocking" relays	7. Motor starter holding coil is not energized	7. Locate open control and determine cause. See individual control instructions
8. Broken or sheared coupling	8. Motor runs but compressor does not	8. Repair or replace. Properly realign
9. Frozen compressor due to locked or damaged mechanism	9. Compressor will not operate	9. Overhaul compressor
10. Suction pressure below cutin setting of low-pressure cutout switch	10. Open contacts on low-pressure switch. Suction pressure below cutin setting	10. Check for loss of refrigerant. Repair leak and recharge
11. Discharge pressure above cutin setting of high-pressure cutout switch	11. Open contacts on high-pressure switch. Discharge pressure above cutin setting	11. Check condenser cooling water, fouled condenser, overcharged system
12. Oil pressure failure control switch has cutout	12. System will restart by resetting oil pressure failure control switch	12. Check oil level, oil pressure, wiring, and control for faulty control

Trouble: Compressor "Short-cycles"

1. Intermittent contact in electric control circuit	1. Normal operation except too frequent stopping and starting	1. Repair or replace faulty electrical control

TABLE 7-18   Troubleshooter's Guide to Refrigeration Problems
(Continued)

Trouble: Compressor "Short-cycles" (contd)

Probable cause	Symptoms	Recommended action
2. Low-pressure controller differential set too close	2. Normal operation except too frequent stopping and starting	2. Reset differential in accordance with proper job conditions
3. Leaky liquid line solenoid valve	3. Valve may hiss when closed. Also temperature change in refrigerant line through valve	3. Repair or replace
4. Dirty or iced evaporator	4. Reduced air flow a. dirty air filters b. broken fan belt c. fan belt tension improperly adjusted	4. Clean or defrost evaporator. Check filters and fan drive
5. Faulty condensing	5. Excessively high discharge pressure	5. Check for water failure or evaporative condenser trouble
6. Overcharge of refrigerant or noncondensable gas	6. High discharge pressure	6. Remove excess refrigerant or purge noncondensable gas
7. Lack of refrigerant	7. Normal operation except too frequent stopping and starting on low-pressure control switch	7. Repair refrigerant leak and recharge
8. Water regulating valve inoperative or restricted by dirt, or water temperature too high	8. High discharge pressure	8. Clean or repair water valve
9. Water piping restricted or supply water pressure too low	9. High discharge pressure	9. Determine cause and correct
10. Restricted liquid line strainer	10. Suction pressure too low and frosting at strainer	10. Clean strainer
11. Faulty motor	11. Motor starts and stops rapidly	11. Repair or replace faulty motor
12. Fouled shell-and-tube condenser	12. Compressor cuts off on high-pressure cutout	12. Clean condenser tubes
13. Faulty operation of evaporative condenser	13. Compressor cuts off on high-pressure cutout a. no water b. spray nozzles clogged c. water pump not operating d. coil surface dirty e. air inlet or outlet obstructed f. fan not operating	13. Determine cause and correct a. fill with water b. clean spray nozzles c. repair faulty pump d. clean coil e. remove obstruction f. repair

TABLE 7-18　Troubleshooter's Guide to Refrigeration Problems (*Continued*)

Trouble: Compressor Runs Continuously		
Probable cause	**Symptoms**	**Recommended action**
1. Excessive load	1. High temperature in conditioned area	1. Check for excessive fresh air or infiltration. Check for inadequate insulation of space
2. Thermostat controlling at too low a temperature	2. Low temperature in conditioned area	2. Reset or repair thermostat
3. "Welded" contacts on electrical control in motor starter circuit	3. Low temperature in conditioned space	3. Repair or replace faulty control
4. Lack of refrigerant	4. Bubbles in sightglass	4. Repair leak and charge
5. Overcharge of refrigerant	5. High discharge pressure	5. Purge or remove excess
6. Leaky valves in compressor	6. Compressor noisy or operating at abnormally low discharge pressure or abnormally high suction pressure	6. Overhaul compressor
7. Solenoid stop valve stuck open or held open by manual lift stem	7. Air-conditioned space too cold	7. Repair valve or restore to automatic operation

Trouble: Compressor Loses Oil		
1. Insufficient oil charge	1. Oil level too low	1. Add sufficient amount of proper compressor oil
2. Clogged strainers or valves	2. Oil level gradually drops	2. Clean or repair and replace
3. Loose expansion valve or remote bulb	3. Excessively cold suction	3. Provide good contact between remote bulb and suction line
4. Liquid flooding back to compressor	4. Excessively cold suction. Noisy compressor operation	4. Readjust superheat setting or check remote bulb contact
5. Short-cycling	5. Too frequent starting and stopping of compressor	5. Defrost; check pressure cutout
6. Crankcase fittings leak oil	6. Oil around compressor base and low crankcase oil level	6. Repair oil leak and add proper refrigerant oil

Trouble: Compressor Is Noisy		
1. Loose compressor drive coupling	1. Coupling bolts loose	1. Tighten coupling and check alignment
2. Lack of oil	2. Compressor cuts out on oil failure control	2. Add oil
3. Dry or scored seal	3. Squeak or squeal when compressor runs	3. Check oil level

TABLE 7-18   Troubleshooter's Guide to Refrigeration Problems (Continued)

Trouble: Compressor Is Noisy (contd)

Probable cause	Symptoms	Recommended action
4. Internal parts of compressor broken	4. Compressor knocks	4. Overhaul compressor
5. Liquid "flood back"	5. Abnormally cold suction line. Compressor knocks	5. Check and adjust superheat. Valve may be too large or remote bulb loose on suction line. Air entering evaporator too cold for complete evaporation of liquid.
6. Dirty water-regulating valve, too high water pressure or intermittent water pressure	6. Water valve chatters or hammers	6. Clean water-regulating valve. Install air chamber ahead of valve
7. Expansion valve stuck in open position	7. Abnormally cold suction line. Compressor knocks	7. Repair or replace
8. Compressor or motor loose on base	8. Compressor or motor jumps on base	8. Tighten motor or compressor hold-down bolts

Trouble: System Short of Capacity

Probable cause	Symptoms	Recommended action
1. Flash gas in liquid line	1. Expansion valve hisses	1. Add refrigerant
2. Clogged strainer or solenoid stop valve	2. Temperature change in refrigerant line through strainer or solenoid stop valve	2. Clean or replace
3. Ice or dirt on evaporator	3. Reduced air flow	3. Clean coil or defrost
4. Expansion valve stuck or obstructed	4. Short-cycling or continuous running	4. Repair or replace expansion valve
5. Excess pressure drop in evaporator	5. Superheat too high	5. Check superheat and reset thermostatic expansion valve
6. Improper superheat adjustment	6. Short-cycling or continuous running	6. Adjust expansion valve. Check superheat and reset thermostatic expansion valve
7. Expansion valve improperly sized	7. Short-cycling or continuous running	7. Replace with correct valve

Trouble: Discharge Pressure Too High

Probable cause	Symptoms	Recommended action
1. Too little or too warm condenser water	1. Excessively warm water leaving condenser	1. Provide adequate cool water, adjust water-regulating valve
2. Fouled tubes in shell-and tube condenser	2. Excessively cool water leaving condenser	2. Clean tubes

TABLE 7-18   Troubleshooter's Guide to Refrigeration Problems
(Continued)

Trouble: Discharge Pressure Too High (contd)		
Probable cause	Symptoms	Recommended action
3. Improper operation of evaporative condenser	3. Low air or spray water volume. Scaled surface	3. Correct air or water flow. Clean coil surface
4. Air or noncondensable gas in system	4. Exceptionally hot condenser and excessive discharge pressure	4. Purge
5. Overcharge of refrigerant	5. Exceptionally hot condenser and excessive discharge pressure	5. Remove excess or purge

Trouble: Discharge Pressure Too Low		
1. Too much condenser water	1. Excessively cold water leaving condenser	1. Adjust water-regulating valve
2. Lack of refrigerant	2. Bubbles in sightglass	2. Repair leak and charge
3. Broken or leaky compressor discharge valves	3. Suction pressure rises faster than 5 lb/min after pressure shutdown	3. Remove head, examine valves, replace faulty ones
4. Leaky relief bypass valve	4. Low discharge pressure and high suction pressure	4. Inspect valve to determine if replacement is necessary

Trouble: Suction Pressure Too High		
1. Excessive load on evaporator	1. Compressor runs continuously	1. Check for excessive fresh air or infiltration, poor insulation of spaces
2. Overfeeding of expansion valve	2. Abnormally cold suction line. Liquid flooding to compressor	2. Regulate superheat setting expansion valve, see remote bulb OK on suction line
3. Expansion valve stuck open	3. Abnormally cold suction line. Liquid flooding to compressor	3. Repair or replace valve
4. Expansion valve too large	4. Abnormally cold suction line. Liquid flooding to compressor	4. Check valve rating, replace if necessary
5. Broken suction valves in compressor	5. Noisy compressor	5. Remove head, examine valves, repair faulty ones

Trouble: Suction Pressure Too Low		
1. Lack of refrigerant	1. Bubbles in sightglass	1. Repair leak, then charge system
2. Light load on evaporator	2. Compressor short-cycles	2. Not enough refrigerant
3. Clogged liquid-line strainer	3. Temp. change in refrigerant line through strainer or solenoid stop valve	3. Clean strainer

TABLE 7-18   Troubleshooter's Guide to Refrigeration Problems
(Continued)

Trouble: Suction Pressure Too Low *(contd)*

Probable cause	Symptoms	Recommended action
4. Expansion valve power assembly has lost charge	4. No flow of refrigerant through valve	4. Replace expansion valve power assembly
5. Obstructed expansion valve	5. Loss of capacity	5. Clean valve or replace if necessary
6. Contacts on control thermostat stuck on closed position	6. Conditioned space too cold	6. Repair thermostat or replace if necessary
7. Compressor capacity control range set too low	7. Compressor short-cycles	7. Reset compressor capacity control range
8. Expansion valve too small	8. Lack of capacity	8. Check valve rating table for correct sizing and replace if necessary
9. Too much pressure drop through evaporator	9. Too high superheat	9. Check for plugged external equalizer

SOURCE: S. M. Elonka and Q. W. Minich, *Standard Refrigeration and Air Conditioning Questions and Answers,* McGraw-Hill, New York, 1973. Used with permission of publisher.

TABLE 7-19   Definition of Psychrometrics Terms

Term	Definition
Atmospheric pressure	Atmospheric pressure variations occur because of site specific elevation above or below sea level.
Chemical dehumidifying	This process usually is an absorption process, using silica gel or a similar moisture absorbent agent to obtain very low humidities.
Cooling and dehumidifying	When the process of sensible cooling is continued until the air is cooled to its dew point temperature.
Density and volume	Refers to weight of moist air, with units of pounds of dry air per cubic foot. Volume is the reciprocal of density.
Dew point	The temperature at which water vapor reaches saturation in the atmosphere and starts to condense.
Effects of altitude	The difference in pressure due to altitude is a significant factor above approximately 2000 ft above sea level. At elevations higher, use high altitude charts when appropriate.

TABLE 7-19   Definition of Psychrometrics Terms (*Continued*)

Term	Definition
Enthalpy	Refers to the heat content of moist air, in Btu per pound of dry air.
Evaporation cooling	A process where water vapor is added to the air stream or flow by adiabatic evaporation; i.e., no heat is added to or subtracted from the system.
Humidifying	The process of adding moisture to the air stream.
Humidity ratio	The quantity of water in the air stated as a ratio, pounds of water per pound of dry air, and often called *specific humidity*.
Mixing	The blending of two or more air streams into one.
Psychrometrics	Particular branch of thermodynamics devoted to the technical study of mixtures of air and water vapor; generally described as moist air.
Psychrometric charts	A primary and essential tool for HVAC design. These charts are simply a graphical expression of the various psychrometric properties.
Psychrometric properties	Basic properties of the air/water vapor mixtures which are used in designing HVAC systems.
Psychrometric tables	Tables of psychrometric properties used in calculating and designing HVAC systems.
Relative humidity	Expresses the relationship of the amount of moisture in the air to the amount the air could hold if saturated at that dry bulb temperature.
Saturated	When the air is saturated, the wet bulb, dry bulb, and dew point temperatures are all equal.
Sensible heating and cooling	*Sensible,* as applied to heating or cooling, means that no moisture is added or subtracted as air temperature increases or decreases.
State points	Any point or exact location on a psychrometric chart can be a *state point*.

TABLE 7-19   Definition of Psychrometrics Terms *(Continued)*

Term	Definition
Temperature—dry bulb	When not otherwise noted, the temperature measured with a typical thermometer is known as *dry bulb* temperature (DB).
Temperature— wet bulb	*Wet-bulb* temperature is measured by a thermometer with a wet cloth sock wrapped around the bulb. The air being measured is blown across the sock allowing moisture to evaporate.

NOTES

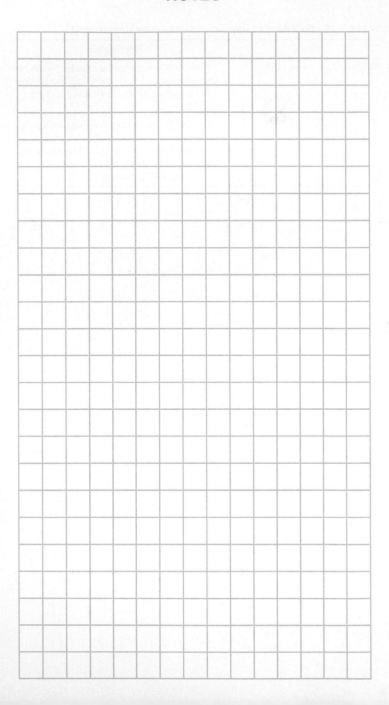

Piping Design

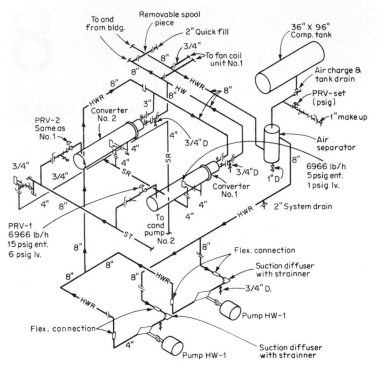

FIGURE 8-1 Hot-water heating piping diagram. (*Source: A. M. Khashab, Heating, Ventilating and Air Conditioning Systems Estimating Manual, McGraw-Hill, New York, 1977. Used with permission.*)

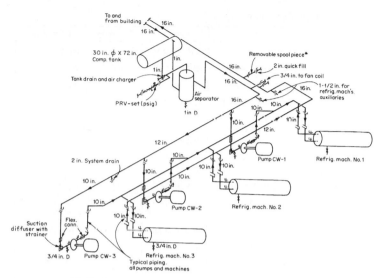

FIGURE 8-2   Chilled-water piping diagram. (*Spool* is a piece of pipe flanged at both ends.) (*Source: A. M. Khashab,* Heating, Ventilating and Air Conditioning Systems Estimating Manual, *McGraw-Hill, New York, 1977. Used with permission.*)

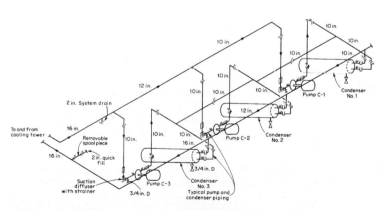

FIGURE 8-3   Condenser water piping diagram. (*Source: A. M. Khashab,* Heating, Ventilating and Air Conditioning Systems Estimating Manual, *McGraw-Hill, New York, 1977. Used with permission.*)

TABLE 8-1   Pipe Schedule

System	Extent	Material	Schedule	Connection
H.P. steam	All	Black steel	80	Welded
H.P. return	2½ in and below	Black steel	80	Screwed
H.P. return	2½ in and up	Black steel	80	Welded
L.P. steam and return	2½ in and below	Yoloy* or black steel	40	Screwed
L.P. steam and return	3 in and up	Yoloy* or black steel	40	Welded
Condenser water	2 in and below	Black steel	40	Threaded and coupled
Condenser water	2½ in and up	Black steel	40	Welded
Chilled water	2 in and below	Black steel	40	Threaded and coupled
Chilled water	2½ in and up	Black steel	40	Welded
Hot-water heating, reheat, and preheat	2 in and below	Black steel	40	Threaded and coupled
Hot-water heating, reheat, and preheat	2½ in and up	Black steel	40	Welded
Indirect drains	All	Galvanized steel	40	Threaded and coupled
City water	All	Type L hard copper	40	Soldered
Vent and relief	All	Black steel	40	Welded
Fuel-oil supply and return piping	All	Black steel	80	Welded
Fuel-oil fill and vent piping	All	Black steel	80	Threaded and coupled
Refrigerant piping	All	ACR copper	80	Soldered
Emergency-generator exhaust pipe	All	Black steel	40	Welded

*Yoloy is a trade name for a nickel-copper-alloy steel pipe material.
SOURCE: A. M. Khashab, *Heating, Ventilating and Air Conditioning Systems Estimating Manual,* McGraw-Hill, New York, 1977. Used with permission.

TABLE 8-2   Fitting Schedule

System	Extent	Material	Schedule	Connection
H.P. steam	All	Steel	80	Welded
H.P. return	All	Steel	80	Welded
L.P. return	2½ in and below	Cast iron	40	Screwed
L.P. return	3 in and up	Steel	40	Welded
Condenser water	2 in and below	Cast iron	40	Screwed
Condenser water	2½ in and up	Steel	40	Welded
Chilled water	2 in and below	Cast iron	40	Screwed
Chilled water	2½ in and up	Steel	40	Welded
Hot-water heating, preheat, and reheat	2 in and below	Cast iron	40	Screwed
Hot-water heating, preheat, and reheat	2½ in and up	Steel	40	Welded
Indirect drains	All	Galvanized malleable	40	Recessed drainage
City water	All	Wrought copper	40	Soldered
Vent and relief	All	Steel	40	Welded
Fuel-oil supply and return piping	All	Cast iron	80	Welded
Fuel-oil fill and vent piping	All	Steel	80	Screwed
Refrigerant piping	All	Wrought copper	80	Soldered

SOURCE: A. M. Khashab, *Heating, Ventilating and Air Conditioning Systems Estimating Manual,* McGraw-Hill, New York, 1977. Used with permission.

TABLE 8-3   Valve Schedule

System	Extent	Material	Connection	Class, lb
H.P. steam and return	2 in and below	All bronze	Screwed	300
H.P. steam and return	2½ in and up	C.I. and bronze	Flanged	300
L.P. steam and return	2 in and below	All bronze	Screwed	125
L.P. steam and return	2½ in and up	C.I. and bronze	Flanged	125
Condenser water	2 in and below	All bronze	Screwed	150
Condenser water	2½ in and up	C.I. and bronze	Flanged	150
Chilled water	2 in and below	All bronze	Screwed	150
Chilled water	2½ in and up	C.I. and bronze	Flanged	150
Hot water	2 in and below	All bronze	Screwed	150
Hot water	2½ in and up	C.I. and bronze	Flanged	150
Indirect drains	All	All bronze	Screwed	150
City water	All	All bronze	Screwed	150
Vent and relief	All	All bronze	Screwed	125
Fuel-oil-tank piping	2 in and below	All bronze	Screwed	250
Fuel-oil-tank piping	2½ in and up	C.I. and bronze	Flanged	250

SOURCE: A. M. Khashab, *Heating, Ventilating and Air Conditioning Systems Estimating Manual,* McGraw-Hill, New York, 1977. Used with permission.

**TABLE 8-4   Recommended Test Pressures
for Piping Systems**

System Classification	Pressure, psig
High-pressure steam and return	150
Low-pressure steam and return	100
Hot-water system	100
Chilled-water system	100
Condenser-water system	150
Fuel-oil system	200

SOURCE: A. M. Khashab, *Heating, Ventilating and Air Conditioning Systems Estimating Manual,* McGraw-Hill, New York, 1977. Used with permission.

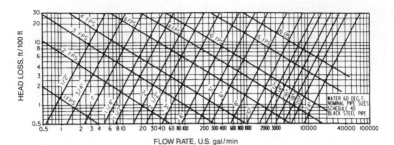

FIGURE 8-4 Friction loss for water in commercial steel pipe (schedule 40). (*Source:* ASHRAE Handbook—1989 Fundamentals. *Reprinted by permission.*)

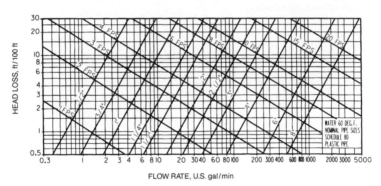

FIGURE 8-5 Friction loss for water in plastic pipe (schedule 80). (*Source:* ASHRAE Handbook—1989 Fundamentals. *Reprinted by permission.*)

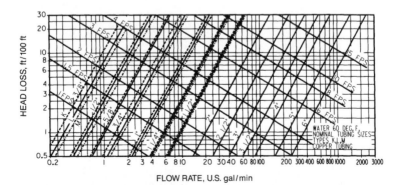

FIGURE 8-6 Friction loss for water in copper tubing (types K, L, M). (*Source:* ASHRAE Handbook—1989 Fundamentals. *Reprinted by permission.*)

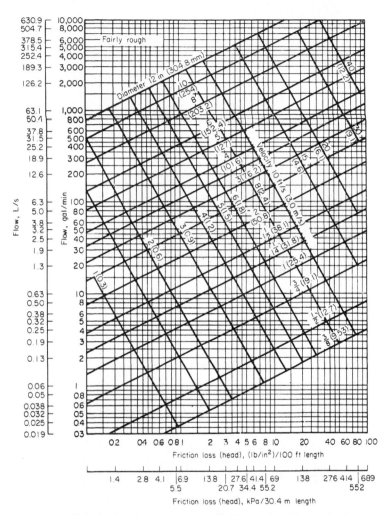

FIGURE 8-7 Friction loss in water piping (fairly rough, older pipe). (*Source: T. G. Hicks,* Power Generation Calculations Reference Guide, *McGraw-Hill, New York, 1987. Used with permission of publisher.*)

TABLE 8-5 Equivalent Length for Fittings

SIZE OF PIPE IN INCHES	GATE VALVE	ST'D ELBOW	RED. COUPLING	SIDE OUTLET "T"	ANGLE VALVE	GLOBE VALVE
			EQUIVALENT LENGTH OF PIPE IN FEET			
1/2	0.3	1.3	1.5	3.0	7.0	14.0
3/4	0.4	1.8	2.0	4.0	10.0	18.0
1	0.5	2.2	2.5	5.0	12.0	23.0
1 1/4	0.6	3.0	3.0	6.0	15.0	29.0
1 1/2	0.8	3.5	3.5	7.0	18.0	34.0
2	1.0	4.3	5.0	8.0	22.0	46.0
2 1/2	1.1	5.0	6.0	11.0	27.0	54.0
3	1.4	6.5	7.0	13.0	34.0	66.0
3 1/2	1.6	8.0	9.0	15.0	40.0	80.0
4	1.9	9.0	10.0	18.0	45.0	92.0
5	2.2	11.0	13.0	22.0	56.0	112.0
6	2.8	13.0	15.0	27.0	67.0	136.0
8	3.7	17.0	20.0	35.0	92.0	180.0
10	4.6	21.0	25.0	45.0	112.0	230.0
12	5.5	27.0	30.0	53.0	132.0	270.0
14	6.4	30.0	35.0	63.0	152.0	310.0

SOURCE: The Trane Company, LaCross, WI. Reproduced with permission.

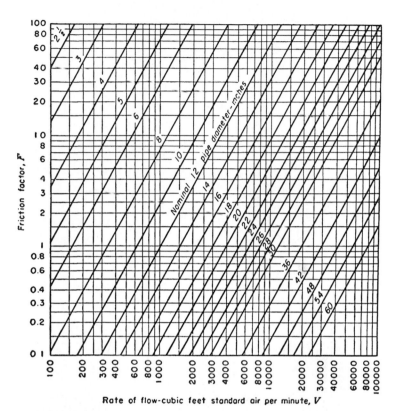

Rate of flow-cubic feet standard air per minute, V

FIGURE 8-8 Loss of air pressure due to pipe friction measured at standard conditions of 14.7 psia and 60°F. (*Source:* Compressed Air and Gas Handbook, *5th ed., 1988. Used with permission from CAGI, 1300 Sumner Ave., Cleveland, Ohio 44115.*)

TABLE 8-6 How to Cut Odd-Angle Elbows

1. MEASURE DISTANCE ON OUTSIDE ARC

	ODD DEGREE LONG RADIUS ELBOWS						
	OUTSIDE ARC						
NOM SIZE	A	B	C	D	E	F	G
2	5/64	3/8	23/32	1 3/32	1 21/32	2 3/4	3 9/32
2½	3/32	7/16	29/32	1 11/32	2 1/32	3 3/8	4 1/16
3	7/64	9/16	1 1/8	1 5/8	2 15/32	4 3/32	4 29/32
3½	1/8	5/8	1 9/32	1 29/32	2 27/32	4 3/4	5 11/16
4	9/64	23/32	1 7/16	2 3/32	3 1/4	5 13/32	6 15/32
5	3/16	29/32	1 25/32	2 11/16	4 1/32	6 23/32	8 1/16
6	7/32	1 1/16	2 5/32	3 7/32	4 27/32	8 1/16	9 21/32
8	9/32	1 7/16	2 27/32	4 9/32	6 13/32	10 11/16	12 13/16
10	11/32	1 25/32	3 9/16	5 11/32	8	13 11/32	16
12	7/16	2 1/8	4 1/4	6 3/8	9 9/16	15 31/32	19 9/32
14	1/2	2 7/16	4 7/8	7 5/16	11	18 5/16	22
16	9/16	2 13/16	5 19/32	8 3/8	12 9/16	20 15/16	25 1/8
18	5/8	3 1/8	6 9/32	9 7/16	14 1/8	23 9/16	28 9/32
20	11/16	3 1/2	7	10 15/32	15 23/32	26 1/8	31 13/32
22	3/4	3 27/32	7 11/16	11 17/32	17 9/32	28 3/16	34 9/16
24	27/32	4 3/16	8 3/8	12 9/16	18 27/32	31 13/32	37 11/16
26	29/32	4 17/32	9 3/32	13 3/8	20 3/16	34 1/32	40 27/32
30	1 1/32	5 1/4	10 15/32	15 3/4	23 3/8	39 1/4	47 1/8
34	1 3/32	5 29/32	11 27/32	17 13/16	26 3/16	44 17/32	53 3/8
36	1 7/32	6 1/4	12 17/32	18 7/8	28 1/32	47	56 17/32
42	1 7/16	7 5/16	14 5/8	22	32 31/32	54 31/32	65 15/16

ODD DEGREE LONG RADIUS ELBOWS

NOM SIZE	INSIDE ARC						
	AA	BB	CC	DD	EE	FF	GG
2	1/32	5/32	5/16	15/32	23/32	1 3/16	1 7/16
2 1/2	3/64	3/16	13/32	19/32	29/32	1 1/2	1 13/16
3	3/64	1/4	1/2	23/32	1 3/32	1 13/16	2 5/32
3 1/2	1/16	9/32	9/16	27/32	1 9/32	2 1/8	2 9/16
4	1/16	5/16	21/32	31/32	1 15/32	2 7/16	2 15/16
5	5/64	13/32	13/16	1 1/4	1 27/32	3 3/32	3 23/32
6	3/32	1/2	1	1 1/2	2 7/32	3 23/32	4 15/32
8	1/8	11/16	1 11/32	2	3 1/32	5 1/32	6 1/32
10	5/32	27/32	1 11/16	2 17/32	3 25/32	6 5/16	7 9/16
12	7/32	1	2 1/32	3 1/16	4 9/16	7 19/32	9 1/8
14	1/4	1 7/32	2 7/16	3 21/32	5 1/2	9 5/32	11
16	9/32	1 13/32	2 13/16	4 3/16	6 9/32	10 15/32	12 5/8
18	5/16	1 9/16	3 1/8	4 23/32	7 1/16	11 25/32	14 1/8
20	11/32	1 3/4	3 1/2	5 1/4	7 27/32	13 3/32	15 11/16
22	3/8	1 29/32	3 27/32	5 3/4	8 5/8	14 3/8	17 9/32
24	13/32	2 3/32	4 3/16	6 9/32	9 7/16	15 11/16	18 27/32
26	15/32	2 9/32	4 17/32	6 13/16	10 7/32	17 1/32	20 13/32
30	17/32	2 5/8	5 1/4	7 7/8	11 25/32	19 5/8	23 9/16
34	19/32	2 31/32	5 29/32	8 29/32	13 3/8	22 9/32	26 11/16
36	5/8	2 13/16	6 1/4	9 7/16	14 1/8	23 5/8	28 1/4
42	23/32	3 21/32	7 5/16	10 19/32	16 1/2	26 3/8	32 31/32

SOURCE: Tube Turns, Inc., Louisville, KY. Used with permission.

2. MEASURE DISTANCE ON INSIDE ARC

3. WRAP TAPE AROUND ELBOW AND MARK CUTTING LINE

8-1   Pipe Alignment*

Proper alignment is one of the most important tasks performed by the pipe fitter. If done correctly, welding will be much easier and the piping system will be properly fabricated. If alignment is poor, however, welding will be difficult and the piping system may not function properly.

Many devices are available to aid alignment. Tube Turns manufactures three types of welding rings which not only make alignment easier but also provide the correct gap for welding.

Methods of alignment vary widely throughout the trade. There is no *best* system; any number of methods have proven successful. The following procedures are popular with many craft workers and will enable you to quickly obtain good alignment.

Pipe to pipe.   Move pipe lengths together until bevels are nearly abutted, allowing space for welding gap. Center squares on top of both pipes and move pipe up and down until squares are aligned. Tack weld top and bottom. Repeat procedure by placing squares on side of pipe. Correct alignment by moving pipe left or right. Tack weld each side. (See Fig. 8-9.)

---

* Text and illustrations in this subsection from material provided by Tube Turns, Inc., Louisville, KY. Used with permission.

FIGURE 8-9   Pipe alignment—pipe to pipe.

90° elbow to pipe.   Place fitting bevel in line with bevel of pipe, allowing for welding gap. Tack weld on top. Center square on top of pipe. Center second square on elbow's alternate face. Move elbow until squares are aligned. (See Fig. 8-10.)

Follow procedure described above except squares will cross. To obtain correct 45° angle, align the same numbers on the inside scale of the tilted square. (*Note:* The numbers 4 and 7 are used in Fig. 8-11.)

Alternate method.   Use same procedure to abut pipe and fitting. Center spirit level on pipe. Next, center 45° spirit level on face of elbow and move elbow until 45° bubble is centered. (See Fig. 8-12.)

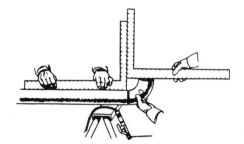

FIGURE 8-10 Pipe alignment —90° elbow to pipe.

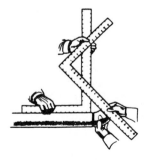

FIGURE 8-11 Pipe alignment— 45° elbow to pipe.

FIGURE 8-12 Pipe alignment—45° elbow to pipe (alternate method).

Tee to pipe. Abut bevels, allowing for welding gap. Tack weld on top. Center square on top of pipe. Place second square on center of branch outlet. Move tee until squares are aligned. (See Fig. 8-13.)

Alternate method. Follow same procedure to abut pipe and fitting. Place square on tee as illustrated in Fig. 8-14. Center rule on top of pipe. Blade of square should be parallel with pipe. Check by measuring with rule at several points along the pipe.

Flange to pipe

Step 1. Abut flange to pipe. Align top two holes of flange with spirit level. Move flange until bubble is centered. Make one tack weld on top. (See Fig. 8-15.)

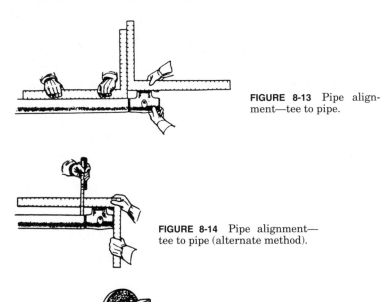

FIGURE 8-13 Pipe alignment—tee to pipe.

FIGURE 8-14 Pipe alignment—tee to pipe (alternate method).

FIGURE 8-15 Pipe alignment—flange to pipe (step 1).

Step 2. Center square on face of flange. Center rule on top of pipe. Move flange until square and pipe are parallel. Tack weld bottom. (See Fig. 8-16.)

Step 3. Center square on face of flange. Center rule on side of pipe and align as in Step 2. Tack both sides.

FIGURE 8-16 Pipe alignment—flange to pipe (step 2).

Many pipe fitters have found this simple jig to be helpful in aligning small diameter pipe and elbows.

It is made from channel iron approximately 3 ft 9 in long. Use ⅛ in × 1½ in for pipe sizes 1¼ in thru 3 in; ⅛ in × ¾ in for sizes 1 in or smaller.

①

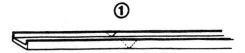

Mark 90° notch on side of channel iron about 9 in from end. Make equal notch on other side.

②

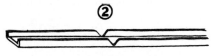

Cut out notches with hack saw.

③

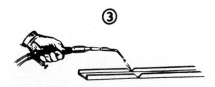

Heat bottom of channel iron between the notches.

FIGURE 8-17 Jig for small-diameter piping. (*Source: Tube Turns, Inc., Louisville, KY. Used with permission.*)

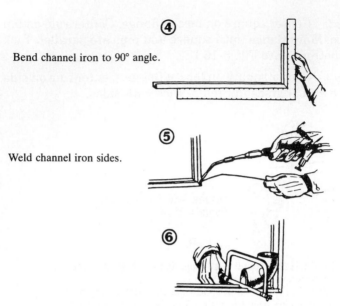

Bend channel iron to 90° angle.

Weld channel iron sides.

Place Tube Turns 90° Long Radius Elbow in jig. Saw half through both sides of channel iron as shown in illustration. Repeat with other size elbows so jig may be used for several sizes.

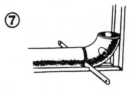

Place used hack saw blade in slot to obtain proper alignment and correct welding gap.

FIGURE 8-17   *(Continued)*

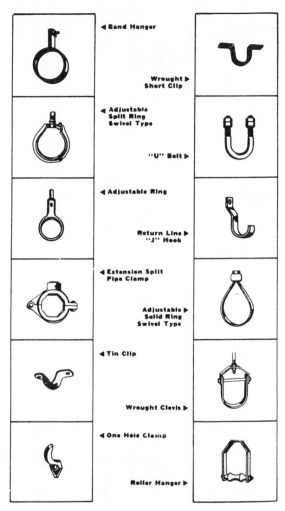

FIGURE 8-18 Typical pipe hangers. (*Source: Tube Turns, Inc., Louisville, KY.*)

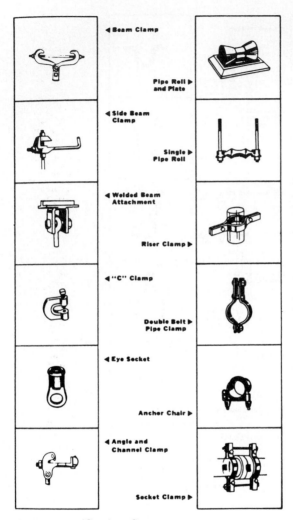

FIGURE 8-18   (*Continued*)

TABLE 8-7   Types of Hangers and Supports

The left portion of this table lists hanger Type Numbers with columns for System Temperature classes:

TYPE No.	CLASS A 120° to 650° Bare	CLASS A Covered	CLASS B 60° thru 119° Bare	CLASS B Covered	CLASS C 59° & down Bare	CLASS C Covered
OVERHEAD SUPPORTS – HORIZONTAL PIPING						
1		●		●		
2						●
3				●		
4	●					
5		●				
6	●					
7		●				
8			●		●	
9						●
10					●	
OVERHEAD SUPPORTS – HORIZ. PIPING – SECONDARY WATER						
11		●		●		●
12		●		●		●
WALL SUPPORTS – HORIZONTAL PIPING						
13		●		●		
14						●
15		●				
16	●					
17		●				
18			●		●	
19			●	●	●	●
20			●	●	●	●
21			●	●		
22			●	●	●	●
WALL SUPPORTS – VERTICAL PIPING						
23			●		●	
24			●		●	
25			●		●	
FLOOR SUPPORTS – HORIZONTAL PIPING						
26	●					
27		●				
28			●		●	
29				●		
30						●
31	●					
32		●				
33			●		●	
34			●			
35						●
FLOOR SUPPORTS – VERTICAL PIPING						
36	●	●	●	●	●	●
37	●	●	●	●	●	●
PRE-ENGINEERED SPRING HANGERS TO BE USED ON LINES SUBJECT TO 1/8" VERTICAL MOVEMENT OR MORE						
38	●	●	●	●	●	●
39	●	●	●	●	●	●
40	●	●	●	●	●	●
41	●	●	●	●	●	●
GUIDES – HORIZONTAL PIPING						
42	●	●	●	●	●	●
43	●	●	●	●	●	●
44						●
GUIDES – VERTICAL PIPING						
45	●		●		●	
46						●
47	●					
48	WATER PROOF SLEEVE – AS PER PLANS & SPECIFICATIONS					
49	UNDERGROUND PIPE HANGER – AS PER PLANS & SPEC.					
50	TUNNEL PIPE STANCHIONS – AS PER PLANS & SPEC.					
	ANCHORS – AS PER PLANS & SPEC.					
	ISOLATION HANGERS – AS PER PLANS & SPEC.					
	ESCUTCHEONS – AS PER PLANS & SPEC.					

TYPE No. 1 6" pipe & smaller — Clevis hanger

TYPE No. 2 6" pipe & smaller — Clevis hanger with welded protection shield

TYPE No. 3 6" pipe & smaller — Clevis hanger with welded protection saddle

TYPE No. 9 8" pipe & larger — Channel trapeze hanger with 360° protection shield

TYPE No. 10 8" pipe & larger — Channel trapeze hanger with protection saddle

TYPE No. 11 6" pipe & smaller — Clevis roller hanger with 360° protection shield

TYPE No. 17 all pipe sizes — Wall bracket with adj. roller chair and protection saddle

TYPE No. 18 4" pipe & smaller — Offset J hook

TYPE No. 19 4" pipe & smaller — Straight J hook

TYPE No. 25 4" pipe & smaller — Hook plate with straps

TYPE No. 26 all pipe sizes — Pipe roll stand

TYPE No. 27 all pipe sizes — Pipe roll stand with protection saddle

TYPE No. 33 all pipe sizes — Adj. pipe stanchion with U-bolt

TYPE No. 34 all pipe sizes — Adj. pipe stanchion with U bolt and protection saddle

TYPE No. 35 all pipe sizes — Adj. pipe stanchion with U bolt and 360° protection shield

TYPE No. 41 all pipe sizes — Variable spring hanger with double bolt riser clamp for vertical piping

TYPE No. 42 all pipe sizes — Slide pipe guide

TYPE No. 43 all pipe sizes — Cylinder pipe guide solid or split with spider clamp

TYPE No. 49 — Underground pipe hanger

TYPE No. 50 — Tunnel pipe stanchions

Spot insert

Continuous slotted insert

SOURCE: A. M. Khashab, *Heating, Ventilating and Air Conditioning Systems Estimating Manual,* McGraw-Hill, New York, 1977. Used with permission.

TABLE 8-7 Types of Hangers and Supports (*Continued*)

TYPE No. 4 6" pipe or smaller	TYPE No. 5 6" pipe & smaller	TYPE No. 6 8" pipe & larger	TYPE No. 7 6" pipe & larger	TYPE No. 8 4" pipe & larger
Clevis roller hanger	Clevis roller hanger with protection saddle	Two-rod roller hanger	Two-rod roller hanger with protection saddle	Channel trapeze hanger
TYPE No. 12 8" pipe & larger	TYPE No. 13 all pipe sizes	TYPE No. 14 all pipe sizes	TYPE No. 15 all pipe sizes	TYPE No. 16 all pipe sizes
Two-rod roller hanger with 360° protection shield	Wall bracket with U-bolt	Wall bracket with U-bolt and 360° protection shield	Wall bracket with U-bolt and protection saddle	Wall bracket with adjustable roller chair
TYPE No. 20 4" pipe & smaller	TYPE No. 21 4" pipe & smaller	TYPE No. 22 8" pipe & smaller	TYPE No. 23	TYPE No. 24 4" pipe & smaller
Welded hook plate	Adj. hook plate with J-hooks	Offset clamp for horizontal piping	Wall brackets with riser clamp	Pipe strap
TYPE No. 28 all pipe sizes	TYPE No. 29 all pipe sizes	TYPE No. 30 all pipe sizes	TYPE No. 31 all pipe sizes	TYPE No. 32 all pipe sizes
Pipe saddle support	Adjustable pipe stand with protection saddle	Adjustable pipe stand with	Adj. pipe stanchion roller	Adj. pipe stanchion roller with protection saddle
TYPE No. 36 all pipe sizes	TYPE No. 37 all pipe sizes	TYPE No. 38 all pipe sizes	TYPE No. 39 6" pipe & smaller	TYPE No. 40 6" pipe & larger
Adj. base elbow support	Riser clamp	Variable spring hanger	FOR PIPE ATTACHMENT SEE TYPE 1 THRU 5 & 11 Variable spring hanger for horizontal piping	FOR PIPE ATTACHMENT SEE TYPE 6 THRU 10 & 12 Variable spring hanger for horizontal piping
TYPE No. 44 all pipe sizes	TYPE No. 45 all pipe sizes	TYPE No. 46 all pipe sizes	TYPE No. 47 all pipe sizes	TYPE No. 48
Cylinder pipe guide solid or split	Guide clamp	Guide clamp with 360° protection shield	Guide clamp with wear bars	Waterproof sleeve
Attachment to structure	Attachment to structure			
Welded beam attachment / Welded beam attachment with bolt & nut	Beam clamp / Adj. anchor rod for fireproofed beams	Supplementary steel	Gang hanger	Typical vertical pipe anchors
Isolation hangers	Escutcheons			

NOTES

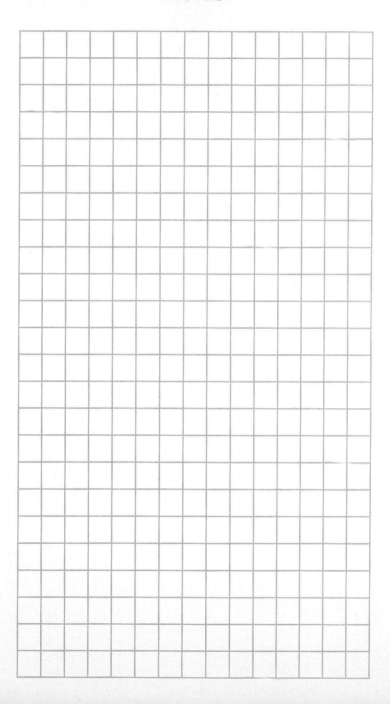

NOTES

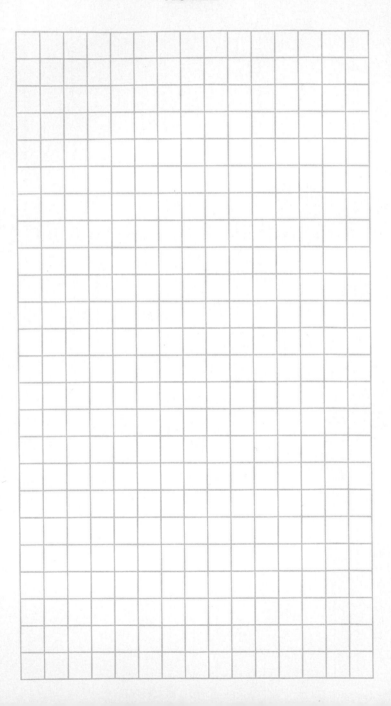

Air Distribution Design

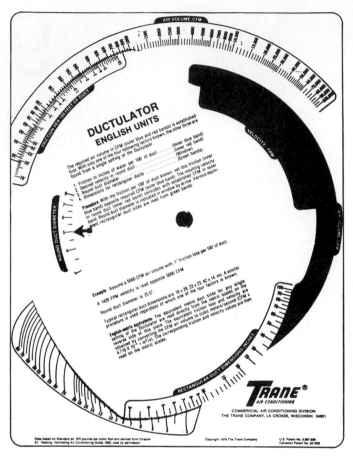

FIGURE 9-1 Ductulator—USCS (English) units. The Ductulator is an accurate, easy-to-use calculating device that aids in the layout of air-handling systems, sizing of ducts, and checking of existing duct systems. With a single setting, the answers to any duct sizing problem may be read directly from the face of the Ductulator. Conversion of units, USCS to metric, or vice versa, is accomplished by setting up a problem on the scales of one unit system, reversing sides of the Ductulator, and reading the converted values from the scales of the other unit system. (*Reproduced by permission of The Trane Company, LaCrosse, WI.*)

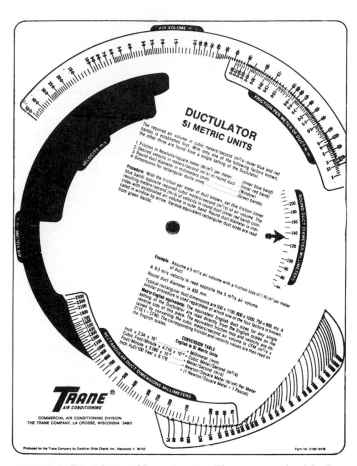

FIGURE 9-2 Ductulator—SI metric units. The reverse side of the Ductulator shown in Fig. 9-1. Its purpose and how to convert from metric to USCS units are the same as given for Fig. 9-1. (*Reproduced by permission of The Trane Company, LaCrosse, WI.*)

TABLE 9-1   Volume of Ventilation Air for Various Buildings

Functional Areas	Ventilation Air	Unit
Auditoriums, churches, dance halls	4–30 10–65 1.5–2	air changes* per hour cfm/occupant cfm/ft²
Barbershops and cafes	7.5	air changes per hour
Bedrooms	1	air changes per hour
Billiards and bowling	6–20	air changes per hour
Classrooms		
Colleges	25–40	cfm/occupant
Schools	30–40	cfm/occupant
Classrooms (schools)	2	cfm/ft²
Corridors	4 ½	air changes per hour cfm/ft²
Dining rooms	4–40 1.5	air changes per hour cfm/ft²
Garages	6–12	air changes per hour
Guest rooms	3–5	air changes per hour
Gymnasiums	12 1.5	air changes per hour cfm/ft²
Halls(residence)	1–3	air changes per hour
Kitchens	4–60 2–4	air changes per hour cfm/ft²
Laboratories	6–20	air changes per hour
Living rooms (residence)	1–2	air changes per hour
Lobbies	3–4	air changes per hour
Locker rooms	2–10 2	air changes per hour cfm/ft²
Lounges	6	air changes per hour
Mechanical rooms	3–12	air changes per hour
Operating rooms	50	cfm/occupant
Projection booths	30 1.5	air changes per hour cfm/ft²
Reading rooms	3–5	air changes per hour
Stores (retail)	6–12 4	air changes per hour cfm/ft²
Toilets		
Private	1–5	air changes per hour
Public	10–30 2	air changes per hour cfm/ft²
Waiting rooms	4–6	air changes per hour

*Air changes is a method of calculating the amount of infiltration or ventilation air in terms of the number of room volumes per hour.

SOURCE: A. M. Khashab, *Heating, Ventilating and Air Conditioning Systems Estimating Manual,* 2d ed., McGraw-Hill, New York, 1977. Used with permission.

TABLE 9-2 Air Diffusion Performance Index (ADPI) Selection Guide

Terminal Device	Room Load Btu/h·ft²	T_{50}/L for Max. ADPI	Maximum ADPI	For ADPI Greater Than	Range of $T_{0.25}/L$
High Sidewall Grilles	30	1.8	68	—	—
	60	1.8	72	70	1.5 – 2.2
	40	1.6	78	70	1.2 – 2.3
	20	1.5	85	80	1.0 – 1.9
Circular Ceiling Diffusers	80	0.8	76	70	0.7 – 1.3
	60	0.8	83	80	0.7 – 1.2
	40	0.8	88	80	0.5 – 1.5
	20	0.8	93	90	0.7 – 1.3
Sill Grille Straight Vanes	80	1.7	61	60	1.5 – 1.7
	60	1.7	72	70	1.4 – 1.7
	40	1.3	86	80	1.2 – 1.8
	20	0.9	95	90	0.8 – 1.3
Sill Grille Spread Vanes	80	0.7	94	90	0.8 – 1.5
	60	0.7	94	80	0.6 – 1.7
	40	0.7	94	—	—
	20	0.7	94	—	—
Ceiling Slot Diffusers (for T_{100}/L)	80	0.3*	85	80	0.3 – 0.7
	60	0.3*	88	80	0.3 – 0.8
	40	0.3*	91	80	0.3 – 1.1
	20	0.3*	92	80	0.3 – 1.5
Light Troffer Diffusers	60	2.5	86	80	<3.8
	40	1.0	92	90	<3.0
	20	1.0	95	90	<4.5
Perforated and Louvered Ceiling Diffusers	11 – 51	2.0	96	90	1.4 – 2.7
				80	1.0 – 3.4

SOURCE: *ASHRAE Handbook—1989 Fundamentals.* Reprinted by permission.

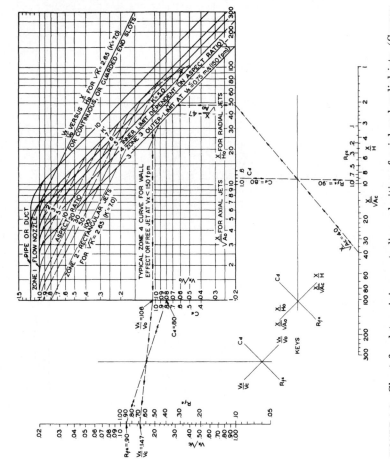

FIGURE 9-3 Chart for determining centerline velocities of axial and radial jets. (*Source: ASHRAE Handbook—1989 Fundamentals. Reprinted by permission.*)

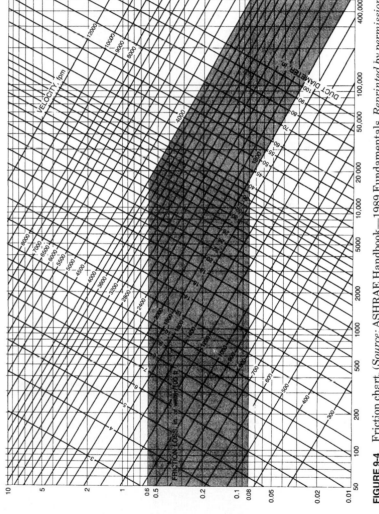

FIGURE 9-4 Friction chart. (*Source: ASHRAE Handbook—1989 Fundamentals. Reprinted by permission.*)

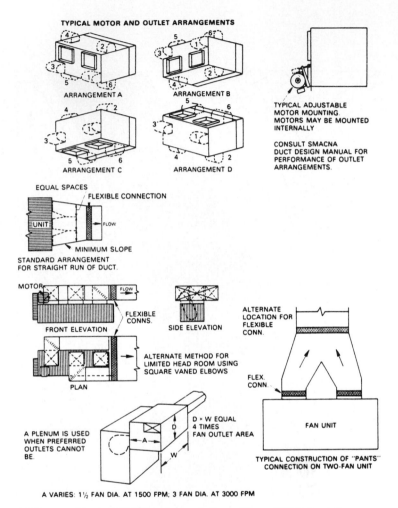

FIGURE 9-5 Typical HVAC unit connections. (*From* HVAC Duct Construction Standards—Metal and Flexible, *1st ed., © 1985 by the Sheet-Metal and Air Conditioning Contractor's National Association. Used with permission of the copyright holder.*)

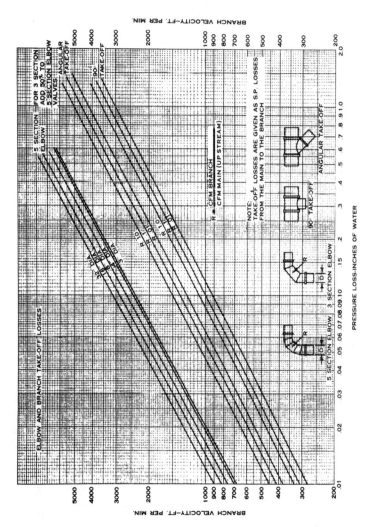

FIGURE 9-6 Elbow and branch takeoff losses. (*Reproduced with permission of The Trane Company, LaCrosse, WI.*)

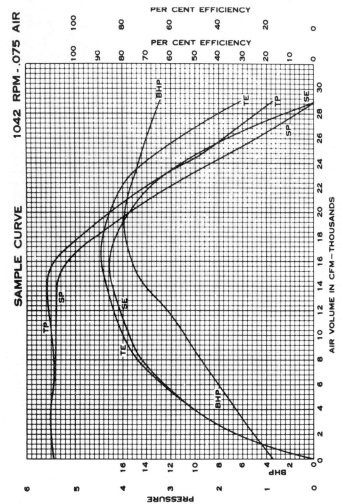

FIGURE 9-7 Performance of an airfoil blade fan. (*Reproduced by permission of The Trane Company, LaCrosse, WI.*)

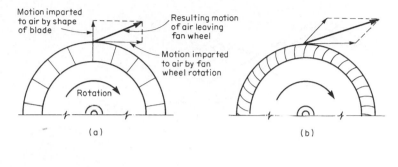

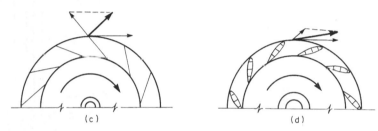

FIGURE 9-8 Types of centrifugal fan blades. (*a*) Radial; (*b*) forward-curved; (*c*) backward-inclined; (*d*) airfoil (forward-inclined). (*Source: A. M. Khashab,* Heating, Ventilating and Air Conditioning Systems Estimating Manual, *McGraw-Hill, New York, 1977. Used with permission.*)

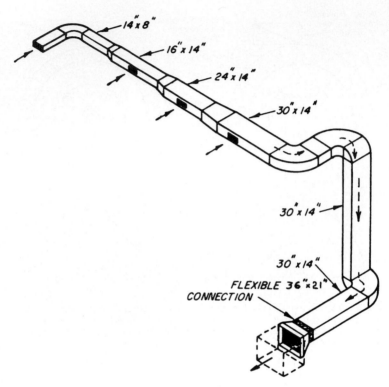

FIGURE 9-9 Isometric view of return air duct.

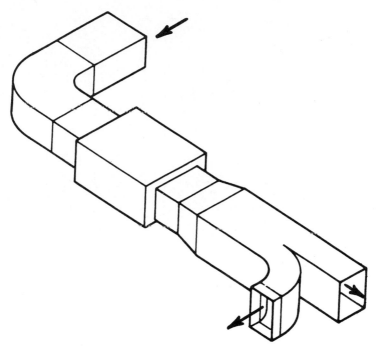

FIGURE 9-10   Isometric view of horizontal mounted furnace ductwork.

NOTES

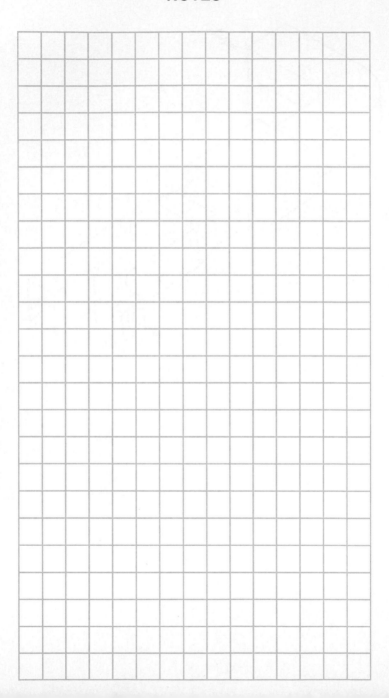

Attic
Ventilation*

* This section was extracted with permission from *Principles of Attic Ventilation,* 4th ed., by Clarke M. Wolfert, P.E., © 1985 by Air Vent, Inc., Peoria, IL.

Until recently there was limited interest in attic ventilation among builders, architects and engineers—to say nothing of the general homeowning public.

Recognition of the importance of effective attic ventilation to solve and prevent problems has grown rapidly, however, and today it is common practice to provide some sort of ventilation openings in attic or ceiling spaces. Vents of one type or another are found on single-family and multi-unit dwellings, manufactured homes, mobile homes and on farm and commercial buildings.

Passive solar housing designs, for example, invariably make use of ample and effective attic or ceiling space ventilation, recognition of its role in lowering energy costs, protecting roof materials, maintaining insulation R-value, and eliminating potential ice-damming problems.

Another significant development affecting attic ventilation requirements is the increasing use of whole house fans to exhaust warm air from living quarters into the attic space. Fan performance—and cooling results—are directly dependent on the amount of net free area provided by attic vents.

Also, the remodeling roofing industry is increasingly aware of the importance of proper ventilation to assure roof shingle durability and performance.

10-1   Why Ventilate?

The two fundamental benefits of an effective attic ventilation system are:

1. A cooler attic in summer

2. A dryer attic in winter

Both benefits result in energy saving, greater homeowner comfort, and higher structural integrity of the dwelling.

Summer

How heat gets in. Summer sunshine causes a buildup of heat in the attic space, created by radiant heat from the sun increasing the roof temperature so that it re-radiates heat into the attic. During the night, the cooler outside air and absence of sunshine permit the attic to re-radiate or conduct stored heat to the atmosphere. However, an unventilated— or poorly ventilated—attic is not able to lose all of its stored heat in this manner. This results in a heat buildup over a period of days, with ever increasing attic temperatures.

How much gets in.   If there is no attic ventilation, a 90° day with full sunshine can heat the roof sheath to 170° or more. Heat radiating from the sheathing down to the attic floor (or ceiling insulation) can raise its temperature to as much as 140°. (See Fig. 10-1.)

Homes today are being built with heavier insulation than ever before. This increases the need for effective attic ventilation. Without it, today's heavier insulation absorbs and holds more of the heat buildup in the attic during the day, making it less likely that all of the heat of the attic will be removed into the cooler nighttime air. Residual stored heat can build up over a period of hot days, maintaining the insulation at a higher temperature and increasing the heat radiation down to the rooms below.

What it does.   Overheated ceiling insulation conducts heat through to the ceiling, and this heat is then radiated down-

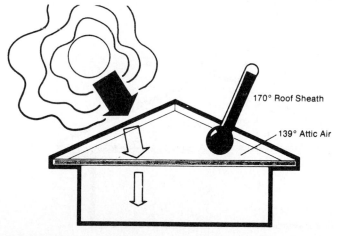

170° Roof Sheath

139° Attic Air

FIGURE 10-1

ward to persons and objects in the rooms below (Fig. 10-2). There are two consequences of this:

1. The home's air conditioning system must operate for longer periods in order to remove this heat.

2. Because it is *radiant* heat, individuals in the home may feel warmer than the air temperature indicates, and their usual reaction is to put the thermostat at a lower setting, further burdening the system and causing higher energy usage. An overheated attic, combined with moisture, can also cause roof shingles to distort and deteriorate.

How to prevent it.   The solution to the problem is attic ventilation. A sufficient volume of ventilation air must be moved through the attic space—under varying conditions of wind force and direction.

Also the ventilation air must move in a uniform pattern along the entire underside of the roof, avoiding isolated non-ventilated "hot spot" areas. The best system is one that:

1. Does not *itself* use electrical power and thus nullify energy savings.

2. Completely ventilates and cools the underside of the roof sheathing, which is the source of heat radiation downward to the attic floor or top of the ceiling insulation.

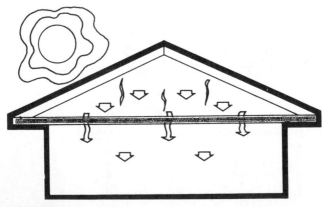

FIGURE 10-2

Winter

How moisture gets in. Winter conditions bring a different type of problem. During the cold months of the year, the air inside the home is warmer and carries more water vapor than the colder, dryer air in the attic. Cooking, laundry, showers, humidifiers and other activities using water contribute to this condition. There is a strong natural force, termed "vapor pressure," that causes water vapor to migrate from high-humidity air or materials to low-humidity air. This migration of water vapor passes through ceilings, insulation and wood—and even successfully circumvents a vapor barrier. It moves into the attic space where it can readily condense into liquid water on the cooler structural members—rafters, trusses, and especially the cold roof sheathing. (See Fig. 10-3.)

Another factor increasing the likelihood of a moisture condensation problem in many homes today is the use of an electrical heating system, more prevalent with the nationwide shortage of gas and oil. Because of the higher cost of heating with electricity, these homes are being built "tighter" and with greater insulation. This helps retain moisture in the home, which is transferred to the attic by the strong driving force—vapor pressure. The moisture problem is heightened by the fact that, in an electrically heated home, outside air is not drawn into the home to replace furnace combustion air. A minimum of dry outside

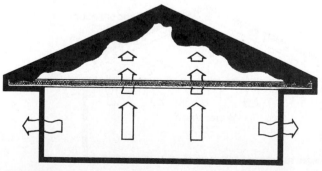

FIGURE 10-3

air enters the home in winter; air within the home acquires a higher moisture content and becomes a more likely source of problems.

What it does. The water vapor moving into the cold attic space in winter can be present in such a volume that it condenses on cold structural parts and drips down to soak ceiling insulation, compressing its volume and reducing its insulating effectiveness. As a consequence, heat loss through the ceiling increases and more energy is required to maintain the residence at a comfortable temperature.

Figure 10-4 shows how condensing moisture within an attic or ceiling space can dampen and compress insulation. Even small amounts of such condensation can have a substantial effect on R-value. Research has shown that, as moisture is added to fiberglass insulation, the R-value is reduced as shown below.

R-19 + 1½% moisture = R-12 (36% loss of R-value)
R-30 + 1½% moisture = R-19 (36% loss of R-value)
R-36 + 1½% moisture = R-23 (36% loss of R-value)

This loss of R-value takes place even without compaction of insulation, which can occur when moisture actually drips onto the insulating material, reducing R-value additionally and proportionately. (10% compaction reduces R-value by 10%, 20% compaction reduces it 20%, etc.)

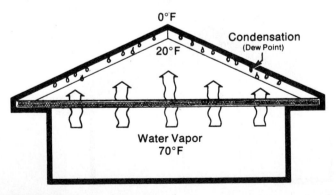

FIGURE 10-4

In addition to its destructive effect upon attic insulation, moisture condensation can lead to many other problems: wood rot, paint peeling, and deterioration of ceiling and roofing materials. It will, in fact, penetrate through the sheathing and will freeze, raise, and rot out the shingles, sheathing, and roofing materials.

How to take it out. As with heat buildup in summer, wintertime attic moisture is removed by ventilation air. (See Fig. 10-5.) The most effective system is one that moves the greatest volume of air through the attic space—under varying conditions of wind force and direction—*and* in a pattern of uniform flow across the underside of the roof sheathing. This is because the roof sheathing is the coldest surface in the attic and the prime location for moisture condensation.

Ice-dams—a special problem. In the northern two-thirds of the U.S. (see Figs. 10-6 and 10-7) there can be a serious problem caused by snow melting on a warm roof surface, usually near the ridge, the water running down to the colder roof overhang and then refreezing into ice. Subsequent melting of roof-snow can then accumulate as a pocket of snow-water that eventually backs up under the shingles to cause major damage in the plateline area.

This damage can appear in the form of soaked and inefficient insulation; stained, cracked, and spalled plaster or

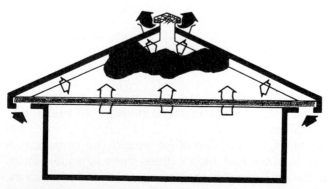

FIGURE 10-5

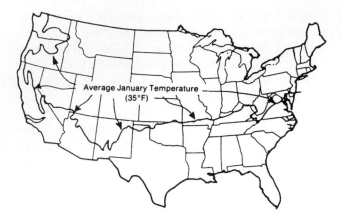

FIGURE 10-6 Map of the United States associating moisture damage susceptibility with temperature.

FIGURE 10-7 U.S. Weather Bureau map and data of mean annual total snowfall with 6- to 8-in line delineated.

sheetrock; damp, odorous, and rotting wall cavities; and stained, blistered, and peeling wall paint, both inside and outside the home.

In some cases, the buildup of ice weight can cause structural damage to the roof. Other times there is a snow or ice fall from the roof, damaging gutters, facia and shrubbery, and creating other problems. (See Fig. 10-8.)

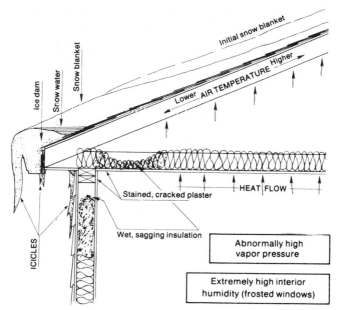

FIGURE 10-8 This sketch of the ice-dam problem identifies both the ice dam and its damage. Of course, all the damage illustrated may (or may not) occur at any one instance. The damages illustrated here are far more common and costly than is generally acknowledged.

A growing problem. Ice-dam problems have become more frequent in recent years because of present construction methods and standards along with concern over energy conservation. Typical American houses built before 1930 were constructed with relatively steeply pitched roofs, with "open" (spaced) sheathing, and mostly with wood shingles. These shingles had gaps between them which served to ventilate attics and to cool the roofdeck. In addition, most pre-'30's houses had no insulation in either ceilings or walls because there was little interest in conserving fuel.

On these older homes, the excessive attic heat supported rapid flows of snow-water downroof to the cooler eave projection and beyond. When air temperatures were less than extremely cold and attic heat was in plentiful supply, rapid melting of roof-snow quickly cleared the roof of snow—with only a gutterful or eave-line of ice and icicles. These houses flowed comparatively rapidly and copiously with snow-water; their roofs cleared fairly rapidly; snow-water penetrations were of comparatively short duration; and massive eave-ice usually was all that remained of the roof-snow until the next snowfall.

Because of concern for energy conservation, modern homes and pre-'30's homes which have been "modernized" now must meet certain standards, codes, and recommendations for insulation and ventilation. However, these standards frequently represent minimum requirements, and as a result, warm, if not hot, attics are an undesirable characteristic of millions of American homes. Roof-snow blanketing on modern houses, in contrast with older homes, melts slowly; beneath the cover water pools may threaten penetration in the house for periods of weeks or months.

Solution—a cold roof.   Though a number of factors enter into the formation of ice-dams, the fundamental problem is undesirable attic heat which results in a warm roof surface. The solution is to maintain a cold roof.

The effectiveness of cold-surface snow behavior to avoid massive ice formations is exhibited everywhere in winterized snow country. There are no massive ice formations on felled timber, rock outcrops, woodpiles, picnic tables, sheds, and unheated buildings. In homes, continuous removal of attic air to maintain the roofdeck at atmospheric air temperatures requires minimizing the amount of heat entering the attic, together with free movement of air from eave inlets to uproof (ridge) outlets. In short, it means insulating room ceilings far more than customary to minimize heat losses and attic temperatures, and ventilating profusely at all eaves and ridge for a natural flow of air to sweep out the warmed attic air.

It is difficult to stipulate the amount of insulation which will be sufficient to adequately contain waste heat because of the innumerable variables in residential designs and requirements. In general, though, a home is not adequately protected until insulation approaches 10 to 12 inches or an R-value of 38.

It is important to emphasize that adequate insulation alone will not prevent ice-dams; it must be done in conjunction with adequate ventilation.

Ventilation is important. When it comes to preventing ice-dams, it is impossible to have too much ventilation. Ventilation should be equally distributed between inlet and outlet areas. Overall, the goal is to have a sweep of cooler outside air entering through the soffit which moves along the ridge rafters and exits through a ridge vent near the peak of the house.

There are other approaches to ventilation, including individual soffit vents, individual roof louvers, and power vents. In general, however, these do not perform as well as the continuous ventilation method.

Eliminating a building code requirement. The problems caused by ice damming have been recognized in the universal building code, which requires an additional course of roofing on homes built in regions where ice-dams may form.

The requirement is as follows: When the roof has a slope of $\frac{4}{12}$ or more and an overhang of less than 36 inches, a course of either 90-lb mineral surfaced roll roofing or 50-lb smooth roll roofing must be installed over the single underlay, extending from the edge of the roof to a minimum of 12 inches beyond the interior face of the wall. For slopes less than $\frac{4}{12}$ and with any amount of overhang, the extra course must extend 24 inches beyond the interior wall face and shall be double and cemented.

This is a substantial additional expense that can be avoided with a cold roof system of ventilation. Builders can obtain a code variance eliminating the requirement, based upon their use of ridge and soffit venting to assure cold-roof performance throughout the winter months, preventing the ice-dam formation.

10-2   How Ventilation Works

What the wind does

There are two natural forces that can move air through an attic. They are the pressure caused by wind striking against the building and the thermal effect of warm air rising within the attic.

There is also an artificial force that can move the air—a ventilating fan powered by electricity.

Of the three forces, *the wind is the most constant and is often the strongest.* As it moves against and around the residential structure it creates areas of positive and negative pressure. When these are known and understood, intake vents can be placed in areas of positive pressure and exhaust vents placed in areas of negative pressure in order to assure a continuous air flow.

Wind force

The force of the wind blowing against the side or end of a house creates a positive pressure area at the point of contact. It then flares out, jumping in a vertical or horizontal direction—or both—depending on the configuration of the building at the point of impact. Within this "jump" area a negative pressure is created. (See Fig. 10-9.)

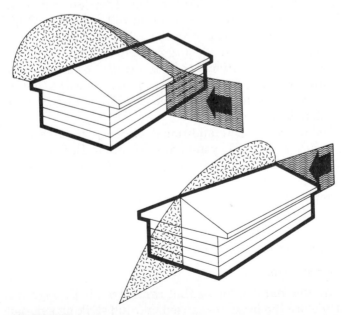

FIGURE 10-9

Wind direction

While the wind appears to maintain a general direction, such as easterly or westerly, it does in fact shift direction frequently, often by as much as 20 or 30 degrees. This shift in direction can change a positive pressure area on a residential structure to a negative pressure area, or vice versa. (See Fig. 10-10.) A vent that was an intake can, a few minutes later, become an exhaust vent. An exhaust vent can become an intake vent, with the undesirable effect of permitting rain and snow infiltration—or reducing ventilation to only a small portion of the attic.

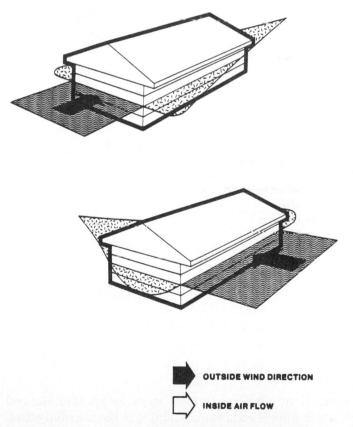

➡ **OUTSIDE WIND DIRECTION**

⇨ **INSIDE AIR FLOW**

FIGURE 10-10

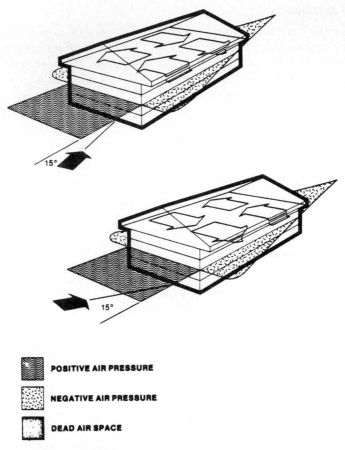

POSITIVE AIR PRESSURE

NEGATIVE AIR PRESSURE

DEAD AIR SPACE

FIGURE 10-10   (*Continued*)

Locating exhaust venting at that portion of the attic structure that is in a negative pressure area *regardless of wind direction* is the first objective for a ventilation system that will be continuously effective.

Thermal effect

In addition to wind movement, there is another natural force which affects attic ventilation. It is the thermal effect, caused by the well known principle that "warm air rises."

Air in the attic is usually warmer than the outside air—during daytime hours—and tends to rise and exhaust at the ridge if there is venting there to allow it. (See Fig. 10-11.)

Also, the air within the attic circulates by convection, warm air rising to the top and cooler air sinking to the attic floor. Exhaust venting at the ridge and intake venting at the soffit assists in this natural movement.

Thermal effect is not a major force in a natural ventilation system, however. The wind effect is far greater. Even a small breeze provides more air movement through a correctly vented attic than thermal effect.

In winter, especially, today's heavier ceiling insulation greatly diminishes heat loss to the attic, resulting in attic air and outside air being nearly the same temperature, for minimal thermal effect.

The inertia of air

An extremely important factor in maintaining good attic ventilation is the inertia of moving air. Air in motion acquires a momentum, similar to moving water or any moving object, which tends to keep it moving in the same direction.

A correctly designed attic ventilation system keeps air moving in the same direction. This makes the inertia of air work to maintain a constant air flow. In effect, it provides more ventilation with less wind.

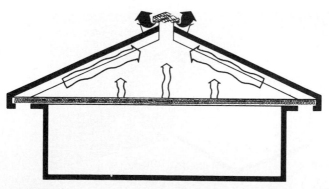

FIGURE 10-11

Conversely, air not in motion tends to remain so and requires more force to begin movement than it does to maintain movement once it has started.

Worse, if vents are placed so that, as the wind changes direction, the air flow through the attic is reversed, very little ventilation will take place. If every few minutes an intake vent becomes an exhaust vent, and vice versa, there is no effective and continuing air flow but only a back-and-forth pulsating movement with little air entering or leaving the attic.

The inertia of air is best utilized when intake vents remain intake vents and exhaust vents remain exhaust vents, regardless of wind force or direction.

High and low balance

Specifications for an attic ventilation system often include the requirement that 50% of the vent area be located high in the attic and 50% low. This is intended for balance, so that there will be equal vent areas for air flow into and out of the attic.

It is better to specify that the net free area of vents located in a positive wind pressure area be approximately equal to the net free area of vents located in a negative pressure area. (See Fig. 10-12.) Only in a ridge vent system are these positions always "low" for intake and "high" for exhaust.

If a ventilation system is not balanced in this manner, the effective ventilation rate is reduced to the air flow through the smaller of the two vent areas.

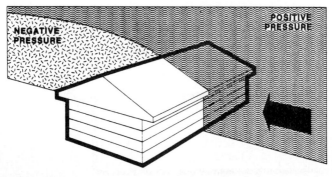

FIGURE 10-12

Underside of roof sheathing

While an attic ventilation system may move air throughout all parts of the attic, there is actually only one area that needs to be bathed with moving air—the underside of the roof sheathing. (See Fig. 10-13.)

Since heat buildup is caused by a radiation effect from the roof sheathing downward to the attic floor, or to the top of the insulation, cooling the roof sheathing effectively cancels this effect and maintains a desirable attic floor temperature.

The winter problem of condensation is also solved by ventilating the underside of the roof sheathing. The roof is the coldest structure in the attic, being in contact with outside air. Water vapor readily condenses there, and by ventilating all parts of the undersurface of the roof, vapor combines with the cold, dry air and is carried out of the attic before it can condense and cause problems.

A ventilation system that brings outside air across the attic floor, or top of ceiling insulation, rather than across the roof sheathing, not only cannot prevent condensation but removes heat from the insulation, adding to heat loss from the residence. It is far better to let the insulation remain as warm as possible and let the roof sheathing be cold, but well ventilated.

How rain and snow get in

Another sometimes troublesome source of moisture in the attic is the infiltration of rain and snow through vents. It occurs

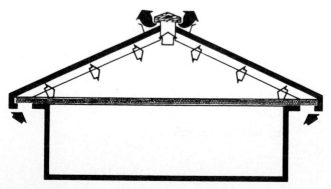

FIGURE 10-13

when intake air enters through the vent, and if wind conditions cause air to enter at high velocity, rain or snow may enter with it in spite of louvers designed to prevent infiltration.

The quantity of this infiltration can be sufficient, at times, to dampen ceiling insulation and even leak through to the ceiling below.

The problem is eliminated or at least minimized when the ventilation system is designed so that air intake is confined to those vents located where rain or snow infiltration is unlikely—such as in the soffit. Also, such vents should be placed adjacent to—or even outboard of—the facia. This provides for any moisture infiltration to either drain back out through the vent or be confined to the soffit, away from insulation, where it can evaporate without causing damage.

Figure 10-14 shows accepted application of continuous soffit vent in conventional construction. Figures 10-15 and 10-16 illustrate a method of combining a drip edge and intake venting on either a soffitless or soffited roof. Figure 10-17 shows placement of regularly spaced eave vents in the soffit.

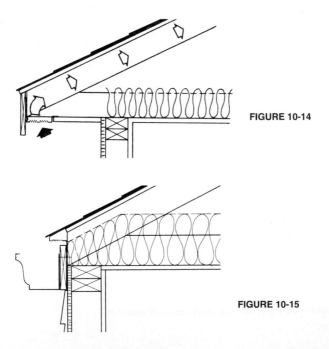

FIGURE 10-14

FIGURE 10-15

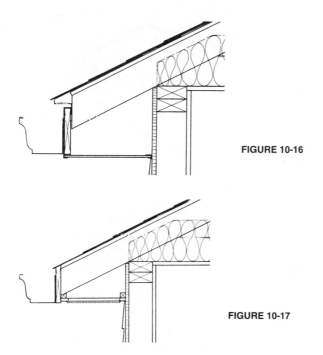

FIGURE 10-16

FIGURE 10-17

10-3   Types of Attic Ventilation

Soffit vents

A common method of ventilating attic space is with soffit vents, usually but not always in combination with other vents in the roof or gable ends. The soffit is an excellent location for venting, particularly for intake venting, because it is less exposed to rain and snow. Also, the wind is parallel to the vent regardless of direction. With soffit venting on both sides of the structure, there is always an equal amount of venting in a positive pressure area as there is in a negative pressure area, regardless of wind direction. Thus a soffit-vent-only system is always in balance.

The problem with a soffit-vent-only system is that ventilation is confined to the attic floor. There is no air movement into the upper part of the attic, and consequently very little moisture removal. There is no air movement due to thermal effect. There is virtually no air movement across the underside of the roof sheathing. (See Fig. 10-18.)

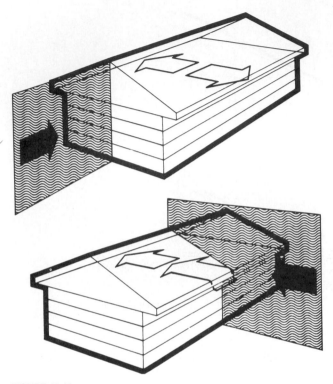

FIGURE 10-18

Roof vents—alone and combined
with soffit vents

Roof vents, while they can be installed near the ridge and thus provide an escape route for overheated air, have a number of disadvantages. If used with no other type of venting, the pattern of air circulation is very small, confined to the space immediately surrounding the vent. (See Fig. 10-19.) Air will enter through some vents and exhaust through others, depending on wind pressure areas on the roof. Weather infiltration can be a problem under these circumstances.

The usual installation combines roof vents with soffit vents. The air flow pattern is not uniform (see Fig. 10-20), with large portions of the roof sheathing unventilated. Also, since roof vents provide only 40 to 80 in^2 of net free area apiece, it is vir-

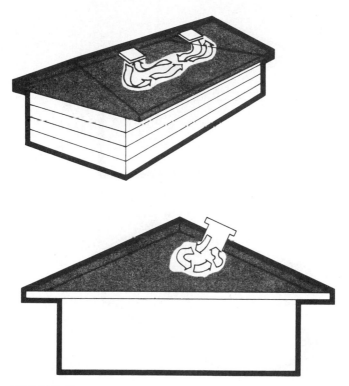

FIGURE 10-19

tually impossible to have enough roof vents to balance the soffit vents. Air usually enters one soffit and flows out the other
soffit, with some finding its way upward to the roof vents.

Gable and end wall louvers—
alone and combined with soffit vents

The gable or end wall louver, whether triangular or rectangular in shape, offers a ventilation air pattern as limited as
the roof vent. When there are louvers in each gable—and no
soffit venting—the ventilated area depends upon wind direction. Figure 10-21 shows what happens when the wind is
perpendicular to the ridge. Each louver acts as both intake
and exhaust, and only the ends of the attic receive any air
circulation. Figure 10-22 shows the flow of air when the

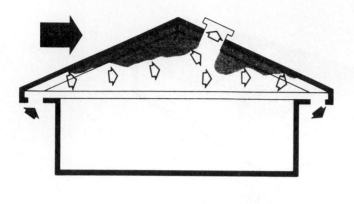

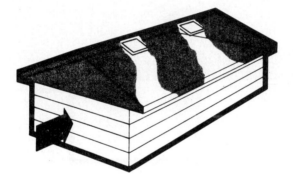

FIGURE 10-20

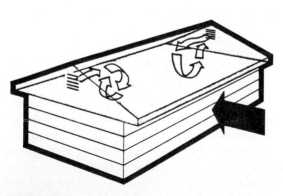

FIGURE 10-21

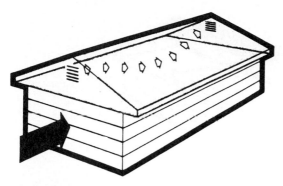

FIGURE 10-22

wind is parallel to the ridge— in at one louver, dipping down to the attic floor in the middle, then up and out the other louver. The roof sheathing is almost entirely unventilated.

With soffit vents added, the air flow is changed in that there is movement across the attic floor—similarly to the pattern of soffit vents used alone. As Figs. 10-23 and 10-24 indicate, gable end louvers and soffit vents largely operate independently of each other instead of combining into an effective system.

Turbine vents

A roof vent with a turbine wheel mounted on top of it continues to act as a roof vent, with the same pattern of air flow

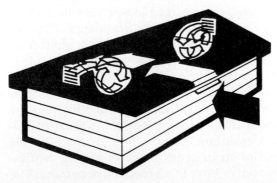

FIGURE 10-23

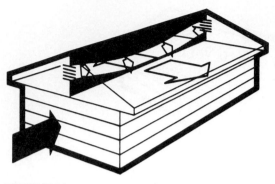

FIGURE 10-24

as shown in Figs. 10-19 and 10-20. At low wind speeds, when the movement of ventilation air into and out of the attic is most needed, the turbine vent is no better or worse than an ordinary roof vent. It has been shown in tests that air may actually enter through the turbine vent, depending on wind pressure conditions. Air, rain, and snow may enter, even with the turbine wheel turning, just as it would with a regular roof louver.

Power fans

Vents in the roof or gable ends employing a powered fan to move air out of the attic are of questionable value. In the first place, they cannot save energy. The power required to operate the fan exceeds the power saved in air conditioning the residence, compared to any fixed louver ventilation system meeting FHA Minimum Property Standards. Depending on their location, air moves in a direct path or shaft from the intake vent—such as in the soffits—to the fan, creating a narrow, restricted pattern. (See Figs. 10-25 and 10-26.)

If the power fan is controlled by a thermostat switch, there is no fan operation during the winter months when air circulation over the roof sheathing is needed in order to prevent moisture condensation. Thus a humidistat must be added to the installation in order to effect winter ventilation, which in any event occurs in an ineffective pattern. The need for electrical installation adds to the high initial cost

FIGURE 10-25

FIGURE 10-26

of a powered vent, with maintenance and repair costs also a factor.

Ridge vent combined with soffit vents

There is one attic ventilation system that effectively uses the two natural forces of wind pressure and thermal effect to continuously and uniformly ventilate the entire underside of the roof sheathing. It is the combination of a continuous ridge vent and an equal net free area of soffit venting, half of it on each side of the structure.

Figure 10-27 shows the air flow pattern of a ridge vent system. Intake is at the soffit. Air moves along the roof sheathing and exhausts at the ridge. *This occurs regardless of wind directions.* Since air flow is always in the same direction, it creates an inertia of constant air movement that is a principal reason a ridge vent system is so efficient. Figure 10-28 shows the importance of the baffle on AV Ridge Filtervent.™ When wind is perpendicular to the ridge, or coming at an angle, it strikes the baffle and jumps over the top of the ridge. A venturi action is created, causing both the windward and lee sides of the ridge vent to be in a negative pressure area. This pulls air out of the attic space.

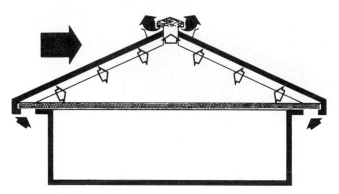

FIGURE 10-27

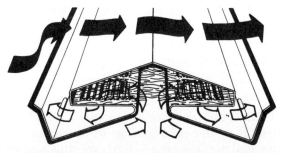

FIGURE 10-28

As illustrated in Fig. 10-9, if the wind is coming against the end of the house it jumps upward and over the ridge, causing a negative pressure area. Thus, regardless of wind direction, ridge vent is always in negative pressure and always acts as an exhaust vent.

When there is little or no wind, thermal effect acts to maintain air circulation across the underside of the roof sheathing. Warm air rises to the ridge and exhausts through the ridge vent and is replaced by cooler air entering at the soffit. No other attic ventilation method utilizes thermal effect as well as a ridge vent system.

A ridge vent system brings a uniform flow of air across the underside of the roof sheathing for the entire length of the attic, leaving no unventilated areas of the sheathing. *No other system provides this type of air flow pattern.*

Contemporary roof designs

Certain residential designs create roofs without a traditional ridge, the top of the roof meeting a vertical wall either at its peak or at some point prior to the peak. (See Fig. 10-29.) Products are available to provide continuous ventilation along the top of the roof in these cases and consist of one-half of a ridge vent in a configuration to fit the roof.

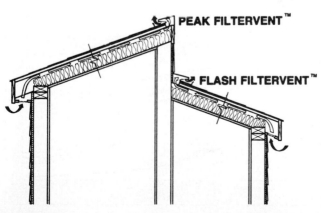

FIGURE 10-29

How vent systems compare in effective air flow

The preceding descriptions cover the general air flow *patterns* of the main types of attic vents. The other important consideration in judging a ventilation system is the *quantity* of air the vent system moves through the attic space under given wind conditions.

Comprehensive tests have been conducted by university laboratories and private industry measuring the cubic feet of air per minute (CFM) moving into and out of an attic for each of the vent systems shown in the three graphs in Fig. 10-30. Test results are expressed as CFM per square inch of vent area and are shown for varying wind velocities. Both winter and summer conditions are shown.

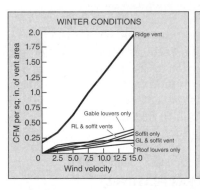

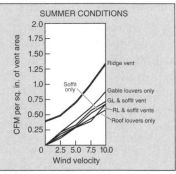

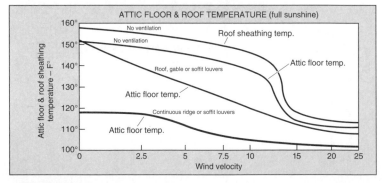

FIGURE 10-30

It is readily seen that a ridge vent system far exceeds other systems in moving more effective ventilating air through the attic space.

Attic floor temperature reduction with various vent systems

Because there is a wide variance in the pattern of air flow and quantity of flow between a ridge vent system and all other systems, it is logical to expect that there will be a corresponding difference in the reduction of attic floor temperatures.

Test results bear this out, as shown in Fig. 10-30. Note that with wind velocity at zero, *only the ridge vent system reduces* attic floor temperature from the 150°+ condition when there is no wind. This is due to thermal effect, the warm air rising along the underside of the roof sheath and exhausting at the ridge, being replaced by lower temperature air at the soffit.

Also, at all wind velocities *the ridge vent system maintains a lower attic floor temperature than other systems,* even when wind exceeds 25 mph.

10-4   How Much Attic Vent Is Required?

In past years there was little knowledge available to help builders determine how much vent area to provide in a residential attic. Builders tended to select one or another type of vent and installed it without aiming for a specific amount of moisture removal or a specific temperature reduction. They were just providing "ventilation."

FHA Minimum Property Standards

So limited has been the information on and concern for proper vent area that the Federal Housing Administration's Minimum Property Standards for one and two living units states:

> Attics and spaces between roof and top floor ceiling may have a free ventilating area of ⅟₃₀₀ of the horizontal area when (a) a vapor barrier having a transmission rate not exceeding one

perm is installed on the warm side of the ceiling or (b) at least 50
per cent of the required ventilating area is provided with fixed
louvers located in the upper portion of the space to be ventilated
(at least 3 ft above eave or cornice vents) with the remainder of
the required ventilation provided by eave or cornice vents.

This ratio of ⅟₃₀₀ translates into slightly less than one half
square inch of vent area for each square foot of attic floor. It
is barely adequate and presumes that the vent system is
properly balanced, with equal net free area in the high vents
and the low vents. Frequently, however, this is not the case.
There are many homes being built today with no consideration for this balance, and, in fact, even less total vent net
free area than the standards require.

With today's greater insulation requirements and a continuing need for energy conservation, both the total amount
and placement of attic venting is critically important.

Recommended rate of air flow

Before determining how much vent area is required for a
given attic space it is necessary to have a specific rate of air
flow as an objective. Tests at the University of Illinois
revealed that attic temperature is reduced by 44.5% when
air flow is 1.5 ft³/min per square foot of attic floor area. This
will provide a floor temperature reduction of 24°F. This
reduction is true regardless of ambient air temperature and
was true on all of the tests, with ambient air temperature
ranging from 85 to 110°F. The university reported that some
further reduction of temperature occurs as air flow is
increased to 2.0 ft³/min per square foot of floor area, but it is
clear from Fig. 10-31 that most of the removable heat is gone
once the ventilation rate reaches 1.5 ft³/min per square foot.

*A ridge vent system (ridge vent plus soffit vents) provides
this rate of air flow when there is 1.5 in² of net free area per
square foot of attic floor. No other vent system can do this.*

Summer requirement vs. winter requirement

While it is true that the summer requirement for a ventilation air flow sufficient to prevent heat buildup is greater

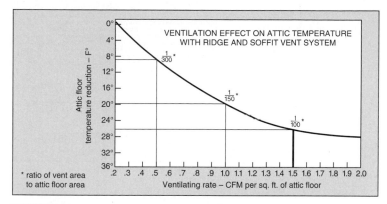

FIGURE 10-31

than the winter requirement to remove moisture, there is an increasing need for higher air flow rates in winter months. The electric utility industry is recommending ceiling insulation factors as high as R-36, and with heavier insulation the attic air is kept at a lower temperature. This lower temperature air absorbs less water vapor, so that there is a greater tendency for moisture to condense on the roof sheathing, rafters, and other structural parts.

In fact, attic temperature may even be low enough to cause escaping residential moisture to freeze in the upper portion of ceiling insulation, thawing at a later time to dampen the insulation.

For these reasons, a good vapor barrier is recommended in order to minimize moisture migration to the insulation and the attic space, along with effective attic ventilation to remove water vapor contained in the air. Rate of water vapor removed is shown in Fig. 10-32.

Builders should use a vent system which will remove at least 3.0 to 4.0 grains of moisture per hour, when using R-19 or heavier ceiling insulation. That requires a ventilation rate of 0.5 ft³/min per square foot of attic floor area. This rate can be accomplished with a ridge vent system that provides 1 square inch of net free area per square foot of attic floor. (This is double the FHA Minimum Property Standard.)

However, if good summer heat removal as well as good winter moisture removal is desired, a ridge vent system pro-

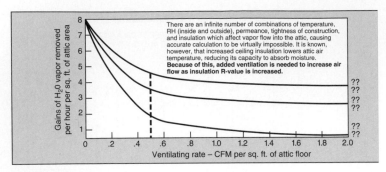

FIGURE 10-32

viding 1.5 in^2 of net free area per square foot of attic floor should be used and will furnish 1.5 ft^3/min of air flow— ample for both summer and winter conditions.

Required vent area for various systems

It was illustrated in graphs earlier that there is a difference in the rate of air flow through various vent systems. Using this data, Table 10-1 has been prepared showing the amount of net free area each vent system must have in order to provide both summer and winter air flow. Note that less is required with a ridge vent system. In some cases, other systems require so much net free area that installation of that much venting is virtually impossible.

TABLE 10-1   Net Free Vent Area Required for Summer and Winter Air Flow

VENT AREA REQUIRED	SUMMER 1.5 CFM per sq. ft. attic area	WINTER 0.5 CFM per sq. ft. attic area
	Sq. in. of net free vent area per sq. ft. of attic floor area	Sq. in. of net free vent area per sq. ft. of attic floor area
VENT SYSTEM		
Roof Louvers	3.3	11.1
Soffit Louvers	2.9	6.5
Gable Louvers	2.4	5.1
Roof Louvers and Soffit Vents	3.1	6.0
Gable Louvers and Soffit Vents	2.7	6.0
Ridge Filtervent and Soffit Vents	1.5	1.0

Ridge-and-soffit system meets all needs

The clear conclusion drawn from the foregoing information is that the most cost-effective method of meeting code requirements is with a ridge vent and soffit vent system. At the same time it provides the ventilation system that removes the most heat from the attic space in summer and the most moisture in winter.

10-5   Ventilating for the Whole House Fan

Formerly a popular cooling device before central air conditioning was commonplace, the large ceiling-mounted fan (Fig. 10-33) is again becoming popular as a viable alternative to air conditioning, particularly on summer days when heat and humidity are less than maximum. The fan reduces energy cost, providing a cooling method that uses approximately $\frac{1}{10}$ the power requirement of a 3-ton central air conditioning system.

How it cools

A whole house fan discharges air from the living space into the attic, cooling the home two ways: (1) When outside air is cooler than inside air, the fan draws the cooler air into the house to provide immediate relief to occupants and reduce inside air temperature. (2) The flow of air removes heat from the walls and objects in the home.

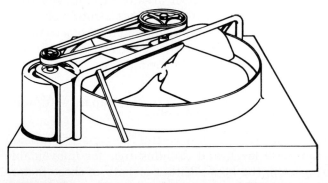

FIGURE 10-33

These cooling opportunities usually occur at night and in the early morning hours, often reducing temperatures enough so that daytime air conditioning is not required.

Two essential considerations

In designing a whole house fan installation, there are two essential considerations:

1. Fan size; i.e., how much air must be circulated to cool the house.

2. Exhaust ventilation; i.e., how much and what type of attic ventilation must be present to handle the exhaust capacity of the fan.

Item No. 1 above depends upon climate and user preference. In very hot and humid conditions, a net change of air once a minute appears to be satisfactory. In less extreme conditions, such as in more northern states, an air change every two minutes provides ample comfort. (It should be noted that some experts believe that only half of this air movement is sufficient.)

Item No. 2 relates directly to achieving the rated air movement of the fan and will be covered later in this section.

Determining correct fan size

The following example illustrates how to determine correct fan size.

Assuming a house with a floor area of 1250 ft^2 and standard 8-ft. ceiling heights, which provides a total volume of 10,000 ft^3. To produce an air flow equivalent to one complete air change every two minutes, the requirement would be a fan of 5,000-ft^3/min capacity. The formula is:

$$\frac{\text{Volume of home in ft}^3}{\text{No. of air changes desired per min}} = \text{fan capacity (ft}^3/\text{min)}$$

Two factors which may influence the fan capacity decision are: (a) fan noise level, and (b) allowance for unavoidable dead air space in the home. Either or both of these factors could reduce the CFM requirement for the fan.

Whole house fans are available from a large number of manufacturers in a range of sizes and capacities. Table 10-2 lists some typical house sizes and the correct fan size for each, to achieve a net change of air of once a minute. In this table fan capacity is shown in cubic feet of air flow per minute against an assumed static pressure of 0.1 inch of water pressure. This is the industry standard and represents the normal resistance encountered when there is unrestricted air flow out of the attic space. The final column of the table shows the net free area of attic ventilation required if the rated fan capacity is to be achieved.

Why sufficient attic ventilation is critical

Often overlooked in whole house fan installations is the provision of sufficient net free area of attic ventilation to permit the fan to operate at its rated capacity. The three- or four-bladed whole house fan is designed for moving a high volume of air against a very low back pressure. As stated earlier, the standard is 0.1 inch of static water pressure, and as back pressure increases beyond this amount, the fan volume capacity drops sharply.

In the curve shown in Fig. 10-34 the fan has a rated capacity of 5375 ft^3/min at 0.1 inches of water pressure. (0.1 inch of water is the equivalent of x lbs/in^2, which is a very small force.) If the back pressure of this particular fan is increased to 0.2 inches of water (i.e., $2 \times$ lbs/in^2), the fan capacity is

TABLE 10-2   Fan Sizing for Houses

House area, ft^2	House volume,* ft^3	Fan size, ft^3/min	Blade size (dia.), in	NFAVA,[†] ft^2
750	6000	3000	20	4.00
1000	8000	4000	24	5.33
1250	10,000	5000	24	6.66
1500	12,000	6000	30	8.00
1750	14,000	7000	30	9.33
2000	16,000	8000	36	10.66

* 8-ft ceiling height assumed.
[†] Net Free Attic Ventilation Area required to accommodate whole house fan exhaust.

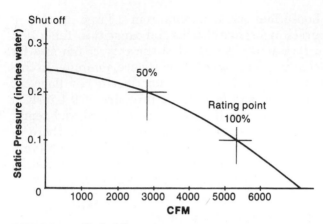

FIGURE 10-34 Typical fan curve.

reduced 50%; and if the back pressure is 0.25 inches of water (i.e., $2.5 \times$ lbs/in^2), the fan ceases to move air.

Whole house fan energy saving calculations are based on having the fan operate at rated capacity. From the above it is easy to see that even a small amount of back pressure eliminates the effectiveness of a whole house fan. Back pressure comes principally from inadequate exhaust venting—inadequate in both amount and placement.

The static pressure of 0.1 inch of water pressure is equivalent to the resistance caused by attic vents and the turning of the air stream as the air flows through the attic. *Forcing a fan to operate against a pressure higher than its rated capacity not only reduces the volume of air flow but can create excessive fan noise and rapid deterioration of the fan motor due to overheating.* Operation can be interrupted due to circuit breaker tripping, or the motor heater controls will cause erratic operation.

Determining NFAVA

To determine the necessary net free attic ventilation area required for a whole house fan installation, the simple and recommended method is to use the industry standard rule of allowing approximately 1.33 ft^2 of net free attic vent area for every 1,000 ft³/min of fan capacity. Following this rule, the required NFAVA for various fan sizes is shown in Table 10-2.

Which type of attic ventilation is best?

Very few homes have enough net free area of attic ventilation to satisfy the requirements for a whole house fan. In almost all cases, attic venting must be added. For this, the *type* of attic vents selected is an important consideration.

Because their design dictates their location in the attic, vents are categorized into two general types: high and low vents. The best ventilation for all purposes is always a combination of high and low vents, preferably balanced so that the net free area of the high vents is equal to the net free area of the low vents. It should be recognized, however, that in many existing homes and some new ones, this balance is not possible and the required net free area must be placed in the best alternative locations.

Another consideration is the desirability of having multiple exhaust vents to minimize the effect of the wind blowing against the vent and increasing static pressure. This can be a significant factor, for a *15 mph wind creates a pressure of 0.1 inch (water pressure).*

If this is being applied against the *only* attic vent, it effectively doubles the resistance that the fan must work against. Whenever possible, vents should be placed so that they enhance, rather than inhibit, the operation of the fan.

Important also is the selection of attic vents that will not admit snow and rain. This occurs when the vent becomes an intake opening for outside air under certain wind conditions. An example would be gable end louvers at each end of the attic. Often air enters at one louver and exits at the other. Another example would be a residence without soffit venting and with several high vents in the gables or roof. Air will exhaust through one or more high vents and enter through other high vents, depending on vent placement.

Table 10-3 lists the types of attic ventilators currently being marketed, along with their net free vent areas.

Table 10-3 clearly shows that the most practical and effective method of providing increased net free attic ventilation

TABLE 10-3   Attic Ventilators

Type of vent	Net free attic vent area, ft^2
High Vents	
Ridge Filtervent®*	
One 8-ft piece	1.00
Typical Installations:	
24 ft (3 pieces)	3.00
32 ft (4 pieces)	4.00
40 ft (5 pieces)	5.00
Roof Louvers	
One roof louver	0.35
Typical Installations:	
2 roof louvers	0.70
3 roof louvers	1.05
4 roof louvers	1.40
5 roof louvers	1.75
Turbine Vent	
One turbine vent	0.78
Typical Installation	
2 turbine vents	1.56
Rectangular Gable Vents	
Size	
14 in × 24 in	1.09
18 in × 24 in	1.42
24 in × 30 in	2.42
Triangular Gable Vents	
Base Length	
30 in	0.57
48 in	1.00
72 in	1.09
96 in	1.91
120 in	2.89
Low Vents	
Unscreened AVI Eave Vent	
16 in × 8 in	0.5
Screened Eave Vents	
16 in × 8 in	0.36
16 in × 4 in	0.15
Strip Vent and Drip Edge Vent	
One 8-ft piece	0.5
Perforated Aluminum Soffit	
1 ft^2	0.1
Lanced Aluminum Soffit	
1 ft^3	0.025–0.05

*Ridge Filtervent is a patented product of Air Vent Inc.

area is with Ridge Filtervent™. Not only does Ridge Filter-vent™ provide the necessary net opening at the lowest cost, but it is the only vent system that *always* aids attic exhaust by utilizing the principle of warm air rising and the lowering of air pressure immediately above the ridge as air flow is lifted by the baffles, creating a venturi effect. Thus Ridge Filtervent™ works with the whole house fan, not against it.

Four points to remember

1. Whole house fans represent a cost-effective alternative to air conditioning, providing comfort at a fraction of air conditioning's operating cost.

2. The correct size fan can be determined, based on house size and climate conditions.

3. Adequate net free area of attic ventilation must be provided to assure that the fan will move its rated volume of air.

4. Lack of sufficient attic ventilation will prevent adequate cooling, increase fan noise, shorten fan life, and cancel the cost efficiency of a whole house fan system.

NOTES

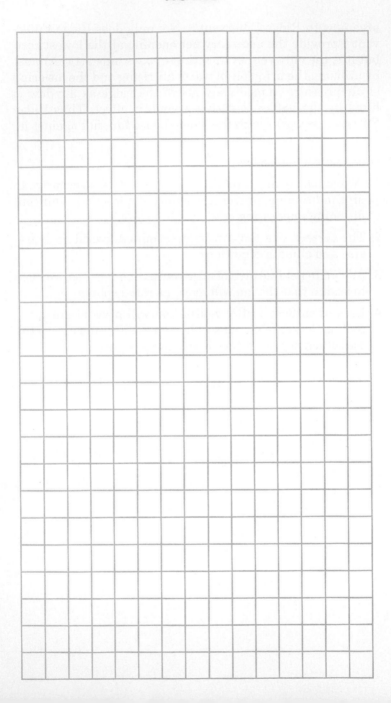

Sheet Metal
Duct Design

Rule 1. Elbow lengths, square throat,
        square heel
   Exterior dimensions of elbow
   establish its length.
   All dimensions are to be aggregated
   to next foot or half-foot.

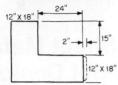

Example elbow, to be computed:

$$2'0" = 2'0"$$
$$+\ 12" = 1'0"$$
$$+\ 12" = 1'0"$$
$$+\ 15" = 1'3"$$
$$= \overline{5'3"}$$
$$= 5'6"$$

Rule 2. Elbows, radius throat, radius heel
   Use throat measurement plus cheek
   size in both directions.
   Then add for total length.
   Throat measurement 12" plus cheek 18" = 30"
   Throat measurement 24" plus cheek 18" = 42"

                          Total $\overline{72"}$ =
                                6' length

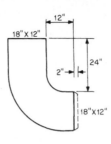

Rule 3. Duct lengths: duct perimeters
   (a) Minimum perimeters shall be 4 ft.
       (Any ducts whose perimeters
       measure less than 4 ft shall be
       considered as having a perimeter of
       4 ft because below this perimeter
       labor remains constant.
   (b) Lengths shall be computed to
       next foot or half-foot.
   (c) 24 ga. is minimum gauge used for
       weight calculations.

Example:
   (a) Duct shown as actual perimeter of
       44" to be computed at 48" = 4'.
   (b) Length 7'10" to be computed at 8'.
       Result: 8' – 12" X 12" Duct.

Rule 4. Sets or raises
   (a) The length of a duct set, extending
       in any direction, shall be considered
       on the extension plus the set.
   (b) The aggregate length shall be
       computed to next foot or half-foot.

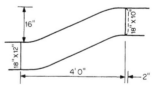

Example: Set to be computed
   4'0" + 16" =
   4'0" + 1'4" =
   5'4" = 5'6" of 18" X 12" duct

FIGURE 11-1   Ductwork takeoff rules. (*Source: A. M. Khashab,* Heating, Ventilating and Air Conditioning Systems Estimating Manual, *McGraw-Hill, New York, 1977. Used with permission.*)

Rule 5. Transformations

In computing the square footage of a transformation, reducing duct from one size to another, the computation shall be made on the basis of the larger dimensions.

Example 1:

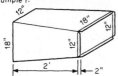

Above transformation to be computed as 2'–18" X 18" duct

Example 2.

Above transformation to be computed as 2'–18" X 12" duct

Rule 6. Collars carrying registers, grilles and diffusers are to be figured a minimum of 1' in length.

Example:

Collar length 10" = 1'
Collar length 3" = 1'

FIGURE 11-1 (*Continued*)

Rule 7. Square foot-duct

(a) Compute perimeter and adjust to next half-foot.
(b) Compute length and adjut to next half-foot.
(c) Calculate footage.

Example: Duct 14" X 12" X 3'10" long

(a) Perimeter 52" = 4' 4" = 4' 6"
(b) Length 3'10" = 4'
(c) Computation 4' X 4 1/2' = 18 ft^2

Rule 8. Square foot

Fire dampers
Louvers
Multiblade dampers
Ducturns

(a) Adjust each dimension to next half-foot.
(b) Multiply for area and adjust to next square foot.
(c) Minimum of 3 ft^2 for any sundry items.

Examples: Fire damper 40" X 32"
= 3' 4" X 2' 8"
= 3 1/2' X 3'
= 10 1/2 ft^2
= 11 ft^2

Use minimum of 12" for any side in computation which is under 12".

Example:
36" X 7" = 36" X 12" = 3 ft^2

TABLE 11-1   Surface Areas of Round Ducts

Diameter, in	Area, ft²/ft	Diameter, in	Area, ft²/ft	Diameter, in	Area, ft²/ft
4	1.05	16	4.19	28	7.33
5	1.31	17	4.45	29	7.59
6	1.57	18	4.72	30	7.85
7	1.83	19	4.98	31	8.11
8	2.09	20	5.24	32	8.38
9	2.36	21	5.50	33	8.65
10	2.62	22	5.76	34	8.91
11	2.88	23	6.02	35	9.17
12	3.15	24	6.28	36	9.43
13	3.40	25	6.54	37	9.69
14	3.67	26	6.80	38	9.95
15	3.93	27	7.07	39	10.21
				40	10.47

SOURCE: A. M. Khashab, *Heating, Ventilating and Air Conditioning Systems Estimating Manual,* McGraw-Hill, New York, 1977. Used with permission.

TABLE 11-2 Surface Areas of Rectangular Ducts

Width + Depth, in	Area, ft²/ft	Width + Depth, in	Area, ft²/ft	Width + Depth, in	Area, ft²/ft
10	1.67	47	7.83	84	14.00
11	1.83	48	8.00	85	14.17
12	2.00	49	8.17	86	14.34
13	2.17	50	8.34	87	14.50
14	2.34	51	8.50	88	14.67
15	2.50	52	8.67	89	14.83
16	2.67	53	8.83	90	15.00
17	2.83	54	9.00	91	15.17
18	3.00	55	9.17	92	15.34
19	3.17	56	9.34	93	15.50
20	3.34	57	9.50	94	15.67
21	3.50	58	9.67	95	15.83
22	3.67	59	9.83	96	16.00
23	3.83	60	10.00	97	16.17
24	4.00	61	10.17	98	16.34
25	4.17	62	10.34	99	16.50
26	4.34	63	10.50	100	16.67
27	4.50	64	10.67	101	16.83
28	4.67	65	10.83	102	17.00
29	4.83	66	11.00	103	17.17
30	5.00	67	11.17	104	17.34
31	5.17	68	11.34	105	17.50
32	5.34	69	11.50	106	17.67
33	5.50	70	11.67	107	17.83
34	5.67	71	11.83	108	18.00
35	5.83	72	12.00	109	18.17
36	6.00	73	12.17	110	18.34
37	6.17	74	12.34	111	18.50
38	6.34	75	12.50	112	18.67
39	6.50	76	12.67	113	18.83
40	6.67	77	12.83	114	19.00
41	6.83	78	13.00	115	19.17
42	7.00	79	13.17	116	19.34
43	7.17	80	13.34	117	19.50
44	7.34	81	13.50	118	19.67
45	7.50	82	13.67	119	19.83
46	7.67	83	13.83	120	20.00

SOURCE: A. M. Khashab, *Heating, Ventilating and Air Conditioning Systems Estimating Manual,* McGraw-Hill, New York, 1977. Used with permission.

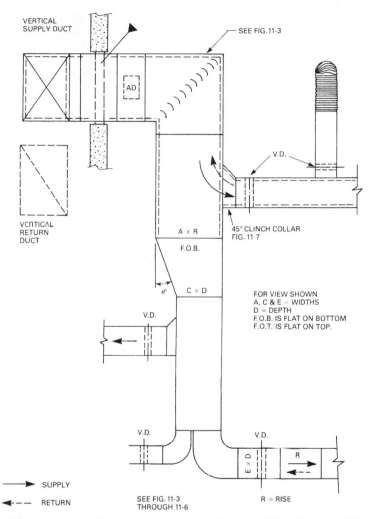

FIGURE 11-2 Typical supply or return duct. (*Courtesy: Sheet Metal and Air Conditioning Contractors National Association, Inc., Chantilly, Virginia, from* HVAC Duct Construction Standards, Metal and Flexible, *1985. Used with permission from SMACNA copyright holder.*)

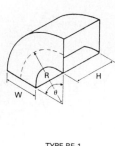

TYPE RE 1
RADIUS ELBOW
(CENTERLINE R $= \dfrac{3W}{2}$ = STD RADIUS)

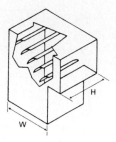

TYPE RE 2
SQUARE THROAT ELBOW
WITH VANES

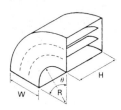

TYPE RE 3
RADIUS ELBOW
WITH VANES

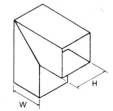

TYPE RE 4
SQUARE THROAT ELBOW
WITHOUT VANES
(1000 FPM MAXIMUM VELOCITY)

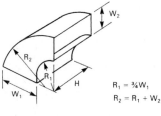

TYPE RE 5
DUAL RADIUS ELBOW

$R_1 = \tfrac{3}{4}W_1$
$R_2 = R_1 + W_2$

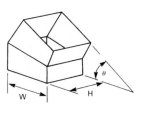

MITERED ELBOW

BEAD, CROSSBREAK AND REINFORCE FLAT SURFACES AS IN STRAIGHT DUCT

FIGURE 11-3 Rectangular elbows. (*Courtesy: Sheet Metal and Air Conditioning Contractors National Association, Inc., Chantilly, Virginia, from* HVAC Duct Construction Standards, Metal and Flexible, *1985. Used with permission from SMACNA copyright holder.*)

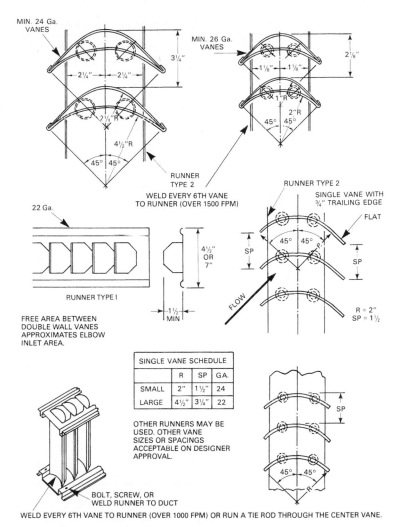

FIGURE 11-4 Vanes and vane runners. (*Courtesy: Sheet Metal and Air Conditioning Contractors National Association, Inc., Chantilly, Virginia, from* HVAC Duct Construction Standards, Metal and Flexible, *1985. Used with permission from SMACNA copyright holder.*)

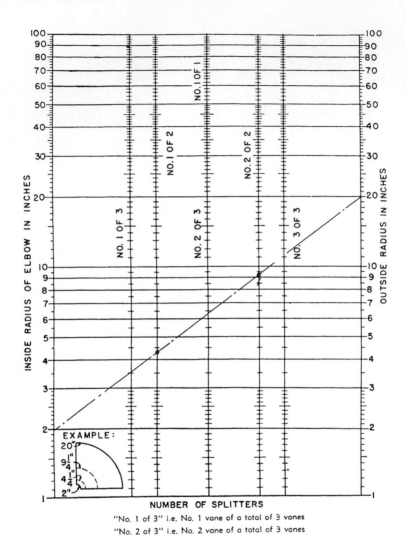

NUMBER OF SPLITTERS

"No. 1 of 3" i.e. No. 1 vane of a total of 3 vanes
"No. 2 of 3" i.e. No. 2 vane of a total of 3 vanes

FIGURE 11-5 Number of short radius vanes. (*Courtesy: Sheet Metal and Air Conditioning Contractors National Association, Inc., Chantilly, Virginia, from* HVAC Duct Construction Standards, Metal and Flexible, *1985. Used with permission from SMACNA copyright holder.*)

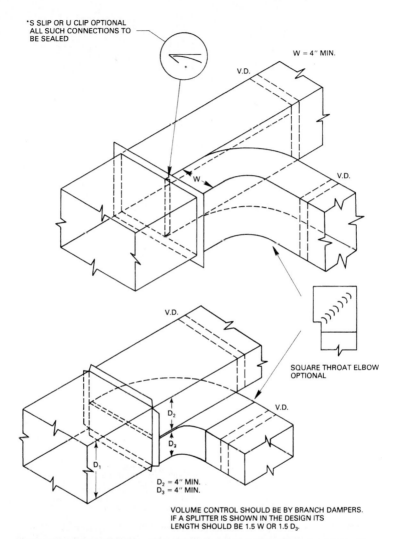

FIGURE 11-6 Parallel flow branches. (*Courtesy: Sheet Metal and Air Conditioning Contractors National Association, Inc., Chantilly, Virginia, from* HVAC Duct Construction Standards, Metal and Flexible, *1985. Used with permission from SMACNA copyright holder.*)

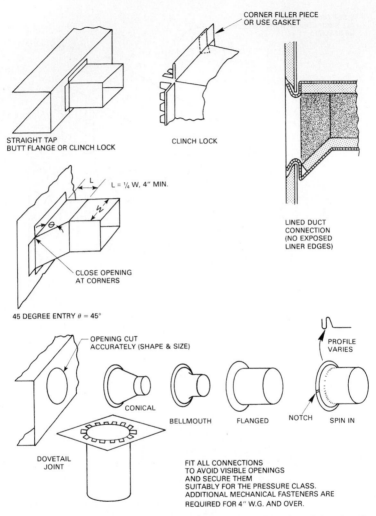

FIGURE 11-7 Branch connections. (*Courtesy: Sheet Metal and Air Conditioning Contractors National Association, Inc., Chantilly, Virginia, from* HVAC Duct Construction Standards, Metal and Flexible, *1985. Used with permission from SMACNA copyright holder.*)

OFFSETS 2 AND 3 AND TRANSITIONS MAY HAVE EQUAL OR UNEQUAL INLET
AND OUTLET AREAS. TRANSITIONS MAY CONVERT DUCT PROFILES TO
ANY COMBINATION FOR RECTANGULAR, ROUND OR FLAT OVAL SHAPES.

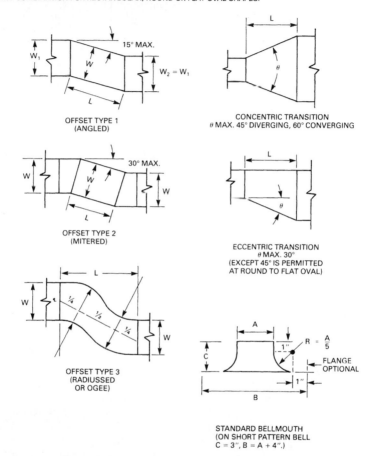

OFFSET TYPE 1
(ANGLED)

CONCENTRIC TRANSITION
θ MAX. 45° DIVERGING, 60° CONVERGING

OFFSET TYPE 2
(MITERED)

ECCENTRIC TRANSITION
θ MAX. 30°
(EXCEPT 45° IS PERMITTED
AT ROUND TO FLAT OVAL)

OFFSET TYPE 3
(RADIUSSED
OR OGEE)

STANDARD BELLMOUTH
(ON SHORT PATTERN BELL
C = 3″, B = A + 4″.)

FIGURE 11-8   Offset and transitions. (*Courtesy: Sheet Metal and Air Conditioning Contractors National Association, Inc., Chantilly, Virginia, from* HVAC Duct Construction Standards, Metal and Flexible, *1985. Used with permission from SMACNA copyright holder.*)

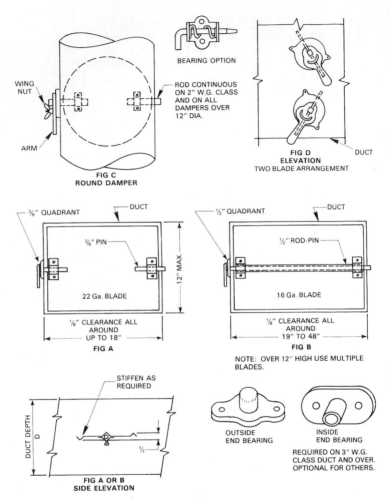

FIGURE 11-9 Volume dampers—single blade type. (*Courtesy: Sheet Metal and Air Conditioning Contractors National Association, Inc., Chantilly, Virginia, from* HVAC Duct Construction Standards, Metal and Flexible, *1985. Used with permission from SMACNA copyright holder.*)

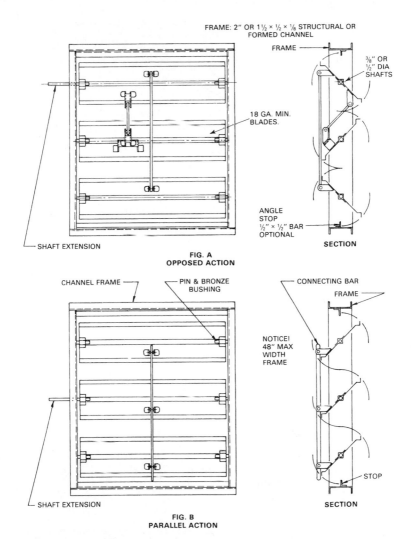

FRAME: 2" OR 1½ × ½ × ⅛ STRUCTURAL OR FORMED CHANNEL

FRAME

⅜" OR ½" DIA SHAFTS

18 GA. MIN. BLADES.

ANGLE STOP ½" × ½" BAR OPTIONAL

SECTION

SHAFT EXTENSION

FIG. A
OPPOSED ACTION

CHANNEL FRAME

PIN & BRONZE BUSHING

CONNECTING BAR

FRAME

NOTICE! 48" MAX WIDTH FRAME

STOP

SHAFT EXTENSION

SECTION

FIG. B
PARALLEL ACTION

FIGURE 11-10 Multiblade volume dampers. (*Courtesy: Sheet Metal and Air Conditioning Contractors National Association, Inc., Chantilly, Virginia, from* HVAC Duct Construction Standards, Metal and Flexible, *1985. Used with permission from SMACNA copyright holder.*)

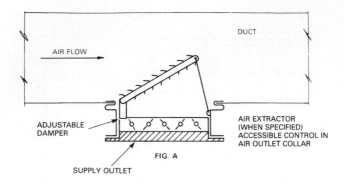

ADJUSTABLE DAMPER

DUCT

AIR FLOW

AIR EXTRACTOR
(WHEN SPECIFIED)
ACCESSIBLE CONTROL IN
AIR OUTLET COLLAR

FIG. A

SUPPLY OUTLET

GRIDS AND DAMPERS MUST NOT DAMAGE LINERS.

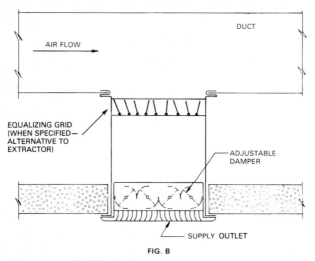

DUCT

AIR FLOW

EQUALIZING GRID
(WHEN SPECIFIED—
ALTERNATIVE TO
EXTRACTOR)

ADJUSTABLE
DAMPER

SUPPLY OUTLET

FIG. B

NOTE:
REGISTER CONTAINS VOLUME
CONTROL AT GRILLE. GRILLE
HAS NO VOLUME CONTROL.

GRILLE FLANGES MUST
COVER DUCT FLANGES.

FIGURE 11-11 Grille and register connections. (*Courtesy: Sheet Metal and Air Conditioning Contractors National Association, Inc., Chantilly, Virginia, from* HVAC Duct Construction Standards, Metal and Flexible, *1985. Used with permission from SMACNA copyright holder.*)

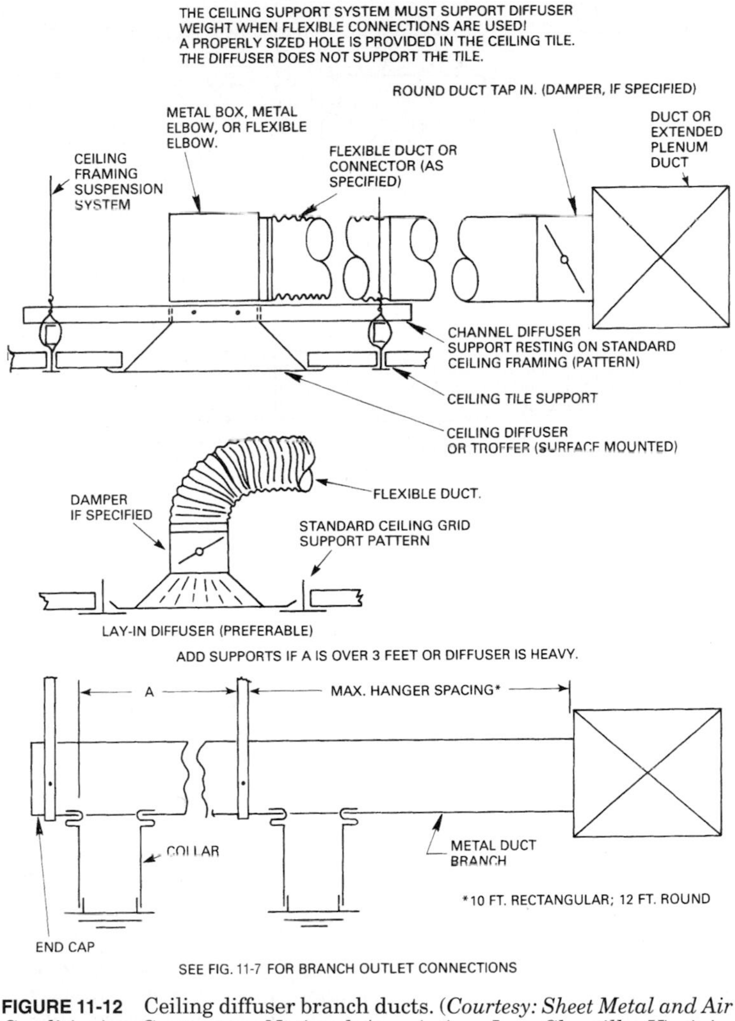

THE CEILING SUPPORT SYSTEM MUST SUPPORT DIFFUSER
WEIGHT WHEN FLEXIBLE CONNECTIONS ARE USED!
A PROPERLY SIZED HOLE IS PROVIDED IN THE CEILING TILE.
THE DIFFUSER DOES NOT SUPPORT THE TILE.

ROUND DUCT TAP IN. (DAMPER, IF SPECIFIED)

METAL BOX, METAL
ELBOW, OR FLEXIBLE
ELBOW.

FLEXIBLE DUCT OR
CONNECTOR (AS
SPECIFIED)

DUCT OR
EXTENDED
PLENUM
DUCT

CEILING
FRAMING
SUSPENSION
SYSTEM

CHANNEL DIFFUSER
SUPPORT RESTING ON STANDARD
CEILING FRAMING (PATTERN)

CEILING TILE SUPPORT

CEILING DIFFUSER
OR TROFFER (SURFACE MOUNTED)

DAMPER
IF SPECIFIED

FLEXIBLE DUCT.

STANDARD CEILING GRID
SUPPORT PATTERN

LAY-IN DIFFUSER (PREFERABLE)

ADD SUPPORTS IF A IS OVER 3 FEET OR DIFFUSER IS HEAVY.

A

MAX. HANGER SPACING*

COLLAR

METAL DUCT
BRANCH

*10 FT. RECTANGULAR; 12 FT. ROUND

END CAP

SEE FIG. 11-7 FOR BRANCH OUTLET CONNECTIONS

FIGURE 11-12 Ceiling diffuser branch ducts. (*Courtesy: Sheet Metal and Air Conditioning Contractors National Association, Inc., Chantilly, Virginia, from* HVAC Duct Construction Standards, Metal and Flexible, *1985. Used with permission from SMACNA copyright holder.*)

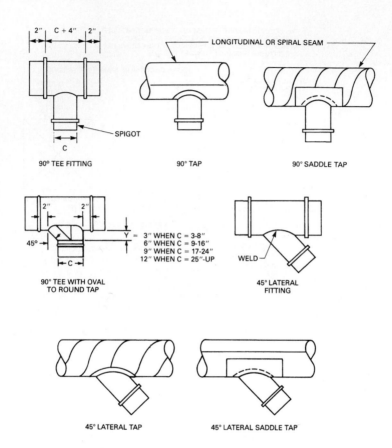

FIGURE 11-13 90° tees and laterals. (*Courtesy: Sheet Metal and Air Conditioning Contractors National Association, Inc., Chantilly, Virginia, from* HVAC Duct Construction Standards, Metal and Flexible, *1985. Used with permission from SMACNA copyright holder.*)

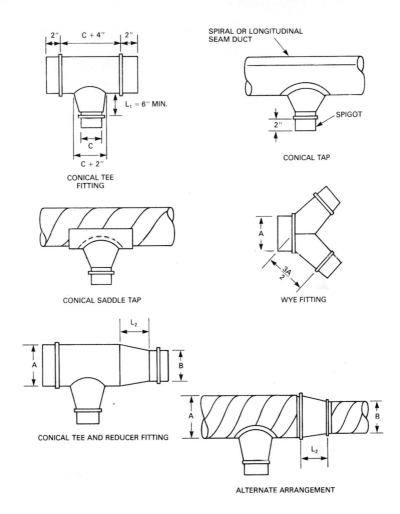

CONICAL TEE FITTING

CONICAL TAP

CONICAL SADDLE TAP

WYE FITTING

CONICAL TEE AND REDUCER FITTING

ALTERNATE ARRANGEMENT

TAPS MAY HAVE SCREWS

$L_2 = A - B$ (4" MIN.)

FIGURE 11-14 Conical tees. (*Courtesy: Sheet Metal and Air Conditioning Contractors National Association, Inc., Chantilly, Virginia, from* HVAC Duct Construction Standards, Metal and Flexible, *1985. Used with permission from SMACNA copyright holder.*)

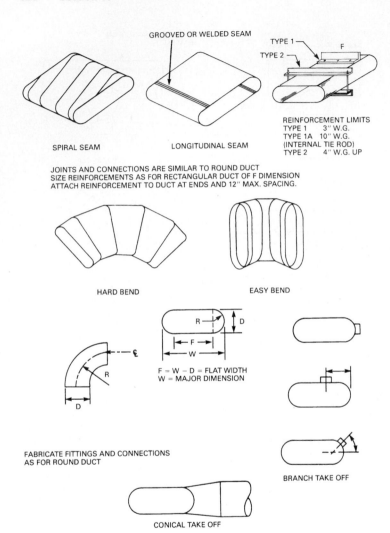

FIGURE 11-15   Flat oval ducts. (*Courtesy: Sheet Metal and Air Conditioning Contractors National Association, Inc., Chantilly, Virginia, from* HVAC Duct Construction Standards, Metal and Flexible, *1985. Used with permission from SMACNA copyright holder.*)

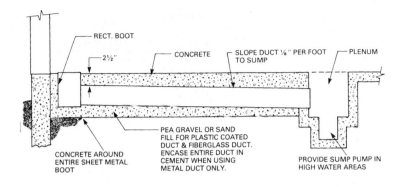

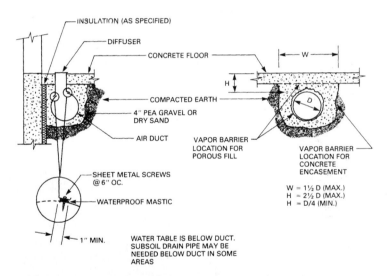

THIS ILLUSTRATION IS A DESIGN GUIDE.

FIGURE 11-16 Typical underslab duct. (*Courtesy: Sheet Metal and Air Conditioning Contractors National Association, Inc., Chantilly, Virginia, from HVAC Duct Construction Standards, Metal and Flexible, 1985. Used with permission from SMACNA copyright holder.*)

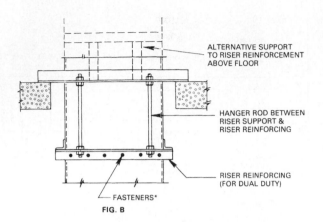

ALTERNATIVE SUPPORT
TO RISER REINFORCEMENT
ABOVE FLOOR

HANGER ROD BETWEEN
RISER SUPPORT &
RISER REINFORCING

RISER REINFORCING
(FOR DUAL DUTY)

FASTENERS*

FIG. B

*MINIMUM NUMBER OF FASTENERS
ON EACH OF TWO SUPPORT BARS

LARGEST DUCT DIM.	MINIMUM NUMBER OF FASTENERS
16" and down	2
17" - 24"	3
over 24"	Largest duct dim. Divided by 8

Locate a fastener within 2" of the duct edges. Locate others at evenly
spaced intervals.

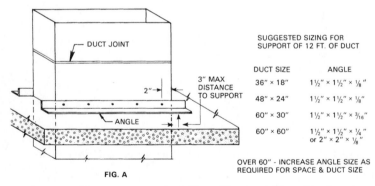

DUCT JOINT

3" MAX
DISTANCE
TO SUPPORT

2"→

ANGLE

SUGGESTED SIZING FOR
SUPPORT OF 12 FT. OF DUCT

DUCT SIZE	ANGLE
36" × 18"	1½" × 1½" × ⅛"
48" × 24"	1½" × 1½" × ⅛"
60" × 30"	1½" × 1½" × 3/16"
60" × 60"	1½" × 1½" × ¼" or 2" × 2" × ⅛"

OVER 60" - INCREASE ANGLE SIZE AS
REQUIRED FOR SPACE & DUCT SIZE

FIG. A

FIGURE 11-17   Riser supports—from floor. (*Courtesy: Sheet Metal and Air
Conditioning Contractors National Association, Inc., Chantilly, Virginia,
from* HVAC Duct Construction Standards, Metal and Flexible, *1985. Used
with permission from SMACNA copyright holder.*)

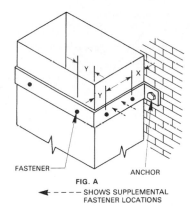

FASTENER

ANCHOR

FIG. A

← – – – SHOWS SUPPLEMENTAL
FASTENER LOCATIONS

FIG. A—SUGGESTED SIZING

DUCT SIZE	BAND
18″ × 12″	1½″ × 16 Ga.
24″ × 20″	1″ × ⅛″

DUCT GAGE	ALLOWABLE LOAD PER FASTENER*
28, 26	25 lb
24, 22, 20	35 lb
18, 16	50 lb

*WELD, BOLT OR NO. 8 SCREW (MIN.).
DEVIATION PERMITTED BY OTHER
ANALYSIS. X = 1″, Y = 2″; ADD OTHERS TO
ACCOMMODATE LOAD. MINIMUM OF 3 ON 24″
WIDTH AND UP. ADD ALONG SIDES NEAREST
ANCHORS.

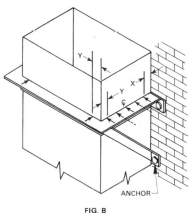

ANCHOR

FIG. B

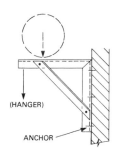

(HANGER)

ANCHOR

SEE KNEE BRACKET TABLES
IN THE ROUND INDUSTRIAL STDS.

FIG. B—SUGGESTED SIZING

DUCT SIZE	ANGLE
30″ × 12″	1″ × 1″ × ⅛″
36″ × 18″	1″ × 1″ × ⅛″
42″ × 24″	1¼″ × 1¼″ × ⅛″
48″ × 30″	1¼″ × 1¼″ × ⅛″

NOTES: 1. BRACKETS ARE SIZED FOR 12 FEET OF DUCT, MAXIMUM.
2. LOCATE DUCTS AGAINST WALL OR MAXIMUM OF 2″ AWAY FROM WALL.
3. EACH WALL ANCHOR SHALL SATISFY THE FOLLOWING CRITERIA UNLESS OTHER
ANALYSIS IS MADE:
A. TENSILE LOAD = ⅜ × DUCT WEIGHT; SAFETY FACTOR OF 4.
B. SHEAR LOAD × ½ × DUCT WEIGHT; SAFETY FACTOR OF 4.

FIGURE 11-18 Supports from wall. (*Courtesy: Sheet Metal and Air Conditioning Contractors National Association, Inc., Chantilly, Virginia, from* HVAC Duct Construction Standards, Metal and Flexible, *1985. Used with permission from SMACNA copyright holder.*)

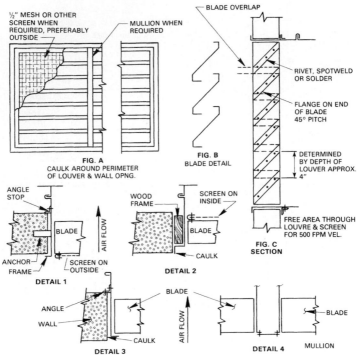

FIGURE 11-19 Louvers and screens. (*Courtesy: Sheet Metal and Air Conditioning Contractors National Association, Inc., Chantilly, Virginia, from* HVAC Duct Construction Standards, Metal and Flexible, *1985. Used with permission from SMACNA copyright holder.*)

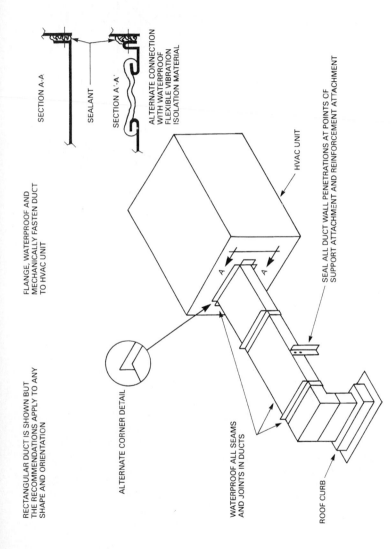

RECTANGULAR DUCT IS SHOWN BUT THE RECOMMENDATIONS APPLY TO ANY SHAPE AND ORIENTATION

FLANGE, WATERPROOF AND MECHANICALLY FASTEN DUCT TO HVAC UNIT

SECTION A-A

SEALANT

SECTION A'-A'

ALTERNATE CONNECTION WITH WATERPROOF FLEXIBLE VIBRATION ISOLATION MATERIAL

HVAC UNIT

SEAL ALL DUCT WALL PENETRATIONS AT POINTS OF SUPPORT ATTACHMENT AND REINFORCEMENT ATTACHMENT

ALTERNATE CORNER DETAIL

WATERPROOF ALL SEAMS AND JOINTS IN DUCTS

ROOF CURB

FIGURE 11-20 Rooftop duct installation. (*Courtesy: Sheet Metal and Air Conditioning Contractors National Association, Inc., Chantilly, Virginia, from HVAC Duct Construction Standards, Metal and Flexible, 1985. Used with permission from SMACNA copyright holder.*)

333

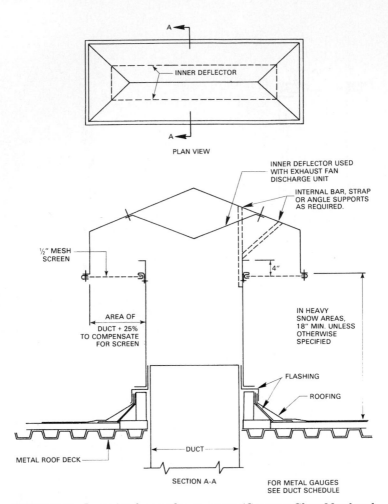

FIGURE 11-21 Large intake or exhaust covers. (*Courtesy: Sheet Metal and Air Conditioning Contractors National Association, Inc., Chantilly, Virginia, from* HVAC Duct Construction Standards, Metal and Flexible, *1985. Used with permission from SMACNA copyright holder.*)

NOTES

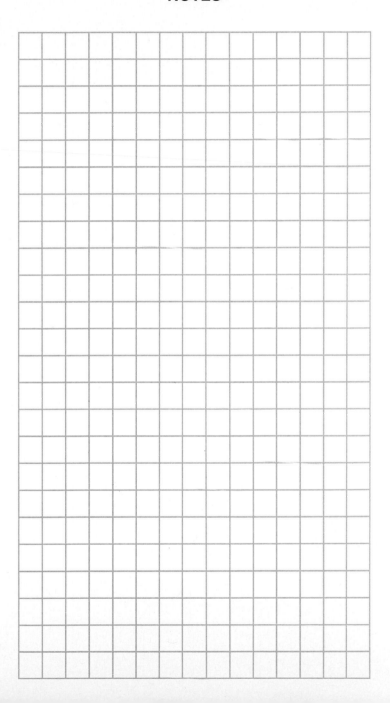

NOTES

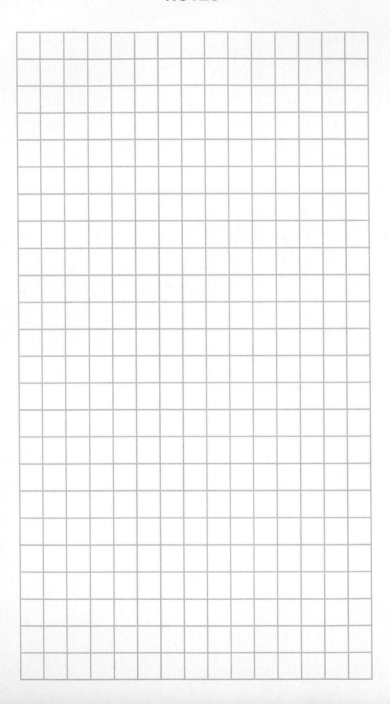

Energy
Conservation

TABLE 12-1   Potential Energy Conservation Checklist

BUILDINGS

1 Add insulation to attic spaces

2 Insulate top of basement walls

3 Weatherstrip around doors and openable windows

4 Insulate backplate of convectors, radiators, electric baseboard with foil faced rigid fiberglass

5 Caulk around exterior doors and windows

6 Close up unused doors during heating season where possible

7 Insert rigid insulation in unused windows during heating season

8 Insulate outside walls of pipe tunnels to at least 2 feet below grade

9 Install automatic closers on exterior doors

10 Install astragals to close opening between double doors

11 Install vestibules, especially at exterior doors used frequently

12 Install storm doors on infrequently-used outside doors

13 Install storm windows where single-glazing now exists

14 Install awnings over windows in air-conditioned spaces

15 Install drapes and/or venetian blinds in windows

16 Install two-inch styrofoam insulation on top of flat roofs

17 Apply epoxy resin over porous exterior masonry walls

18 Add insulation to the inside of exterior masonry walls

19 Add insulation to the outside of exterior masonry walls and cover with stucco, siding, etc.

20 Install insulating window coverings

21 Excavate around building perimeter and insulate exterior surface of basement walls

22 Close up and insulate unneeded windows permanently

HEATING SYSTEMS

1 Install automatic night temperature setback controls

2 Provide locking covers on thermostats

3 Change range of thermostats to a wider band (dead band between heat and cool)

4 Install self-contained thermostat control valves on uncontrolled radiation

5 Replace existing thermostats with new dual temperature thermostats controlled by a 7-day time clock

6 Improve boiler or furnace efficiency (test for air/fuel mix-oxygen test)

7 Install automatic stack dampers on gas-fired furnaces and domestic hot water heaters

8 Insulate hot air supply ducts and outside air ducts up to the damper in furnace rooms

9 Add insulation to surface of boilers

10 Insulate bare domestic hot water piping and hot water heaters

TABLE 12-1   Potential Energy Conservation Checklist (*Continued*)

11 Insulate bare steam condensate return lines

12 Add insulation to existing steam lines

13 Add resilient edge seals to outside air dampers

14 Seal air leaks around boiler access doors to combustion chamber

15 Install automatic indoor-outdoor control to maintain lowest hot water boiler temperature necessary for space heating needs

16 Install an automatic combustion air damper

17 Install turbulators in firetube boilers after obtaining the boiler manufacturers approval

18 Install replacement burners that burn either natural gas or fuel oil

19 Replace gas pilots with electric or electronic pilots

20 Install a "shell-head" adapter at the end of the gun on oil burners firing No. 2 fuel oil

21 Install glass doors on fireplace openings

VENTILATING SYSTEMS

1 Disconnect interlock between ventilation supply fan and exhaust fan, and install a manual timer switch to control the exhaust fan

2 Provide a manual positioning switch (or potentiometer) for outside air dampers on ventilation supply air

3 Consider installing destratification fans in the high ceiling area of the building

4 Reduce ventilation rate to code minimum

5 Seal with duct tape or caulk all joints in ductwork systems

6 Reduce toilet exhaust to code minimum

7 Provide light switch control of the toilet exhaust fan

8 Install timers to automatically turn off exhaust fans

9 Install smoke removal units in meeting rooms

10 Set the minimum position on unit ventilator outside air dampers to fully closed

11 Replace gravity backdraft dampers in exhaust fan or relief system with insulated powered dampers

12 Replace existing outside air dampers with insulating dampers

13 Install a baffle in the kitchen hood

PLUMBING SYSTEMS

1 Install water dams in toilet water closets

2 Reduce urinal flushing

3 Install 7-day time clock to control domestic hot water circulating pump

4 Install 7-day time clock and controls to reduce the temperature of domestic hot water heater that is not required

TABLE 12-1   Potential Energy Conservation Checklist (*Continued*)

5 Install separate domestic hot water heater for summer use

6 Relocate electric domestic hot water heater closer to use area when possible

ELECTRICAL SYSTEMS

1 Replace selected incandescent bulbs with fluorescent circular adapter units

2 Replace incandescent light fixtures with fluorescent light fixtures

3 Install photocells to automatically shut off outdoor or vestibule security lighting when adequate daylight is available

4 Install additional switches to control lighting in unused areas

5 Install dimmers to reduce lighting level when possible

6 Install manual wall timers in selected rooms to automatically turn off lights

7 Insall a time clock to shut off refrigerated drinking fountains during unoccupied periods

8 Install a timer and disconnect lights on refrigerated vending machines

9 Replace outdoor security lighting with sodium vapor fixtures

10 Replace incandescent gymnasium, banquet hall, meeting room lighting with mercury vapor, metal halide, or sodium vapor fixtures

SOURCE: R. O. Parmley, *HVAC Field Manual,* McGraw-Hill, New York, 1988. Used with permission of publisher.

SIZE, GROSS SQ. FT. _____

AREA COOLED_____ AREA HEATED_____

TYPE(S) OF OCCUPANCY: (% OR SQ. FT.)

Office_____ (Other)_____

Warehouse_____ (Other)_____

Manufacturing_____ (Other)_____

Retail_____

Lobbies & Mall_____
 (Enclosed)

BUILDING USE AND OCCUPANCY

Fully Occupied: (50% or more of normal)

Weekdays (Hours)_____ to _____

Weekends (Hours)_____ to _____

    _____ to _____ Sunday

    _____ to _____ Holidays

Remarks: Describe below if occupancy differs for different floors, areas, buildings: _____

_____

_____

LIGHTING SURVEY

1. Interior Lighting Type_____ Watts/Ft² Offices_____

 _____ Other _____

 Total Install KW_____

 On-Off from Breaker Panel?_____

 Wall Switches?_____ Control Switching?_____

 Operating Schedule_____

 _____

2. Exterior Lighting Type_____ Sq. Ft. Served: _____

 Total KW_____ Foot-Candles:_____

Remarks: _____

_____

Operating Schedule: _____

FIGURE 12-1 Space conditioning equipment and schedules. (*From* Energy Conservation with Comfort, *2d ed.,* © *1979 by Honeywell. Used with permission of the copyright holder.*)

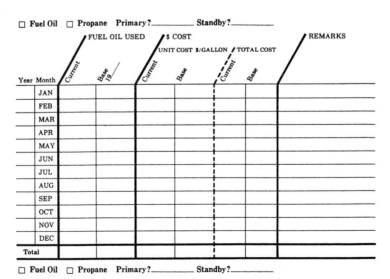

FIGURE 12-2 Gas consumption tabulation. (*From* Energy Conservation with Comfort, *2d ed.,* © *1979 by Honeywell. Used with permission of the copyright holder.*)

FIGURE 12-3 Fuel oil and propane consumption tabulation. (*From* Energy Conservation with Comfort, *2d ed.* © *1979 by Honeywell. Used with permission of the copyright holder.*)

☐ Electricity

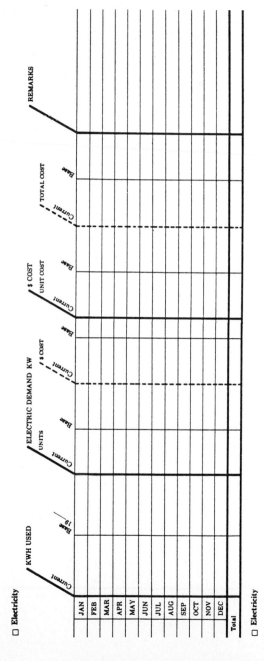

☐ Electricity

FIGURE 12-4 Electricity consumption tabulation. (*From* Energy Conservation with Comfort, 2d ed., © 1979 by Honeywell. Used with permission of the copyright holder.)

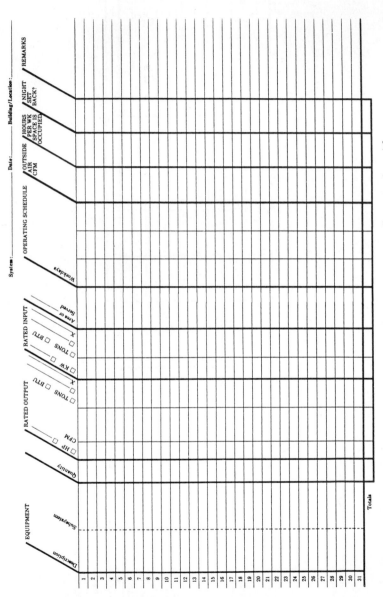

FIGURE 12-5 Building survey—equipment list tabulation. (*From Energy Conservation with Comfort, 2d ed.,* © 1979 by Honeywell. Used with permission of the copyright holder.*)

TABLE 12-2   Energy Equivalents

One	Is Equal to	One	Is Equal to
Btu (mean)	778.104 foot-pounds	Foot-pound	1.3557 joules
U.S. horsepower-hour	2544.65 Btu	Btu (mean)	1054.90 joules
Kilowatthour	3412.66 Btu	Watthour	3600.00 joules
Btu (mean)	252.0 calories	Calorie	4.186 joules

SOURCE: A. M. Khashab, *Heating, Ventilating, and Air Conditioning Systems Estimating Manual,* McGraw-Hill, New York, 1977. Used with permission.

TABLE 12-3   Power Equivalents

One	Is Equal to	One	Is Equal to
U.S. horsepower	550 foot-pounds per second	U.S. horsepower	745.70 watts
U.S. horsepower	0.70685 Btu per second	Metric horsepower	735.50 watts
U.S. horsepower	76.04 kilogram-meters per second	Boiler horsepower	9809.50 watts
Boiler horsepower	33,475 Btu per hour	Btu per hour	0.2929 watt
Ton of refrigeration	12,000 Btu per hour	Ton of refrigeration	3516.8 watts
Metric horsepower	75 kilogram-meters per second	Calorie per second	4.184 watts

SOURCE: A. M. Khashab, *Heating, Ventilating, and Air Conditioning Systems Estimating Manual,* McGraw-Hill, New York, 1977. Used with permission.

NOTES

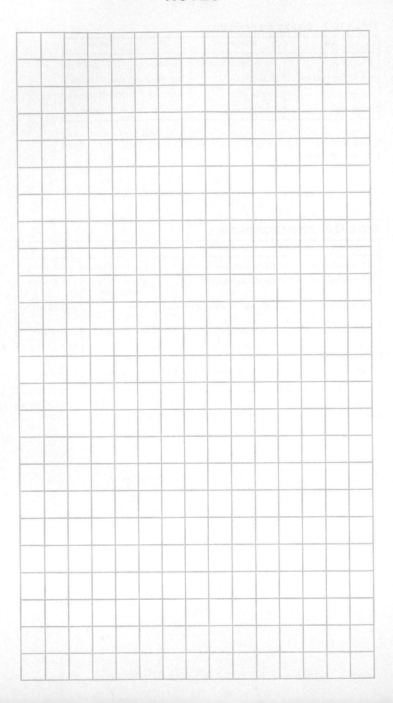

Section

13

Climatic Data

TABLE 13-1 Climatic Conditions for the United States

Note: AP = airport; AFB = air force base; CO = office location within an urban area.

Col. 1	Col. 2		Col. 3		Col. 4	Winter,[d] °F Col. 5		Summer,[e] °F Col. 6			Col. 7	Col. 8		
	Lati- tude[b]		Longi- tude[b]		Eleva- tion[c]	Design Dry-Bulb		Design Dry-Bulb and Mean Coincident Wet-Bulb			Mean Daily	Design Wet-Bulb		
State and Station	°	'	°	'	Ft	99%	97.5%	1%	2.5%	5%	Range	1%	2.5%	5%
ALABAMA														
Alexander City	33	0	86	0	660	18	22	96/77	93/76	91/76	21	79	78	78
Anniston AP	33	4	85	5	599	18	22	97/77	94/76	92/76	21	79	78	78
Auburn	32	4	85	3	730	18	21	96/77	93/76	91/76	21	79	78	78
Birmingham AP	33	3	86	5	610	17	21	96/74	94/75	92/74	21	78	77	76
Decatur	34	4	87	0	580	11	16	95/75	93/74	91/74	22	78	77	76
Dothan AP	31	2	85	2	321	23	27	94/76	92/76	91/76	20	80	79	78
Florence AP	34	5	87	4	528	17	21	97/74	94/74	92/74	21	78	77	76
Gadsden	34	0	86	0	570	16	20	96/75	94/75	92/74	22	78	77	76
Huntsville AP	34	4	86	4	619	11	16	95/75	93/74	91/74	23	78	77	76
Mobile AP	30	4	88	2	211	25	29	95/77	93/77	91/76	18	80	79	78
Mobile CO	30	4	88	1	119	25	29	95/77	93/77	91/76	16	80	79	78
Montgomery AP	32	2	86	2	195	22	25	96/76	95/76	93/76	21	79	79	78
Selma-Craig AFB	32	2	87	0	207	18	26	97/78	95/77	93/77	21	81	80	79
Talladega	33	3	86	1	565	18	22	97/77	94/76	92/76	21	79	78	78
Tuscaloosa AP	33	1	87	4	170r	20	23	98/75	96/76	94/76	22	79	78	77
ALASKA														
Anchorage AP	61	1	150	0	90	-23	-18	71/59	68/58	66/56	15	60	59	57
Barrow (S)	71	2	156	5	22	-45	-41	57/53	53/50	49/47	12	54	50	47
Fairbanks AP (S)	64	5	147	4	436	-51	-47	82/62	78/60	75/59	24	64	62	60
Juneau AP	58	2	134	3	17	-4	1	74/60	70/58	67/57	15	61	59	58
Kodiak	57	3	152	3	21	10	13	69/58	65/56	62/55	10	60	59	58
Nome AP	64	3	165	3	13	-31	-27	66/57	62/55	59/54	10	58	56	55
ARIZONA														
Douglas AP	31	3	109	3	4098	27	31	98/63	95/63	93/63	31	70	69	68
Flagstaff AP	35	1	111	4	6973	-2	4	84/55	82/55	80/54	31	61	60	59
Fort Huachuca AP (S)	31	3	110	3	4664	24	28	95/62	92/62	90/62	27	69	68	67

Station	Lat °	Lat '	Long °	Long '	Elev									
Kingman AP	35	2	114	0	3446	18	25	103/65	100/64	97/64	30	70	69	69
Nogales	31	2	111	0	3800	28	32	99/64	96/64	94/64	31	71	70	69
Phoenix AP(S)	33	3	112	0	1117	31	34	109/71	107/71	105/71	27	76	75	75
Prescott AP	34	4	112	3	5014	4	9	96/61	94/60	92/60	30	66	65	64
Tuscon AP (S)	32	1	111	0	2584	28	32	104/66	102/66	100/66	26	72	71	71
Winslow AP	35	0	110	4	4880	5	10	97/61	95/60	93/60	32	66	65	64
Yuma AP	32	4	114	4	199	36	39	111/72	109/72	107/71	27	79	78	77
ARKANSAS														
Blytheville AFB	36	0	90	0	264	10	15	96/78	94/77	91/76	21	81	80	78
Camden	33	4	92	5	116	18	23	98/76	96/76	94/76	21	80	79	78
El Dorado AP	33	1	92	5	252	18	23	98/76	96/76	94/76	21	80	79	78
Fayetteville AP	36	0	94	1	1253	7	12	97/72	94/73	92/73	23	77	76	75
Fort Smith AP	35	2	94	2	449	12	17	101/75	98/76	95/76	24	80	79	78
Hot Springs	34	3	93	1	535	17	23	101/77	97/77	94/77	22	80	79	78
Jonesboro	35	5	90	4	345	10	15	96/78	94/77	91/76	21	81	80	78
Little Rock AP (S)	34	4	92	1	257	15	20	99/76	96/77	94/77	22	80	79	78
Pine Bluff AP	34	1	92	0	204	16	22	100/78	97/77	95/78	22	81	80	80
Texarkana AP	33	3	94	0	361	18	23	98/76	96/77	93/76	21	80	79	78
CALIFORNIA														
Bakersfield AP	35	2	119	0	495	30	32	104/70	101/69	98/68	32	73	71	70
Barstow AP	34	5	116	5	2142	26	29	106/68	104/68	102/67	37	73	71	70
Blythe AP	33	4	114	3	390	30	33	112/71	110/71	108/70	28	75	75	74
Burbank AP	34	1	118	2	699	37	39	95/68	91/68	88/67	25	71	70	69
Chico	39	5	121	5	205	28	30	103/69	101/68	93/67	36	71	70	68

[a]Table 13.1 was prepared by ASHRAE Technical Committee 4.2, Weather Data, from data compiled from official weather stations where hourly weather observations are made by trained observers.

[b]Latitude, for use in calculating solar loads, and longitude are given to the nearest 10 minutes. For example, the latitude and longitude for Anniston, Alabama, are given as 33°34 and 85°55, respectively, or 33°40 and 85°50.

[c]Elevations are ground elevations for each station. Temperature readings are generally made at an elevation of 5 ft above ground, except for locations marked r, indicating roof exposure thermometer.

[d]Percentage of winter design data shows the percent of the 3-month period, December through February.

[e]Percentage of summer design data shows the percent of 4-month period, June through September.

349

TABLE 13-1 Climatic Conditions for the United States (Continued)

Col. 1	Col. 2	Col. 3	Col. 4	Col. 5		Col. 6			Col. 7	Col. 8		
				Winter, °F		Summer, °F						
	Lati-	Longi-	Eleva-	Design Dry-Bulb		Design Dry-Bulb and Mean Coincident Wet-Bulb			Mean Daily	Design Wet-Bulb		
State and Station	tude[b]	tude[b]	tion[c] Ft	99%	97.5%	1%	2.5%	5%	Range	1%	2.5%	5%
	° ′	° ′										
Concord	38 0	122 1	195	24	27	100/69	97/68	94/67	32	71	70	68
Covina	34 1	117 5	575	32	35	98/69	95/68	92/67	31	73	71	70
Crescent City AP	41 5	124 1	50	31	33	68/60	65/59	63/58	18	62	60	59
Downey	34 0	118 1	116	37	40	93/70	89/70	86/69	22	72	71	70
El Cajon	32 4	117 0	525	42	44	83/69	80/69	78/68	30	71	70	68
El Centro AP (S)	32 5	115 4	-30	35	38	112/74	110/74	108/74	34	81	80	78
Escondido	33 0	117 1	660	39	41	89/68	85/68	82/68	30	71	70	69
Eureka/ Arcata AP	41 0	124 1	217	31	33	68/60	65/59	63/58	11	62	60	59
Fairfield- Travis AFB	38 2	122 0	72	29	32	99/68	95/67	91/66	34	70	68	67
Fresno AP (S)	36 5	119 4	326	28	30	102/70	100/69	97/68	34	72	71	70
Hamilton AFB	38 0	122 3	3	30	32	89/68	84/66	80/65	28	72	69	67
Laguna Beach	33 3	117 5	35	41	43	83/68	80/68	77/67	18	70	69	68
Livermore	37 4	122 0	545	24	27	100/69	97/68	93/67	24	71	70	68
Lompoc, Vandenburg AFB	34 4	120 3	552	35	38	75/61	70/61	67/60	20	63	61	60
Long Beach AP	33 5	118 1	34	41	43	83/68	80/68	77/67	22	70	69	68
Los Angeles AP (S)	34 0	118 2	99	41	43	83/68	80/68	77/67	15	70	69	68
Los Angeles CO (S)	34 0	118 1	312	37	40	93/70	89/70	86/69	20	72	71	70
Merced-Castle AFB	37 2	120 3	178	29	31	102/70	99/69	96/68	36	72	71	70
Modesto	37 4	121 0	91	28	30	101/69	98/68	95/67	36	71	70	69
Monterey	36 4	121 5	38	35	38	75/63	71/61	68/61	20	64	62	61
Napa	38 2	122 2	16	30	32	100/69	96/68	92/67	30	71	69	68
Needles AP	34 5	114 4	913	30	33	112/71	110/71	108/70	27	75	75	74
Oakland AP	37 4	122 1	3	34	36	85/64	80/63	75/62	19	66	64	63
Oceanside	33 1	117 2	30	41	43	83/68	80/68	77/67	13	70	69	68
Ontario	34 0	117 36	995	31	33	102/70	99/69	96/67	36	74	72	71

Note: AP = airport; AFB = air force base; CO = office location within an urban area.

Station	Lat °	Lat ′	Long °	Long ′	Elev	99%	97½%	Summer DB/WB 1%	Summer DB/WB 2½%	Summer DB/WB 5%	MDR	WB 1%	WB 2½%	WB 5%
Obhard	34	1	119	1	43	36	34	83/66	80/64	77/63	19	70	68	67
Palmdale AP	34	4	118	2	2517	22	18	103/65	101/65	98/64	35	69	67	66
Palm Springs	33	5	116	4	411	35	32	112/71	110/70	108/70	35	76	74	73
Pasadena	34	1	118	1	864	35	32	98/69	95/68	92/67	29	73	71	70
Petaluma	38	1	122	4	27	29	26	94/68	90/66	87/65	31	72	70	68
Pomona CO	34	0	117	5	871	30	28	102/70	99/69	95/68	36	74	72	71
Redding AP	40	3	122	1	495	31	29	105/68	102/67	100/66	32	71	69	68
Redlands	34	0	117	1	1318	33	31	102/70	99/69	96/68	33	74	72	71
Richmond	38	0	122	2	55	36	34	85/64	80/63	75/62	17	66	64	63
Riverside-March AFB (S)	33	5	117	2	1511	32	29	100/68	98/68	95/67	37	72	71	70
Sacramento AP	38	3	121	3	17	32	30	101/70		94/69	36	72	71	70
Salinas AP	36	4	121	4	74	32	30	74/61	70/60	67/59	24	62	61	59
San Bernardino, Norton AFB	34	1	117	1	1125	33	31	102/70	99/69	96/68	38	74	72	71
San Diego AP	32	4	117	4	19	44	42	83/69	80/69	78/68	12	71	70	68
San Fernando	34	1	118	3	977	39	37	95/68	91/68	88/67	38	71	70	69
San Francisco AP	37	4	122	2	8	38	35	82/64	77/63	73/62	20	65	64	62
San Francisco CO	37	5	122	3	52	40	38	74/63	71/62	69/61	14	64	62	61
San Jose AP	37	2	122	0	70r	36	34	85/66	81/65	77/64	26	68	67	65
San Luis Obispo	35	2	120	4	315	35	33	92/69	88/70	84/69	26	73	71	70
Santa Ana AP	33	4	117	5	115r	39	37	89/69	85/68	82/68	28	71	70	69
Santa Barbara MAP	34	3	119	2	10	36	34	81/67	77/66	75/65	24	68	67	66
Santa Cruz	37	0	122	3	125	38	35	75/63	71/61	68/61	28	64	62	61
Santa Maria AP (S)	34	5	120	0	238	33	31	81/64	76/63	73/62	23	65	64	63
Santa Monica CO	34	0	118	4	57	43	41	83/68	80/68	77/67	16	70	69	68
Santa Paula	34	2	119	0	263	35	33	90/68	86/67	84/66	36	71	69	68
Santa Rosa	38	3	122	3	167	29	27	99/68	95/67	91/66	34	70	68	67
Stockton AP	37	5	121	3	28	30	28	100/69	97/68	94/67	37	71	70	68
Ukiah	39	1	122	1	620	29	27	99/69	95/68	91/67	40	70	68	67
Visalia	36	2	119	1	354	30	28	102/70	100/69	97/68	38	72	71	70
Yreka	41	4	122	4	2625	17	13	95/65	92/64	89/63	38	67	65	64
Yuba City	39	1	121	4	70	31	29	104/68	101/67	99/66	36	71	69	68
COLORADO														
Alamosa AP	37	3	105	5	7536	-16	-21	84/57	82/57	80/57	35	62	61	60
Boulder	40	0	105	2	5385	8	-2	93/59	91/59	89/59	27	64	63	62

TABLE 13-1 Climatic Conditions for the United States (Continued)

Col. 1	Col. 2	Col. 3	Col. 4	Winter,[d] °F Col. 5		Summer,[e] °F Col. 6 Design Dry-Bulb and Mean Coincident Wet-Bulb			Col. 7	Col. 8 Design Wet-Bulb		
State and Station	Lati- tude[b] ° '	Longi- tude[b] ° '	Eleva- tion[c] Ft	Design Dry-Bulb 99%	97.5%	1%	2.5%	5%	Mean Daily Range	1%	2.5%	5%

Note: AP = airport; AFB = air force base; CO = office location within an urban area.

State and Station	Lat	Long	Elev	99%	97.5%	1%	2.5%	5%	Range	1%	2.5%	5%
Colorado												
Springs AP	38	104	6173	−3	2	91/58	88/57	86/57	30	63	62	61
Denver AP	39	104	5283	−5	1	93/59	91/59	89/59	28	64	63	62
Durango	37	107	6550	−1	4	89/59	87/59	85/59	30	64	63	62
Fort Collins	40	105	5001	−10	−4	93/59	91/59	89/59	28	64	63	62
Grand Junction AP (S)	39	108	4849	2	7	96/59	94/59	92/59	29	65	64	63
Greeley	40	104	4648	−11	−5	96/60	94/60	92/60	29	65	64	63
La Junta AP	38	103	4188	−3	3	100/68	98/68	95/67	31	72	70	69
Leadville	39	106	10177	−8	−4	84/52	81/51	78/50	30	56	55	54
Pueblo AP	38	104	4639	−7	0	97/61	95/61	92/61	31	67	66	65
Sterling	40	103	3939	−7	−2	95/62	93/62	90/62	30	67	66	65
Trinidad AP	37	104	5746	−2	3	93/61	91/61	89/61	32	66	65	64
CONNECTICUT												
Bridgeport AP	41	73	7	6	9	86/73	84/71	81/70	18	75	74	73
Hartford,												
Brainard Field	41	72	15	3	7	91/74	88/73	85/72	22	77	75	74
New Haven AP	41	73	6	3	7	88/75	84/73	82/72	17	76	75	74
New London	41	72	60	5	9	88/73	85/72	83/71	16	76	75	74
Norwalk	41	73	37	6	9	86/73	84/71	81/70	19	75	74	73
Norwich	41	72	20	3	7	89/75	86/73	83/72	18	76	75	74
Waterbury	41	73	605	−4	2	88/73	85/71	82/70	21	75	74	72
Windsor Locks,												
Bradley Field (S)	42	72	169	0	4	91/74	88/72	85/71	22	76	75	73
DELAWARE												
Dover AFB	39	75	38	11	15	92/75	90/75	87/74	18	79	77	76
Wilmington AP	39	75	78	10	14	92/74	89/74	87/73	20	77	76	75
DISTRICT OF COLUMBIA												
Andrews AFB	38	76	279	10	14	92/75	90/74	87/73	18	78	76	75
Washington												
National AP	38	77	14	14	17	93/75	91/74	89/74	18	78	77	76

FLORIDA

Station	Lat °	Lat '	Long °	Long '	Elev	Winter 99%	Winter 97½%	Summer 1%	Summer 2½%	Summer 5%	Daily Range	WB 1%	WB 2½%	WB 5%
Belle Glade	26	4	80	4	16	41	44	92/76	91/76	89/76	16	79	78	78
Cape Kennedy AP	28	3	80	3	16	35	38	90/78	88/78	87/78	15	80	79	78
Daytona Beach AP	29	1	81	0	31	32	35	92/78	90/77	88/77	15	80	79	79
Fort Lauderdale	26	0	80	1	13	42	46	92/78	91/78	90/78	15	80	79	79
Fort Myers AP	26	4	81	5	13	41	44	93/78	92/78	91/77	18	80	79	79
Fort Pierce	27	3	80	2	10	38	42	91/78	90/78	89/78	15	80	79	78
Gainesville AP (S)	29	4	82	4	155	28	31	95/77	93/77	92/77	18	79	79	79
Jacksonville AP	30	5	81	4	24	29	32	96/77	94/77	92/77	19	79	79	79
Key West AP	24	3	81	4	6	55	57	90/78	90/78	89/78	9	80	79	79
Lakeland CO (S)	28	0	82	0	214	39	41	93/76	91/76	89/76	17	79	78	78
Miami AP (S)	25	5	80	2	7	44	47	91/77	90/77	89/77	15	79	79	78
Miami Beach CO	25	5	80	1	9	45	48	90/77	89/77	88/77	10	79	79	78
Ocala	29	1	82	1	86	31	34	95/77	93/77	92/76	18	80	79	78
Orlando AP	28	3	81	2	106r	35	38	94/76	93/76	91/76	17	79	78	78
Panama City, Tyndall AFB	30	0	85	4	22	29	33	92/78	90/77	89/77	14	81	80	79
Pensacola CO	30	3	87	1	13	25	29	94/77	93/77	91/77	14	80	79	79
St. Augustine	29	5	81	2	15	31	35	92/78	89/78	87/78	16	80	79	79
St. Petersburg	28	0	82	4	35	36	40	92/77	91/77	90/76	16	79	79	78
Sanford	28	5	81	2	14	35	38	94/76	92/76	91/76	17	79	78	78
Sarasota	27	2	82	3	30	39	42	93/77	92/77	90/76	17	79	79	78
Tallahassee AP (S)	30	2	84	2	58	27	30	94/77	92/76	90/76	19	79	78	78
Tampa AP (S)	28	0	82	3	19	36	40	92/77	91/77	90/76	17	79	79	78
West Palm Beach AP	26	4	80	1	15	41	45	92/78	91/78	90/78	16	80	79	79

GEORGIA

Station	Lat °	Lat '	Long °	Long '	Elev	Winter 99%	Winter 97½%	Summer 1%	Summer 2½%	Summer 5%	Daily Range	WB 1%	WB 2½%	WB 5%
Albany, Turner AFB	31	3	84	1	224	25	29	97/77	95/76	93/76	20	80	79	78
Americus	32	0	84	2	476	21	25	97/77	94/76	92/75	20	79	78	77
Athens	34	0	83	2	700	17	22	94/74	92/74	90/74	21	78	77	76
Atlanta AP (S)	33	4	84	0	1005	17	22	94/74	92/74	90/73	19	77	76	75
Augusta AP	33	2	82	0	143	20	23	97/77	95/76	93/76	19	80	79	78
Brunswick	31	1	81	3	14	29	32	92/78	89/78	87/78	18	80	79	79
Columbus, Lawson AFB	32	3	85	0	242	21	24	95/76	93/76	91/75	21	79	78	78
Dalton	34	5	85	0	720	17	22	94/76	93/76	91/76	22	79	78	77

TABLE 13-1 Climatic Conditions for the United States (Continued)

	Col. 2		Col. 3		Col. 4	Winter, °F Col. 5		Summer, °F Col. 6			Col. 7	Col. 8		
	Lattude		Longitude		Elevation	Design Dry-Bulb		Design Dry-Bulb and Mean Coincident Wet-Bulb			Mean Daily	Design Wet-Bulb		
Col. 1 State and Station	°	'	°	'	Ft	99%	97.5%	1%	2.5%	5%	Range	1%	2.5%	5%
Dublin	32	3	83	0	215	21	25	96/77	93/76	91/75	20	79	78	77
Gainesville	34	2	83	5	1254	16	21	93/74	91/74	89/73	21	77	76	75
Griffin (S)	33	1	84	2	980	18	22	93/76	90/75	88/74	21	78	77	76
La Grange	33	0	85	0	715	19	23	94/76	91/75	89/74	21	78	77	76
Macon AP	32	4	83	4	356	21	25	96/77	93/76	91/75	22	79	78	77
Marietta, Dobbins AFB	34	0	84	3	1016	17	21	94/74	92/74	90/74	21	78	77	76
Moultrie	31	1	83	4	340	27	30	97/77	95/77	92/76	20	80	79	78
Rome AP	34	2	85	1	637	17	22	94/76	93/76	91/76	23	79	78	77
Savannah-Travis AP	32	1	81	1	52	24	27	96/77	93/77	91/77	20	80	78	78
Valdosta-Moody AFB	31	0	83	1	239	28	31	96/77	94/77	92/76	20	80	79	78
Waycross	31	2	82	2	140	26	29	96/77	94/77	91/76	20	80	79	78
HAWAII														
Hilo AP (S)	19	4	155	1	31	61	62	84/73	83/72	82/72	15	75	74	74
Honolulu AP	21	2	158	0	7	62	63	87/73	86/73	85/72	12	76	75	74
Kaneohe Bay MCAS	21	2	157	0	18	65	66	85/73	84/74	83/74	12	76	76	75
Wahiawa	21	3	158	0	900	58	59	86/73	85/72	84/72	14	75	74	73
IDAHO														
Boise AP(S)	43	3	116	1	2842	3	10	96/65	94/64	91/64	31	68	66	65
Burley	42	3	113	5	4180	-3	2	99/62	95/61	92/66	35	64	63	61
Coeur d'Alene AP	47	5	116	5	2973	-8	-1	89/62	86/61	85/60	31	64	63	61
Idaho Falls AP	43	3	112	0	4730r	-11	-6	89/61	87/61	84/59	38	65	63	61
Lewiston AP	46	2	117	0	1413	-1	6	96/65	93/64	90/63	32	67	66	64
Moscow	46	4	117	0	2660	-7	0	90/63	87/62	84/61	32	65	64	62
Mountain Home AFB	43	0	115	5	2992	6	12	99/64	97/63	94/62	36	66	65	63
Pocatello AP	43	0	112	4	4444	-8	-1	94/61	91/60	89/59	35	64	64	63
Twin Falls AP (S)	42	3	114	3	4148	-3	2	99/62	95/61	92/60	34	64	63	63
ILLINOIS														
Aurora	41	5	88	2	744	-6	-1	93/76	91/76	88/75	20	79	78	76
Belleville, Scott AFB	38	3	89	5	447	1	6	94/76	92/76	89/75	21	79	78	76

Note: AP = airport; AFB = air force base; CO = office location within an urban area.

Bloomington	40	3	89	0	775	-6	-2	92/75	90/74	88/73	21	78	76	75
Carbondale	37	5	89	1	380	2	7	95/77	93/77	90/76	21	80	79	77
Champaign/Urbana	40	0	88	5	743	-3	2	95/75	92/74	90/73	21	78	77	75
Chicago, Midway AP	41	4	87	5	610	-5	0	94/74	91/73	88/72	20	77	75	74
Chicago, O'Hare AP	42	0	87	4	658	-8	-4	91/74	89/74	86/72	15	79	76	75
Chicago CO	41	5	87	4	594	-4	2	94/75	91/74	88/73	21	78	77	75
Danville	40	1	87	4	558	-4	1	93/75	90/74	88/73	21	78	77	75
Decatur	39	5	88	4	670	-3	2	94/75	91/74	88/73	21	78	77	75
Dixon	41	5	89	2	696	-7	-2	93/75	91/75	88/73	23	78	77	75
Elgin	42	0	88	3	820	-7	-2	91/75	88/74	86/73	24	78	77	74
Freeport	42	2	89	2	780	-9	-4	91/74	89/73	87/72	24	78	77	75
Galesburg	41	0	90	2	771	-7	-2	93/75	91/75	83/74	22	79	77	76
Greenville	39	0	89	2	563	-1	4	94/76	92/75	89/74	21	79	78	76
Joliet	41	3	88	1	588	-5	0	93/75	90/74	88/73	20	78	77	75
Kankakee	41	1	87	2	625	-4	-1	93/75	90/73	88/73	21	78	77	75
La Salle/Peru	41	2	89	2	520	-7	-2	93/75	91/75	88/74	22	78	77	76
Macomb	40	3	90	4	702	-5	0	95/76	92/76	89/75	22	79	78	76
Moline AP	41	3	90	3	582	-9	-4	93/75	91/75	88/74	23	78	77	75
Mt Vernon	38	2	88	4	500	0	5	95/76	92/75	39/74	21	79	78	76
Peoria AP	40	4	89	4	652	-8	-4	91/75	89/74	37/73	22	78	76	75
Quincy AP	40	0	91	1	762	-2	3	96/76	93/76	90/76	22	80	78	77
Chanute AFB, Rantoul,	40	2	88	1	740	-4	1	94/75	91/74	89/73	21	78	77	75
Rockford	42	1	89	0	724	-9	-4	91/74	89/73	87/72	24	77	76	74
Springfield AP	39	5	89	4	587	-3	1	94/75	92/74	89/74	21	79	77	76
Waukegan	42	2	87	5	680	-6	-3	92/76	89/74	87/73	21	78	76	75
INDIANA														
Anderson	40	0	85	4	847	0	6	95/76	92/75	89/74	22	79	77	76
Bedford	38	3	86	3	670	0	5	95/76	92/75	89/74	22	79	78	76
Bloomington	39	1	86	1	820	0	5	95/75	92/75	89/74	22	79	78	76
Columbus, Bakalar AFB	39	2	85	5	661	3	7	95/76	92/75	90/74	22	79	78	76
Crawfordsville	40	0	86	5	752	-2	3	94/75	91/74	88/73	22	79	77	76
Evansville AP	38	0	87	5	381	4	9	95/76	93/75	91/75	22	79	78	77
Fort Wayne AP	41	0	85	1	791	-4	1	93/75	89/72	86/72	24	77	75	74
Goshen AP	41	3	85	2	823	-3	1	91/73	89/73	85/72	23	77	75	74
Hobart	41	2	87	3	600	-4	2	91/73	88/73	85/72	21	77	75	74
Huntington	40	4	85	3	802	-4	1	92/73	89/72	87/72	23	77	75	74
Indianapolis AP (S)	39	4	86	2	793	-2	2	92/74	90/74	87/73	22	78	76	75

TABLE 13-1 Climatic Conditions for the United States (Continued)

Col. 1	Col. 2		Col. 3		Col. 4	Winter, °F		Summer, °F						
						Col. 5		Col. 6			Col. 7	Col. 8		
	Lati-tude		Longi-tude		Eleva-tion	Design Dry-Bulb		Design Dry-Bulb and Mean Coincident Wet-Bulb			Mean Daily Range	Design Wet-Bulb		
State and Station	°	'	°	'	Ft	99%	97.5%	1%	2.5%	5%		1%	2.5%	5%
Jeffersonville	38	2	85	5	455	5	10	95/74	93/74	90/74	23	79	77	76
Kokomo	40	3	86	1	790	-4	0	91/74	90/73	88/73	22	77	75	74
Lafayette	40	2	86	5	600	-3	3	94/74	91/73	88/73	22	78	76	75
La Porte	41	4	86	4	810	-3	3	93/74	90/74	87/73	22	78	76	75
Marion	40	3	85	3	791	-4	0	91/74	90/73	88/73	23	77	75	74
Muncie	40	1	85	2	955	-3	2	92/74	90/73	87/73	22	78	76	75
Peru, Grissom AFB	40	4	86	1	804	-6	-1	90/74	88/73	86/72	22	77	75	74
Richmond AP	39	5	84	5	1138	-2	2	92/74	90/74	87/73	22	78	76	75
Shelbyville	39	3	85	5	765	-1	3	93/74	91/74	88/73	22	78	76	75
South Bend AP	41	4	86	2	773	-3	1	91/73	89/73	86/72	22	77	75	74
Terre Haute AP	39	3	87	2	601	-2	4	95/75	92/74	89/73	22	79	77	76
Valparaiso	41	4	87	0	801	-3	3	93/74	90/74	87/73	22	78	76	75
Vincennes	38	4	87	3	420	1	6	95/75	92/74	90/73	22	79	77	76
IOWA														
Ames (S)	42	0	93	4	1004	-11	-6	93/75	90/74	87/73	23	78	76	75
Burlington AP	40	5	91	1	694	-7	-3	94/74	91/75	88/73	22	78	78	75
Cedar Rapids AP	41	5	91	4	863	-10	-5	91/76	88/75	86/74	23	78	77	75
Clinton	41	5	90	1	595	-8	-3	92/75	90/75	87/74	23	78	77	75
Council Bluffs	41	2	95	5	1210	-8	-3	94/76	91/75	88/74	22	78	77	75
Des Moines AP	41	3	93	4	948r	-10	-5	94/75	91/74	88/73	23	78	77	75
Dubuque	42	2	90	2	1065	-12	-7	90/74	88/73	86/72	22	77	75	74
Fort Dodge	42	3	94	1	1111	-12	-7	91/74	88/74	86/72	23	77	75	74
Iowa City	41	4	91	3	645	-11	-6	92/76	89/76	87/74	22	80	78	76
Keokuk	40	2	91	2	526	-5	0	95/75	92/75	89/74	22	79	77	76
Marshalltown	42	0	92	5	898	-12	-7	92/76	90/75	88/74	23	78	77	75
Mason City AP	43	1	93	2	1194	-15	-11	90/74	88/74	85/72	24	77	75	74
Newton	41	4	93	0	946	-10	-5	94/75	91/74	88/73	23	78	77	75
Ottumwa AP	41	1	92	2	842	-8	-4	94/75	91/74	88/73	22	78	77	75

Note: AP = airport; AFB = air force base; CO = office location within an urban area.

Sioux City AP	42	2	96	2	1095	−11	−7	95/74	92/74	89/73	24	78	77	75
Waterloo	42	3	92	2	868	−15	−10	91/76	89/75	86/74	23	78	77	75
KANSAS														
Atchison	39	3	95	1	945	−2	2	96/77	93/76	91/76	23	81	79	77
Chanute AP	34	4	95	3	977	3	7	100/74	97/74	94/74	23	78	77	76
Dodge City AP (S)	37	5	100	0	2594	0	5	100/69	97/69	55/69	25	74	73	71
El Dorado	37	5	96	5	1282	3	7	101/72	98/73	56/73	24	77	76	75
Emporia	38	2	96	5	1209	1	4	100/74	97/74	94/73	25	78	77	76
Garden City AP	38	0	101	3	2882	−1	4	99/69	96/69	94/69	28	74	73	71
Goodland AP	39	2	101	4	3645	−5	0	99/66	96/65	93/66	31	71	70	68
Great Bend	38	2	98	5	1940	0	4	101/73	98/73	95/73	28	78	76	75
Hutchinson AP	38	0	97	5	1524	4	8	102/72	99/72	97/72	28	77	75	74
Liberal	37	0	101	0	2838	2	7	99/68	96/68	34/68	28	73	72	71
Manhattan,														
Fort Riley (S)	39	0	96	5	1076	−1	3	99/75	95/75	92/74	24	78	77	76
Parsons	37	2	95	5	908	5	9	100/74	97/74	94/74	23	79	77	76
Russell AP	38	5	98	5	1864	0	4	101/73	98/73	95/73	29	78	76	75
Salina	38	5	97	4	1271	0	5	103/74	100/74	97/73	26	78	77	76
Topeka AP	39	0	95	4	877	0	4	99/75	96/75	93/74	24	79	78	76
Wichita AP	37	4	97	3	1321	3	7	101/72	98/73	96/73	23	77	76	75
KENTUCKY														
Ashland	38	3	82	4	551	5	10	94/76	91/74	89/73	22	78	77	75
Bowling Green AP	37	0	86	3	535	4	10	94/77	92/75	89/74	21	79	77	76
Corbin AP	37	0	84	1	1175	4	9	94/73	92/73	89/72	23	78	76	75
Covington AP	39	0	84	4	869	1	6	92/73	90/72	88/72	22	77	75	74
Hopkinsville,														
Ft. Campbell	36	4	87	3	540	4	10	94/77	92/75	89/74	21	79	77	76
Lexington AP (S)	38	0	84	4	979	4	8	93/73	91/73	88/72	22	77	76	75
Louisville AP	38	1	85	4	474	5	10	95/74	93/74	90/74	23	79	77	76
Madisonville	37	2	87	3	439	5	10	96/76	93/75	90/75	22	79	78	77
Owensboro	37	5	87	1	420	5	10	97/76	94/75	91/75	23	79	78	77
Paducah AP	37	0	88	4	398	7	12	98/76	95/75	92/75	20	79	78	77
LOUISIANA														
Alexandria AP	31	2	92	2	92	23	27	95/77	94/77	92/77	20	80	79	78
Baton Rouge AP	30	3	91	1	64	25	29	95/77	93/77	92/77	19	80	80	79
Bogalusa	30	5	89	0	103	24	28	95/77	93/77	92/77	19	80	80	79
Houma	29	3	90	4	13	31	35	95/78	93/78	92/77	15	81	80	79

TABLE 13-1 Climatic Conditions for the United States (Continued)

Col. 1	Col. 2		Col. 3		Col. 4	Winter, °F		Summer, °F					
	Lati-tude[b]		Longi-tude[b]		Eleva-tion[c]	Col. 5 Design Dry-Bulb		Col. 6 Design Dry-Bulb and Mean Coincident Wet-Bulb			Col. 7 Mean Daily	Col. 8 Design Wet-Bulb	
State and Station	°	′	°	′	Ft	99%	97.5%	1%	2.5%	5%	Range	1%	2.5%	5%

Note: AP = airport; AFB = air force base; CO = office location within an urban area.

State and Station	°	′	°	′	Ft	99%	97.5%	1%	2.5%	5%	Range	1%	2.5%	5%
Lafayette AP	30	1	92	0	38	26	30	95/78	94/78	92/78	18	81	80	79
Lake Charles AP (S)	30	1	93	1	14	27	31	95/77	93/77	92/77	17	80	79	79
Minden	32	4	93	2	250	20	25	99/77	96/76	94/76	20	79	79	78
Monroe AP	32	3	92	0	78	20	25	99/77	96/76	94/76	20	79	79	78
Natchitoches	31	5	93	0	120	22	26	97/77	95/77	93/77	20	80	79	78
New Orleans AP	30	0	90	2	3	29	33	93/78	92/78	90/77	16	81	80	79
Shreveport AP(S)	32	3	93	5	252	20	25	99/77	96/76	94/76	20	79	79	78
MAINE														
Augusta AP	44	2	69	5	350	−7	−3	88/73	85/70	82/68	22	74	72	70
Bangor, Dow AFB	44	5	68	5	162	−11	−6	86/70	83/68	80/67	22	73	71	69
Caribou AP (S)	46	5	68	0	624	−18	−13	84/69	81/67	78/66	21	71	69	67
Lewiston	44	0	70	1	182	−7	−2	88/73	85/70	82/68	21	74	72	70
Millinocket AP	45	4	68	4	405	−13	−9	87/69	83/68	80/66	22	72	70	68
Portland (S)	43	4	70	2	61	−6	−1	87/72	84/71	81/69	22	74	72	70
Waterville	44	3	69	4	89	−8	−4	87/72	84/69	81/68	22	74	72	70
MARYLAND														
Baltimore AP	39	1	76	4	146	10	13	94/75	91/75	89/74	21	78	77	76
Baltimore CO	39	2	76	5	14	14	17	92/77	89/76	87/75	17	80	78	76
Cumberland	39	4	78	5	945	6	10	92/75	89/74	87/74	22	77	76	75
Frederick AP	39	3	77	3	294	8	12	94/76	91/75	88/74	22	78	77	76
Hagerstown	39	4	77	4	660	8	12	94/75	91/74	89/74	22	77	76	75
Salisbury (S)	38	2	75	3	52	12	16	93/75	91/75	88/74	18	79	77	76
MASSACHUSETTS														
Boston AP (S)	42	2	71	0	15	6	9	91/73	88/71	85/70	16	75	74	72
Clinton	42	2	71	4	398	−2	2	90/72	87/71	84/69	17	75	73	72
Fall River	41	4	71	1	190	5	9	87/72	84/71	81/69	18	74	73	72
Framingham	42	2	71	3	170	3	6	89/72	86/71	83/69	17	74	73	71
Gloucester	42	3	70	4	10	2	5	89/73	86/71	83/70	15	75	74	72
Greenfield	42	3	72	4	205	−7	−2	88/72	85/71	82/69	23	74	73	71
Lawrence	42	4	71	1	57	−6	0	90/73	87/72	84/70	22	76	74	73

Station	Lat °	Lat ′	Long °	Long ′	Elev	Winter 99%	Winter 97.5%	Summer 1%	Summer 2.5%	Summer 5%	Mean Daily Range	WB 1%	WB 2.5%	WB 5%
Lowell	42	3	71	2	90	−4	1	91/73	88/72	85/70	21	76	74	73
New Bedford	41	4	71	0	70	5	9	85/72	82/71	80/69	19	74	73	73
Pittsfield AP	42	3	73	2	1170	−8	−3	87/71	84/70	81/68	23	73	72	70
Springfield, Westover AFB	42	1	72	3	247	−5	0	90/72	87/71	84/69	19	75	73	72
Taunton	41	5	71	1	20	5	9	89/73	86/72	83/70	18	75	74	73
Worcester AP	42	2	71	5	986	0	4	87/71	84/70	81/68	18	73	72	70
MICHIGAN														
Adrian	41	5	84	0	754	−1	3	91/73	88/72	85/71	23	76	75	73
Alpena AP	45	0	83	3	689	−11	−6	89/70	85/70	83/69	27	73	72	70
Battle Creek AP	42	2	85	2	939	1	5	92/74	88/72	85/70	23	76	74	73
Benton Harbor AP	42	1	86	0	649	1	5	91/72	88/72	85/70	20	75	74	73
Detroit	42	2	83	0	633	3	6	91/73	88/72	86/71	20	76	74	73
Escanaba	45	4	87	4	594	−11	−7	87/70	83/69	80/68	17	73	71	69
Flint AP	42	0	83	4	766	−4	1	90/73	87/72	85/71	25	76	74	72
Grand Rapids AP	42	5	85	3	681	1	5	91/72	88/72	85/70	24	75	74	72
Holland	42	5	86	1	612	2	6	88/72	86/71	83/70	22	76	73	72
Jackson AP	42	2	84	2	1003	1	5	92/74	88/72	85/70	23	76	74	73
Kalamazoo	42	1	85	3	930	−1	5	92/74	88/72	85/70	23	76	74	73
Lansing AP	42	4	84	4	852	−3	5	90/73	87/72	84/70	24	75	74	72
Marquette CO	46	3	87	5	677	−12	−8	84/70	81/69	77/66	18	72	70	68
Mt Pleasant	43	4	84	1	796	0	4	91/73	87/72	84/71	24	76	74	72
Muskegon AP	43	1	86	1	627	2	6	86/72	84/70	82/70	21	75	73	72
Pontiac	42	4	83	2	974	0	0	90/73	87/72	85/71	21	76	74	73
Port Huron	43	0	82	3	586	0	0	90/73	87/72	83/71	21	76	74	73
Saginaw AP	43	3	84	1	662	0	0	91/73	87/72	84/71	23	76	74	72
Sault Ste. Marie AP (S)	46	3	84	2	721	−12	−8	84/70	81/69	77/66	23	72	70	68
Traverse City AP	44	4	85	4	618	−3	1	89/72	86/71	83/69	22	75	73	71
Ypsilanti	42	1	83	3	777	1	5	92/72	89/71	86/70	22	75	74	72
MINNESOTA														
Albert Lea	43	4	93	2	1235	−17	−12	90/74	87/72	84/71	24	77	75	73
Alexandria AP	45	5	95	0	1421	−22	−16	91/72	88/72	85/70	24	76	74	72
Bemidji AP	47	3	95	0	1392	−31	−26	88/69	85/69	81/67	24	73	71	69
Brainerd	46	2	94	2	1214	−20	−16	90/73	87/71	84/69	24	75	73	71
Duluth AP	46	5	92	1	1426	−21	−16	85/70	82/68	79/66	22	72	70	68

TABLE 13-1 Climatic Conditions for the United States (Continued)

Note: AP = airport; AFB = air force base; CO = office location within an urban area.

Col. 1	Col. 2		Col. 3		Col. 4	Col. 5		Col. 6			Col. 7	Col. 8		
	Lati-tude[b]		Longi-tude[b]		Eleva-tion[c]	Winter,[d] F Design Dry-Bulb		Summer,[e] F Design Dry-Bulb and Mean Coincident Wet-Bulb			Mean Daily	Design Wet-Bulb		
State and Station	°	'	°	'	Ft	99%	97.5%	1%	2.5%	5%	Range	1%	2.5%	5%
Fairbault	44	2	93	2	1190	−17	−12	91/74	88/72	85/71	24	77	75	73
Fergus Falls	46	0	96	0	1210	−21	−17	91/72	88/72	85/70	24	76	74	72
International Falls AP	48	3	93	2	1179	−29	−25	85/68	83/68	80/66	26	71	70	68
Mankato	44	1	94	0	785	−17	−12	91/72	88/72	85/70	24	77	75	73
Minneapolis/ St Paul AP	44	5	93	1	822	−16	−12	92/75	89/73	86/71	22	77	75	73
Rochester AP	44	0	92	3	1297	−17	−12	90/74	87/72	84/71	24	77	75	73
St Cloud AP (S)	45	4	94	1	1034	−15	−11	91/74	88/72	85/70	24	76	74	72
Virginia	47	3	92	3	1435	−25	−21	85/69	83/68	80/66	23	71	70	68
Willmar	45	1	95	1	1133	−15	−11	91/74	88/72	85/71	24	76	74	72
Winona	44	4	91	4	652	−14	−10	91/75	88/73	85/72	24	77	75	74
MISSISSIPPI Biloxi, Keesler AFB	30	2	89	0	25	28	31	94/79	92/79	90/78	16	82	81	80
Clarksdale	34	1	90	3	178	14	19	96/77	94/77	92/76	21	80	79	78
Columbus AFB	33	4	88	3	224	15	20	95/77	93/77	91/76	22	80	79	78
Greenville AFB	33	3	91	1	139	15	20	95/77	93/77	91/76	21	80	79	78
Greenwood	33	3	90	1	128	15	20	95/77	93/77	91/76	21	80	79	78
Hattiesburg	31	2	89	2	200	24	27	96/78	94/77	92/77	21	81	80	79
Jackson AP	32	2	90	1	330	21	25	97/76	95/76	93/76	21	79	78	78
Laurel	31	4	89	1	264	24	27	96/78	94/77	92/77	21	81	80	79
McComb AP	31	1	90	3	458	21	26	96/77	94/76	92/76	18	80	79	78
Meridian AP	32	5	88	5	294	19	23	97/77	95/76	93/76	22	80	79	78
Natchez	31	4	91	3	168	23	27	96/78	94/78	92/77	21	81	80	79
Tupelo	34	2	88	4	289	14	19	96/77	94/77	92/76	22	80	79	78
Vicksburg CO	32	2	91	0	234	22	26	97/78	95/78	93/77	21	81	80	79
MISSOURI Cape Girardeau	37	1	89	3	330	8	13	98/76	95/75	92/75	21	79	78	77

City	Lat °	Lat ′	Long °	Long ′	Elev								
Columbia AP (S)	39	0	92	2	778	-1	4	97/74	94/74	91/73	22	78	77	76
Farmington AP	37	5	90	3	928	3	8	96/76	93/75	90/74	22	78	77	75
Hannibal	39	4	91	2	489	-2	3	96/76	93/76	90/76	22	80	78	77
Jefferson City	38	4	92	1	640	2	7	98/75	95/75	92/74	23	78	77	76
Joplin AP	37	1	94	3	982	6	10	100/73	97/73	94/73	24	78	77	76
Kansas City AP	39	1	94	4	742	2	6	99/75	96/74	93/74	20	78	77	76
Kirksville AP	40	1	92	4	966	-5	0	96/74	93/74	90/73	24	78	77	76
Mexico	39	1	92	0	775	-1	4	97/74	94/74	91/73	22	78	77	76
Moberly	39	3	92	0	850	-2	3	97/74	94/74	91/73	23	78	77	76
Poplar Bluff	36	5	90	3	322	11	16	98/78	95/76	92/76	22	81	79	78
Rolla	38	0	91	5	1202	3	9	94/77	91/75	89/74	22	78	77	76
St Joseph AP	39	5	95	0	809	-3	2	96/77	93/76	91/76	23	81	79	77
St Louis AP	38	5	90	2	535	2	6	97/75	94/75	91/74	21	78	77	76
St Louis CO	38	4	90	2	465	3	8	98/75	94/75	91/74	18	78	77	76
Sedalia, Whiteman AFB	38	4	93	3	838	-1	4	95/76	92/76	90/75	22	79	78	76
Sikeston	36	5	89	3	318	9	15	98/77	95/76	92/75	21	80	78	77
Springfield AP	37	1	93	2	1265	3	9	96/73	93/74	91/74	23	78	77	75
MONTANA														
Billings AP	45	5	108	3	3567	-15	-10	94/64	91/64	88/63	31	67	66	64
Bozeman	45	5	111	0	4856	-20	-14	90/61	87/60	84/59	32	63	62	60
Butte AP	46	0	112	3	5226r	-24	-17	86/58	83/56	80/56	35	60	58	57
Cut Bank AP	48	4	112	2	3838r	-25	-20	88/61	85/61	82/60	35	64	62	61
Glasgow AP (S)	48	1	106	4	2277	-22	-18	92/64	89/63	85/62	29	68	66	64
Glendive	47	1	104	4	2076	-18	-13	95/66	92/64	89/62	29	69	67	65
Great Falls AP (S)	47	3	111	2	3664r	-21	-15	91/60	88/60	85/59	28	64	62	60
Havre	48	3	109	4	2488	-18	-11	94/65	90/64	87/63	33	68	66	65
Helena AP	46	4	112	0	3893	-21	-16	91/60	88/60	85/59	32	64	62	61
Kalispell AP	48	2	114	2	2965	-14	-7	91/62	87/61	84/60	34	65	63	62
Lewiston AP	47	0	109	3	4132	-22	-16	90/62	87/61	83/60	30	65	63	62
Livingston AP	45	4	110	3	4653	-20	-20	90/61	87/60	84/59	32	63	62	60
Miles City AP	46	3	105	5	2629	-20	-15	98/66	95/66	92/65	30	70	68	67
Missoula AP	46	5	114	1	3200	-13	-6	92/62	88/61	85/60	36	65	63	62
NEBRASKA														
Beatrice	40	2	96	5	1235	-5	-2	99/75	95/74	92/74	24	78	77	76

TABLE 13-1 Climatic Conditions for the United States (Continued)

Col. 1	Col. 2	Col. 3		Col. 4	Winter,° F Col. 5		Summer,° F Col. 6 Design Dry-Bulb and Mean Coincident Wet-Bulb			Col. 7	Col. 8 Design Wet-Bulb		
State and Station	Lati- tude °	Long- itude °	′	Eleva- tion Ft	Design Dry-Bulb 99%	97.5%	1%	2.5%	5%	Mean Daily Range	1%	2.5%	5%
Chadron AP	42	103	0	3300	−8	−3	97/66	94/65	91/65	30	71	69	68
Columbus	41	97	2	1442	−6	−2	98/74	95/73	92/73	25	77	76	75
Fremont	41	96	3	1203	−6	−2	98/75	95/74	92/74	22	78	77	76

Note: AP = airport; AFB = air force base; CO = office location within an urban area.

Col. 1	Col. 2	Col. 3		Col. 4	Col. 5		Col. 6			Col. 7	Col. 8		
State and Station	Lat °	Long °	′	Elev Ft	99%	97.5%	1%	2.5%	5%	MDR	1%	2.5%	5%
Grand Island AP	41	98	2	1841	−8	−3	97/72	94/71	91/71	28	75	74	73
Hastings	40	98	2	1932	−7	−3	97/72	94/71	91/71	27	75	74	73
Kearney	40	99	1	2146	−9	−4	96/71	93/70	90/70	28	74	73	72
Lincoln CO (S)	40	96	5	1150	−5	−2	99/75	95/74	92/74	24	78	77	76
McCook	40	100	4	2565	−6	−2	98/69	95/69	91/69	28	74	72	71
Norfolk	42	97	3	1532	−8	−4	97/74	93/74	90/73	30	78	72	75
North Platte AP (S)	41	100	4	2779	−8	−4	97/69	94/69	90/69	28	74	72	71
Omaha AP	41	95	5	978	−8	−3	94/76	91/75	88/74	22	78	77	75
Scottsbluff AP	41	103	5	3950	−8	−3	95/65	92/65	90/64	31	70	68	67
Sidney AP	41	103	0	4292	−8	−3	95/65	92/65	90/64	31	70	68	67
NEVADA													
Carson City	39	119	5	4675	4	9	94/60	91/59	89/58	42	63	61	60
Elko AP	40	115	5	5075	−10	−2	94/59	92/59	90/58	42	63	62	60
Ely AP (S)	39	114	5	6257	−10	−4	89/57	87/56	85/55	39	60	59	58
Las Vegas AP (S)	36	115	1	2162	25	28	108/66	106/65	104/65	30	71	70	69
Lovelock AP	40	118	3	3900	8	12	98/63	96/63	93/62	42	66	65	64
Reno AP (S)	39	119	5	4404	5	10	95/61	92/60	90/59	45	64	62	61
Reno CO	39	119	5	4490	6	11	96/61	93/60	91/59	45	64	62	61
Tonopah AP	38	117	1	5426	5	10	94/60	92/59	90/58	40	64	62	61
Winnemucca AP	40	117	5	4299	−1	3	96/60	94/60	92/60	42	64	62	61
NEW HAMPSHIRE													
Berlin	44	71	1	1110	−14	−9	87/71	84/69	81/68	22	73	71	70
Claremont	43	72	2	420	−9	−4	89/72	86/70	83/69	24	74	73	71
Concord AP	43	71	2	339	−10	−3	90/72	87/70	84/69	26	74	73	71
Keene	43	72	2	490	−12	−7	90/72	87/70	83/69	24	74	73	71
Laconia	43	71	3	505	−10	−5	89/72	86/70	83/69	25	74	73	71
Manchester, Grenier AFB	43	71	3	253	−8	−3	91/72	88/71	85/70	24	75	74	72
Portsmouth, Pease AFB	43	70	5	127	−2	2	89/73	85/71	83/70	22	75	74	72

State / City	Lat °	Lat ′	Long °	Long ′	Elev	Winter 99%	Winter 97.5%	Summer 1%	Summer 2.5%	Summer 5%	Daily Range	WB 1%	WB 2.5%	WB 5%
NEW JERSEY														
Atlantic City CO	39	3	74	3	11	10	13	92/74	89/74	86/72	18	78	77	75
Long Branch	40	2	74	0	20	10	13	93/74	90/73	87/72	18	78	77	75
Newark AP	40	4	74	1	11	10	14	94/74	91/73	88/72	19	77	76	75
New Brunswick	40	3	74	1	86	6	10	94/74	91/73	86/72	20	77	76	75
Paterson	40	5	74	1	100	6	10	94/74	92/73	88/72	21	77	75	74
Phillipsburg	40	4	75	—	180	9	11	92/73	89/72	86/71	21	76	76	75
Trenton CO	40	1	74	5	144	11	11	91/75	89/74	85/73	19	78	77	75
Vineland	39	3	75	0	95	8	11	91/75	89/74	86/73	19	78	77	75
NEW MEXICO														
Alamogordo, Holloman AFB	32	5	106	1	4070	14	19	98/64	96/64	94/64	30	69	68	67
Albuquerque AP (S)	35	0	106	4	5310	12	16	96/61	94/61	92/61	27	66	65	64
Artesia	35	5	104	3	3375	12	19	103/67	100/67	97/67	30	72	71	70
Carlsbad AP	32	2	104	2	3234	8	19	103/67	100/67	97/67	28	72	71	70
Clovis AP	34	3	103	5	4279	13	13	95/65	93/62	91/65	28	69	68	67
Farmington AP	36	5	108	5	5495	6	6	95/63	93/62	91/61	30	67	65	64
Gallup	35	1	108	1	6465	0	5	90/59	89/58	85/58	32	64	62	61
Grants	35	1	107	1	6520	-1	4	89/59	88/58	85/57	32	64	63	62
Hobbs AP	32	4	103	0	3664	13	18	101/66	99/66	97/66	29	71	70	69
Las Cruces	32	2	107	2	3900	15	20	100/64	98/64	96/64	32	69	68	67
Los Alamos	35	5	106	3	7410	-4	9	89/60	87/60	85/60	34	62	61	60
Raton AP	36	5	104	0	6379	1	1	91/60	89/60	87/60	33	65	64	63
Roswell, Walker AFB	33	2	104	4	3643	13	18	100/66	98/66	96/66	28	71	70	69
Santa Fe CO	35	4	106	2	7045	6	10	90/61	88/61	86/61	30	63	62	61
Silver City AP	32	4	108	1	5373	13	10	95/61	93/60	91/60	30	66	64	63
Socorro AP	34	0	106	5	4617	13	17	97/62	95/62	93/62	30	67	66	65
Tucumcari AP	35	1	103	4	4053	8	13	99/66	97/66	95/65	28	70	69	68
NEW YORK														
Albany AP (S)	42	5	73	5	277	-6	-1	91/73	88/72	85/70	23	75	74	72
Albany CO	42	5	73	3	19	-4	1	91/73	88/72	85/70	20	75	74	72
Auburn	43	5	76	3	715	-3	-2	90/72	87/71	84/70	22	75	73	72
Batavia	43	3	78	1	900	-3	1	90/72	87/71	84/70	22	75	73	72
Binghamton AP	42	0	76	0	1590	-2	2	86/71	83/69	81/68	20	73	72	70
Buffalo AP	43	0	78	0	705r	2	5	88/71	85/70	83/69	21	74	73	72
Cortland	42	4	76	2	1129	-5	0	88/71	85/71	82/70	23	74	73	71
Elmira AP	42	1	76	5	860	-4	1	89/71	86/71	83/70	24	74	73	71
Geneva (S)	42	5	77	0	590	-3	2	90/73	87/71	84/70	22	75	73	72
Glens Falls	43	2	73	2	321	-11	-5	88/72	85/71	82/69	23	74	73	71
Gloversville	43	1	74	1	790	-8	-2	89/72	86/71	83/69	23	75	74	72

TABLE 13-1 Climatic Conditions for the United States (Continued)

Note: AP = airport; AFB = air force base; CO = office location within an urban area.

Col. 1	Col. 2 Latitude °	'	Col. 3 Longitude °	'	Col. 4 Elevation Ft	Winter, F Col. 5 Design Dry-Bulb 99%	97.5%	Summer, F Col. 6 Design Dry-Bulb and Mean Coincident Wet-Bulb 1%	2.5%	5%	Col. 7 Mean Daily Range	Col. 8 Design Wet-Bulb 1%	2.5%	5%
Hornell	42	2	77	4	1325	-4	0	88/71	85/70	82/69	24	74	73	72
Ithaca (S)	42	3	76	3	950	-5	0	88/71	85/71	82/70	24	74	73	71
Jamestown	42	1	79	2	1390	-1	3	88/70	86/70	83/69	20	74	72	71
Kingston	42	0	74	0	279	-3	3	91/73	88/72	85/70	22	76	74	73
Lockport	43	1	78	4	520	4	7	89/74	86/72	84/71	21	76	74	73
Massena AP	45	0	75	0	202r	-13	-8	86/70	83/69	80/68	20	73	72	70
Newburg-Stewart AFB	41	3	74	1	460	-1	4	90/73	88/72	85/70	21	76	74	73
NYC-Central Park (S)	40	5	74	0	132	11	15	92/74	89/73	87/72	17	76	75	74
NYC-Kennedy AP	40	4	73	5	16	12	15	90/73	87/72	84/71	16	76	75	74
NYC-La Guardia AP	40	5	73	5	19	11	15	92/74	89/73	87/72	16	76	75	74
Niagra Falls AP	43	1	79	0	596	4	7	89/74	86/72	84/71	20	76	74	73
Olean	42	1	78	2	1420	-2	2	87/71	84/71	81/70	23	74	73	71
Oneonta	42	3	75	0	1150	-7	-4	86/71	83/69	80/68	24	73	72	70
Oswego CO	43	3	76	3	300	-1	7	86/73	83/71	80/70	20	75	73	72
Plattsburg AFB	44	4	73	3	165	-13	-8	86/70	83/69	80/68	22	73	72	70
Poughkeepsie	41	4	73	5	103	0	6	92/74	89/74	86/72	21	77	75	74
Rochester AP	43	1	77	4	543	1	5	91/73	88/71	85/70	22	75	73	72
Rome-Griffiss AFB	43	3	75	3	515	-11	-5	88/71	85/70	83/69	22	75	73	71
Schenectady (S)	42	5	74	0	217	-4	1	90/73	87/72	84/70	22	75	74	72
Suffolk County AFB	40	5	72	5	57	7	10	86/72	83/71	80/70	16	76	74	73
Syracuse AP	43	1	76	1	424	-3	2	90/73	87/71	84/70	20	75	73	72
Utica	43	1	75	2	714	-12	-6	88/73	85/71	82/70	22	75	73	71
Watertown	44	0	76	0	497	-11	-6	86/73	83/71	81/70	20	75	73	72
NORTH CAROLINA														
Asheville AP	35	3	82	3	2170r	10	14	89/73	87/72	85/71	21	75	74	72
Charlotte AP	35	0	81	0	735	18	22	95/74	93/74	91/74	20	77	76	76
Durham	36	0	78	5	406	16	20	94/75	92/75	90/75	20	78	77	76
Elizabeth City AP	36	2	76	1	10	12	19	93/78	91/77	89/76	18	80	78	78
Fayetteville, Pope AFB	35	1	79	0	95	17	20	95/76	92/76	90/75	20	79	78	77

	Lat°		Long°		Elev	W 99%	W 97.5%	S 1%	S 2.5%	S 5%	Range	WB 1%	WB 2.5%	WB 5%
Goldsboro, Seymour-Johnson AFB	35	2	78	0	88	18	21	94/77	91/76	89/75	18	79	78	77
Greensboro AP (S)	36	1	80	0	887	14	18	93/74	91/73	89/73	21	77	76	75
Greenville	35	4	77	2	25	18	21	93/77	91/76	89/75	19	79	78	77
Henderson	36	2	78	2	510	12	15	95/77	92/76	90/76	20	79	78	77
Hickory	35	4	81	2	1165	14	18	92/73	90/72	88/72	21	75	74	73
Jacksonville	34	5	77	3	24	20	24	92/78	90/78	88/77	18	80	79	78
Lumberton	34	4	79	0	132	18	21	95/76	92/76	90/75	20	79	78	77
New Bern AP	35	1	77	0	17	20	24	92/78	90/78	88/77	18	80	79	78
Raleigh/Durham AP (S)	35	5	78	5	433	16	20	94/75	92/75	90/75	20	78	77	76
Rocky Mount	36		77	5	81	18	21	94/77	91/76	89/75	21	79	78	77
Wilmington AP	34	2	78	0	30	23	26	93/79	91/78	89/77	19	81	80	79
Winston-Salem AP	36	1	80	1	967	16	20	94/74	91/73	89/73	20	76	75	74
NORTH DAKOTA														
Bismarck AP (S)	46	5	100	5	1647	-23	-19	95/68	91/68	88/67	27	73	71	70
Devil's Lake	48		98	5	1471	-25	-21	91/69	88/68	85/66	25	73	71	69
Dickinson AP	46	5	102	5	2595	-21	-17	94/68	90/66	87/65	25	71	69	68
Fargo AP	46	5	96	5	900	-22	-18	92/73	89/71	85/69	25	76	74	72
Grand Forks AP	48	4	97	4	832	-26	-22	91/70	87/70	84/68	25	74	72	70
Jamestown AP	47	2	98	4	1492	-22	-18	94/70	90/69	87/68	26	74	74	71
Minot AP	48	1	101	2	1713	-24	-20	92/68	89/67	86/65	25	72	70	68
Williston	48	3	103	4	1877	-25	-21	91/68	88/67	85/65	25	72	70	68
OHIO														
Akron-Canton AP	41	0	81	3	1210	4	6	89/72	86/71	84/70	21	75	73	72
Ashtabula	42	0	80	5	690	4	9	88/73	85/72	83/71	18	75	74	72
Athens	39	2	82	1	700	0	6	95/75	92/74	90/73	22	78	76	74
Bowling Green	41	3	83	4	675	-2	7	92/73	89/73	86/71	23	76	75	73
Cambridge	40	0	81	4	800	1	7	93/75	90/74	87/73	23	78	76	75
Chillicothe	39	2	83	2	638	0	6	95/75	92/74	90/73	22	78	76	74
Cincinnati CO	39	1	84	1	761	1	6	92/73	90/72	88/72	21	77	75	74
Cleveland AP (S)	41	0	81	5	777r	0	5	91/73	88/72	86/71	22	76	75	73
Columbus AP (S)	40	5	82	1	812	-1	5	92/73	90/73	87/72	24	77	75	74
Dayton AP	39	2	84	2	997	-1	4	91/73	89/72	86/71	20	76	75	73
Defiance	41	0	84	4	700	-1	4	94/74	91/73	88/72	24	77	76	74
Findlay AP	41	0	83	0	797	2	3	92/74	90/73	87/72	24	77	76	74
Fremont	41	2	83	1	600	-3	1	90/73	88/73	85/71	24	76	75	73
Hamilton	39	2	84	3	650	0	5	92/73	90/72	87/71	22	76	75	73
Lancaster	39	4	82	4	920	0	5	93/74	91/73	88/72	23	77	75	74

TABLE 13-1 Climatic Conditions for the United States *(Continued)*

Col. 1	Col. 2 Lati-tude[b]		Col. 3 Longi-tude[b]		Col. 4 Eleva-tion[c] Ft	Winter,[d] F Col. 5 Design Dry-Bulb 99%	97.5%	Summer,[e] F Col. 6 Design Dry-Bulb and Mean Coincident Wet-Bulb 1%	2.5%	5%	Col. 7 Mean Daily Range	Col. 8 Design Wet-Bulb 1%	2.5%	5%
State and Station	°	'	°	'										
Lima	40	4	84	0	860	−1	4	94/74	91/73	88/72	24	77	76	74
Mansfield AP	40	5	82	3	1297	0	5	90/73	87/72	85/72	22	76	74	73
Marion	40	4	83	1	920	0	5	93/74	91/73	88/72	23	77	76	74
Middletown	39	3	84	3	635	0	5	92/73	90/72	87/71	22	76	75	73
Newark	40	1	82	4	825	−1	5	94/73	92/73	89/72	23	77	75	74
Norwalk	41	1	82	4	720	−3	1	90/73	88/73	85/71	22	76	75	73
Portsmouth	38	3	83	0	530	5	10	95/76	92/74	89/73	22	78	77	75
Sandusky CO	41	3	82	4	606	1	6	93/73	91/72	88/71	21	76	74	73
Springfield	40	0	83	5	1020	−1	3	91/74	89/73	87/72	21	77	76	74
Steubenville	40	2	80	4	992	1	5	89/72	86/71	84/70	22	74	73	72
Toledo AP	41	4	83	5	676r	−3	1	90/73	88/73	85/70	25	76	75	73
Warren	41	2	80	5	900	0	6	89/71	87/71	85/70	23	74	74	71
Wooster	40	5	82	0	1030	−1	4	89/72	86/71	84/70	22	75	73	72
Youngstown AP	41	2	80	4	1178	−1	4	88/71	86/71	84/70	23	74	73	71
Zanesville AP	40	0	81	5	881	1	7	93/75	90/74	87/73	23	78	76	75
OKLAHOMA														
Ada	34	5	96	4	1015	10	14	100/74	97/74	95/74	23	77	76	75
Altus AFB	34	4	99	2	1390	11	16	102/73	100/73	98/73	25	77	76	75
Ardmore	34	2	97	1	880	13	17	100/74	98/74	95/74	23	77	76	76
Bartlesville	36	5	96	0	715	6	10	101/73	98/74	95/74	23	77	77	76
Chickasha	35	0	98	0	1085	10	14	101/74	98/74	95/74	24	78	77	76
Enid-Vance AFB	36	2	98	0	1287	9	13	103/74	100/74	97/74	24	79	77	76
Lawton AP	34	3	98	2	1108	12	16	101/74	99/74	96/74	24	78	78	76
Mc Alester	34	5	95	5	760	14	19	99/74	96/75	93/74	23	77	76	75
Muskogee AP	35	5	95	2	610	10	15	101/74	98/75	95/75	23	79	78	77
Norman	35	1	97	3	1109	9	13	99/74	96/74	94/74	24	77	77	75
Oklahoma City AP (S)	35	2	97	4	1280	9	13	100/74	97/74	95/73	23	78	77	76
Ponca City	36	4	97	0	996	5	9	100/74	97/74	94/74	24	77	76	76
Seminole	35	2	96	4	865	11	15	99/74	96/74	94/73	23	77	76	75
Stillwater (S)	36	1	97	0	884	8	13	100/74	96/74	93/74	24	77	76	75
Tulsa AP	36	1	95	5	650	8	13	101/74	98/75	95/75	22	79	78	77
Woodward	36	3	99	3	1900	6	10	100/73	97/73	94/73	26	78	78	75

Note: AP = airport; AFB = air force base; CO = office location within an urban area.

Station	Lat	Long	Elev	99%	97½%	1% DB/MWB	2½% DB/MWB	5% DB/MWB	DR	WB 1%	WB 2½%	WB 5%
OREGON												
Albany	44 4	123 1	224	18	29	92/67	89/66	86/65	31	69	67	66
Astoria AP (S)	46 1	123 5	8	25	29	75/65	71/62	68/61	16	65	63	62
Baker	44 5	117 5	3368	−7	6	92/63	89/61	86/60	30	65	63	61
Bend	44 0	121 2	3599	−3	4	90/62	87/60	84/59	33	64	62	60
Corvallis (S)	44 3	123 2	221	18	22	92/67	89/66	86/65	31	69	67	66
Eugene AP	44 1	123 1	364	17	22	92/67	89/66	86/65	31	69	67	66
Grants Pass	42 3	123 2	925	20	24	99/69	96/68	93/67	33	71	69	68
Klamath Falls AP	42 1	121 4	4091	4	9	90/61	87/60	84/59	36	63	61	60
Medford AP (S)	42 2	122 5	1298	19	23	98/68	94/67	91/66	35	70	68	67
Pendleton AP	45 4	118 5	1492	−2	5	97/65	93/64	90/62	29	66	65	63
Portland AP	45 4	122 4	21	17	23	89/68	85/67	81/65	23	69	67	66
Portland CO	45 3	122 4	57	18	24	90/68	86/67	82/65	21	68	67	66
Roseburg AP	43 1	123 2	505	18	23	93/67	90/66	87/65	30	69	67	66
Salem AP	45 0	123 0	195	18	23	92/68	88/66	84/65	31	69	68	66
The Dalles	45 4	121 1	102	13	19	93/69	89/68	85/66	28	70	68	67
PENNSYLVANIA												
Allentown AP	40 4	75 3	376	4	9	92/73	88/72	86/72	22	76	75	73
Altoona CO	40 2	78 1	1468	0	5	90/72	87/71	84/70	23	74	73	72
Butler	40 4	80 1	1100	−1	6	90/73	87/72	85/71	23	75	74	73
Chambersburg	40 0	77 0	640	4	8	93/75	90/74	87/73	23	77	76	75
Erie AP	42 1	80 1	732	4	9	88/73	85/72	83/71	18	75	74	72
Harrisburg AP	40 1	76 5	335	7	11	94/75	91/74	88/73	21	77	76	75
Johnstown	40 2	78 5	1214	−3	2	86/70	83/70	80/68	23	72	71	70
Lancaster	40 1	80 1	255	8	13	93/75	90/74	87/73	22	77	76	75
Meadville	41 4	80 2	1065	4	8	88/71	85/70	83/69	21	73	72	73
New Castle	39 0	75 1	825	2	7	91/73	90/74	86/71	23	75	74	73
Philadelphia AP	40 5	80 3	7	10	14	93/75	90/74	87/72	21	77	76	75
Pittsburgh AP	40 3	80 2	1137	−1	5	89/72	88/71	84/70	22	74	73	72
Pittsburgh CO	40 2	80 0	749r	3	8	91/72	88/71	86/70	19	74	73	72
Reading CO	40 2	75 5	226	9	13	92/73	89/72	86/72	19	76	75	73
Scranton/Wilkes-Barre	41 2	75 4	940	1	6	90/72	87/71	84/70	19	74	73	72
State College (S)	40 5	76 5	1175	3	7	90/72	87/71	84/70	23	74	73	72
Sunbury	40 5	76 5	480	2	6	92/73	89/72	86/70	22	75	74	73
Uniontown	39 5	79 5	1040	5	10	91/74	88/73	85/72	22	76	75	74
Warren	41 1	79 1	1280	−2	2	89/71	86/71	83/70	24	74	73	73
West Chester	40 0	75 1	440	9	13	92/75	89/74	86/72	20	77	74	73
Williamsport AP	41 1	77 0	527	2	7	92/73	89/72	86/70	23	75	74	73
York	40 0	76 4	390	8	13	94/75	91/74	88/73	22	77	76	75
RHODE ISLAND												
Newport (S)	41 3	71 2	20	5	9	88/73	85/72	82/70	16	76	75	73
Providence AP	41 4	71 3	55	5	9	89/73	86/72	83/70	19	75	74	73

TABLE 13-1 Climatic Conditions for the United States (Continued)

Col. 1	Col. 2 Lat.[b] °	′	Col. 3 Long.[b] °	′	Col. 4 Elevation[c] Ft	Winter, °F Col. 5 Design Dry-Bulb 99%	97.5%	Summer, °F Col. 6 Design Dry-Bulb and Mean Coincident Wet-Bulb 1%	2.5%	5%	Col. 7 Mean Daily Range	Col. 8 Design Wet-Bulb 1%	2.5%	5%
SOUTH CAROLINA														
Anderson	34	3	82	4	764	19	23	94/74	92/74	90/74	21	77	76	75
Charleston AFB (S)	32	5	80	0	41	24	27	93/78	91/78	89/77	18	81	80	79
Charleston CO	32	5	80	0	9	25	28	94/78	92/78	90/77	13	81	80	79
Columbia AP	34	0	81	1	217	20	24	97/76	95/75	93/75	22	79	78	77
Florence AP	34	1	79	4	146	22	25	94/77	92/77	90/76	21	80	79	78
Georgetown	33	2	79	2	14	23	26	92/79	90/78	88/77	18	81	80	79
Greenville AP	34	5	82	1	957	18	22	93/74	91/74	89/74	21	77	76	75
Greenwood	34	1	82	1	671	19	22	93/75	91/74	89/74	21	78	77	76
Orangeburg	33	3	80	5	244	20	24	97/76	95/75	93/75	20	79	78	77
Rock Hill	35	0	81	0	470	19	23	96/75	94/74	92/74	20	78	77	76
Spartanburg AP	35	0	82	0	816	18	22	93/74	91/74	89/74	20	77	76	75
Sumter-Shaw AFB	34	0	80	3	291	22	25	95/77	92/76	90/75	21	79	78	77
SOUTH DAKOTA														
Aberdeen AP	45	3	98	3	1296	−19	−15	94/73	91/72	88/70	27	77	75	73
Brookings	44	2	96	5	1642	−17	−13	95/73	92/72	89/71	25	77	75	73
Huron AP	44	3	98	1	1282	−18	−14	96/73	93/72	90/71	28	77	75	73
Mitchell	43	4	98	0	1346	−15	−10	96/72	93/71	90/70	28	76	74	72
Pierre AP	44	2	100	2	1718r	−15	−10	99/71	95/71	92/69	29	75	74	72
Rapid City AP (S)	44	0	103	0	3165	−11	−7	95/66	92/65	89/65	28	71	69	67
Sioux Falls AP	43	4	96	4	1420	−15	−11	94/73	91/72	88/71	24	76	75	73
Watertown AP	45	0	97	1	1746	−19	−15	94/73	91/72	88/71	26	76	75	73
Yankton	43	0	97	2	1280	−13	−7	94/73	91/72	88/71	25	77	76	74
TENNESSEE														
Athens	33	3	84	4	940	13	18	95/74	92/73	90/73	22	77	76	75
Bristol-Tri City AP	36	3	82	2	1519	9	14	91/72	89/72	87/71	22	75	73	73
Chattanooga AP	35	0	85	1	670	13	18	96/75	93/74	91/74	22	78	77	76
Clarksville	36	4	87	2	470	6	12	95/76	93/74	90/74	21	78	77	76
Columbia	35	4	87	0	690	10	15	97/75	94/74	91/74	21	78	77	76
Dyersburg	36	0	89	3	334	10	15	96/78	94/77	91/76	21	81	80	78
Greenville	35	5	82	5	1320	11	16	92/73	90/72	88/72	22	76	75	74
Jackson AP	35	4	88	0	413	11	16	98/76	95/75	92/75	21	79	78	77
Knoxville AP	35	5	84	0	980	13	19	94/74	92/73	90/73	21	77	76	75
Memphis AP	35	0	90	0	263	13	18	98/77	95/76	93/76	21	80	79	78
Murfreesboro	35	5	86	2	608	9	14	97/75	94/74	91/74	22	78	77	76

Note: AP = airport; AFB = air force base; CO = office location within an urban area.

368

Station	Lat°	′	Long°	′	Elev	99%	97.5%	1%	2.5%	5%	DR	1%	2.5%	5%
Nashville AP (S)	36	1	86	4	577	9	14	97/75	94/74	91/74	21	78	77	76
Tullahoma	35	2	86	1	1075	8	13	96/74	93/73	91/73	22	77	76	75
TEXAS														
Abilene AP	32	3	98	4	1759	15	20	101/71	99/71	97/71	22	75	74	74
Alice AP	27	4	98	0	180	31	34	100/78	98/77	95/77	20	82	81	79
Amarillo AP	35	1	101	4	3607	6	11	98/67	95/67	93/67	26	71	70	70
Austin AP	30	1	97	4	597	24	28	100/74	98/74	97/74	22	78	77	77
Bay City	29	0	96	0	52	29	33	96/77	94/77	92/77	16	80	79	79
Beaumont	30	0	94	0	18	27	31	95/79	93/78	91/78	19	81	80	80
Beeville	28	2	97	4	225	30	33	99/78	97/77	95/77	18	82	81	80
Big Spring AP (S)	32	2	101	2	2537	16	20	100/69	97/69	95/69	26	74	73	72
Brownsville AP (S)	25	5	97	5	16	35	39	94/77	93/77	92/77	18	80	79	79
Bryan AP	30	4	96	2	275	24	29	98/76	96/76	94/76	20	79	78	78
Corpus Christi AP	27	5	97	3	43	31	35	95/78	94/78	92/78	19	80	80	78
Corsicana	32	5	96	3	425	20	25	100/75	98/75	96/75	21	79	78	77
Dallas AP	32	5	96	5	481	18	22	102/75	100/75	97/75	20	78	78	77
Del Rio, Laughlin AFB	29	2	101	1	1072	26	31	100/73	98/73	97/75	24	78	77	76
Denton	33	0	97	1	655	17	22	101/74	99/74	97/73	22	79	77	76
Eagle Pass	28	1	100	0	743	27	32	101/73	99/73	98/73	24	78	78	77
El Paso AP (S)	31	5	106	2	3918	20	24	100/64	98/64	96/64	27	69	68	68
Fort Worth AP (S)	32	5	97	0	544r	17	22	101/74	98/74	97/74	22	78	77	76
Galveston AP	29	2	94	5	5	31	36	90/79	89/79	88/78	10	81	80	80
Greenville	33	0	96	1	575	17	22	101/74	99/74	97/74	21	78	77	76
Harlingen	26	1	97	4	37	35	39	96/77	94/77	93/77	19	80	79	79
Houston AP	29	4	95	1	50	27	32	96/77	94/77	92/77	18	80	79	79
Houston CO	29	4	95	0	158r	28	33	97/77	95/77	93/77	18	80	79	79
Huntsville	30	4	95	4	494	22	27	100/75	98/75	96/75	20	78	77	77
Killeen-Gray AFB	31	0	97	0	1021	20	25	99/73	97/73	95/73	22	77	76	75
Lamesa	32	2	102	4	2965	13	17	99/69	96/69	94/69	26	73	72	71
Laredo AFB	27	3	99	4	503	32	36	102/73	101/73	99/73	23	78	78	77
Longview	32	2	94	4	345	19	24	99/76	97/76	95/76	20	80	79	78
Lubbock AP	33	4	101	4	3243	10	15	98/69	96/69	94/69	26	73	72	71
Lufkin AP	31	1	94	1	286	25	29	99/76	97/76	94/76	20	80	79	78
Mc Allen	26	1	98	1	122	35	39	97/77	95/77	94/77	21	80	79	79
Midland AP (S)	32	0	102	5	2815r	16	21	100/69	98/69	96/69	26	73	72	71
Mineral Wells AP	32	5	98	0	934	17	22	101/74	99/74	97/74	22	78	77	76
Palestine CO	31	5	95	5	580	23	27	100/76	98/76	96/76	20	79	79	78
Pampa	35	3	101	3	3230	7	12	99/67	96/67	94/67	26	71	70	70
Pecos	31	2	103	3	2580	16	21	100/69	98/69	96/69	27	73	72	71
Plainview	34	1	101	4	3400	8	13	98/68	96/68	94/68	26	72	71	70

TABLE 13-1 Climatic Conditions for the United States (Continued)

Note: AP = airport; AFB = air force base; CO = office location within an urban area.

Col. 1	Col. 2		Col. 3		Col. 4	Col. 5		Col. 6			Col. 7	Col. 8		
	Lati-tude		Longi-tude		Eleva-tion	Winter, F Design Dry-Bulb		Summer, F Design Dry-Bulb and Mean Coincident Wet-Bulb			Mean Daily	Design Wet-Bulb		
State and Station	°	'	°	'	Ft	99%	97.5%	1%	2.5%	5%	Range	1%	2.5%	5%
Port Arthur AP	30	0	94		16	27	31	95/79	93/78	91/78	19	81	80	80
San Angelo, Goodfellow AFB	31	2	100	2	1878	18	22	101/71	99/71	97/70	24	75	74	73
San Antonio AP (S)	29	3	98	3	792	25	30	99/72	97/73	96/73	19	77	76	76
Sherman Perrin AFB	33	4	96	4	763	15	20	100/75	98/75	95/74	22	78	77	76
Snyder	32	4	101	1	2325	13	18	100/70	98/70	96/70	26	74	73	72
Temple	31	1	97	2	675	22	27	100/74	99/74	97/74	22	78	77	77
Tyler AP	32	2	95	2	527	19	24	99/76	97/76	95/76	21	80	79	78
Vernon	34	1	99	2	1225	13	17	102/73	100/73	97/73	24	77	76	75
Victoria AP	28	5	97	0	104	29	32	98/78	96/77	94/77	18	82	81	79
Waco AP	31	4	97	1	500	21	26	101/75	99/75	97/75	22	78	78	77
Wichita Falls AP	34	0	98	3	994	14	18	103/73	101/73	98/73	24	77	76	75
UTAH														
Cedar City AP	37	4	113	1	5613	-2	5	93/60	91/60	89/59	32	65	63	62
Logan	41	5	111	5	4775	-3	2	93/62	91/61	88/60	33	65	64	63
Moab	38	5	109	3	3965	6	11	100/60	98/60	96/60	30	65	64	64
Ogden AP	41	1	112	0	4455	1	5	93/63	91/61	88/61	33	66	65	64
Price	39	4	110	5	5580	-2	3	93/60	91/60	89/59	33	65	65	62
Provo	40	1	111	4	4470	1	6	98/62	96/62	94/61	32	66	65	64
Richfield	38	5	112	1	5300	-1	5	93/60	93/60	89/59	34	65	63	62
St George CO	37	1	113	3	2899	14	21	103/65	101/65	99/64	33	70	68	67
Salt Lake City AP (S)	40	5	111	5	4220	3	8	97/62	95/62	92/61	32	66	65	64
Vernal AP	40	3	109	3	5280	-5	0	91/61	89/60	86/59	32	64	63	62
VERMONT														
Barre	44	1	72	3	1120	-16	-11	84/71	81/69	78/68	23	73	71	70
Burlington AP (S)	44	3	73	1	331	-12	-7	88/72	85/70	82/69	23	74	72	71
Rutland	43	3	73	0	620	-13	-8	87/72	84/70	81/69	23	74	72	71
VIRGINIA														
Charlottesville	38	1	78	3	870	14	18	94/74	91/74	88/73	23	77	76	75
Danville AP	36	3	79	2	590	14	16	94/74	92/73	90/73	21	77	76	75
Fredericksburg	38	2	77	3	50	10	14	96/76	93/75	90/74	21	78	77	76
Harrisonburg	38	3	78	5	1340	12	16	93/72	91/72	88/71	23	75	74	73
Lynchburg AP	37	2	79	1	947	12	16	93/74	90/74	88/73	21	77	76	75

Station	Lat.°		Long.°		Elev. ft	Winter 99%	Winter 97.5%	Summer 1% DB/WB	Summer 2.5% DB/WB	Summer 5% DB/WB	Daily Range	WB 1%	WB 2.5%	WB 5%
Norfolk AP	36	5	76	1	26	20	22	93/77	91/76	89/76	18	79	78	77
Petersburg	37	1	77	3	194	14	17	95/76	92/76	90/75	20	79	78	77
Richmond AP	37	3	77	4	162	14	17	95/76	92/76	90/75	21	79	78	77
Roanoke AP	37	2	80	0	1174r	12	16	93/72	91/72	88/71	23	75	74	73
Staunton	38	2	78	5	1480	12	16	93/72	91/72	88/71	23	75	74	73
Winchester	39	1	78	1	750	6	10	93/75	90/74	88/74	21	77	76	75
WASHINGTON														
Aberdeen	47	0	123	5	12	25	28	80/65	77/62	73/61	16	65	63	62
Bellingham AP	48	5	122	3	150	10	15	81/67	77/65	74/63	19	68	65	63
Bremerton	47	3	122	4	162	21	25	82/65	78/64	75/62	20	68	64	63
Ellensburg AP	47	0	120	3	1729	-2	6	94/65	91/64	87/62	34	66	65	63
Everett-Paine AFB	47	5	122	2	598	21	25	80/65	76/64	73/62	20	67	64	63
Kennewick	46	0	119	1	392	5	11	99/68	96/67	92/66	30	70	68	67
Longview	46	1	123	0	12	19	24	88/68	85/67	81/65	30	69	67	66
Moses Lake, Larson AFB	47	1	119		1183	1	7	97/66	94/65	90/63	32	67	66	64
Olympia AP	47	0	122	5	190	16	22	87/66	83/65	79/64	32	67	66	64
Port Angeles	48	1	123	3	99	24	27	72/62	69/61	67/60	18	64	62	61
Seattle-Boeing Field	47	3	122	2	14	21	26	84/68	81/66	77/65	24	69	67	65
Seattle CO (S)	47	4	122	2	14	22	27	85/68	82/66	78/65	19	69	67	65
Seattle-Tacoma AP (S)	47	3	122	2	386	21	26	84/65	80/64	76/62	22	66	64	63
Spokane AP (S)	47	4	117	3	2357	-6	2	93/64	90/63	87/62	28	65	64	62
Tacoma-McChord AFB	47	1	122	2	350	19	24	86/66	82/65	79/63	22	68	66	64
Walla Walla AP	46	1	118	2	1185	0	7	97/67	94/66	90/65	27	68	67	66
Wenatchee	47	2	120	3	634	7	11	99/67	96/66	92/64	32	68	67	65
Yakima AP	46	3	120	3	1061	-2	5	96/65	93/65	89/63	36	68	66	65
WEST VIRGINIA														
Beckley	37	5	81	1	2330	-2	4	83/71	81/69	79/69	22	73	71	70
Bluefield AP	37	2	81	2	2850	-2	4	83/71	81/69	79/69	22	73	71	70
Charleston AP	38	2	81	4	939	7	11	92/74	90/73	87/72	20	76	75	74
Clarksburg	39	2	80	5	977	6	10	92/74	90/73	87/72	21	76	75	74
Elkins AP	38	5	79	0	1970	-1	6	86/72	84/70	82/70	22	74	72	71
Huntington CO	38	2	82	0	565r	6	10	94/76	91/74	89/73	22	78	77	75
Martinsburg AP	39	2	78	3	537	6	10	93/75	90/74	88/74	21	77	76	75
Morgantown AP	39	4	80	4	1245	8	8	90/74	87/73	85/73	21	76	75	73
Parkersburg CO	39	2	81	2	615r	11	11	93/75	90/74	88/73	21	77	76	75
Wheeling	40	1	80	1	659	1	5	89/72	86/71	84/70	21	74	73	72

TABLE 13-1 Climatic Conditions for the United States (Continued)

Col. 1	Col. 2	Col. 3	Col. 4	Col. 5 Winter, F		Col. 6 Summer, F			Col. 7	Col. 8		
				Design Dry-Bulb		Design Dry-Bulb and Mean Coincident Wet-Bulb			Mean	Design Wet-Bulb		
State and Station	Lati-tude° '	Longi-tude° '	Eleva-tion° Ft	99%	97.5%	1%	2.5%	5%	Daily Range	1%	2.5%	5%

Note: AP = airport; AFB = air force base; CO = office location within an urban area.

WISCONSIN

State and Station	Lat	Long	Elev	99%	97.5%	1%	2.5%	5%	Range	1%	2.5%	5%
Appleton	44 2	88 5	742	-14	-9	89/74	86/72	83/71	23	76	74	72
Ashland	46 3	90 5	650	-21	-16	85/70	82/68	79/66	23	72	70	68
Beloit	42 5	89 0	780	-7	-3	92/75	90/75	88/74	24	78	77	75
Eau Claire AP	44 5	91 3	888	-15	-11	92/75	89/73	86/71	23	77	75	73
Fond du Lac	43 5	88 3	760	-12	-8	89/74	86/72	84/71	23	76	74	72
Green Bay AP	44 5	88 1	683	-13	-9	88/74	85/72	83/71	22	76	74	72
La Crosse AP	43 5	91 2	652	-13	-9	91/75	88/73	85/72	22	77	75	74
Madison AP (S)	43 1	89 2	858	-11	-7	91/74	88/73	85/71	22	77	75	73
Manitowoc	44 1	87 4	660	-11	-7	89/74	86/72	83/71	21	76	74	72
Marinette	45 0	87	605	-15	-11	87/73	84/71	82/70	20	75	73	71
Milwaukee AP	43 4	87 5	672	-8	-4	90/74	87/73	84/71	21	76	74	73
Racine	42 4	87 4	640	-6	-2	91/75	88/73	85/72	21	77	75	74
Sheboygan	43 4	87 4	648	-10	-6	89/75	86/73	83/72	20	77	75	73
Stevens Point	43 3	89 3	1079	-15	-11	92/75	89/73	86/71	23	77	74	73
Waukesha	43 0	88 1	860	-9	-5	90/74	87/73	84/71	22	76	75	72
Wausau AP	44 6	89 4	1196	-16	-12	91/74	88/72	85/70	23	76	74	73

WYOMING

State and Station	Lat	Long	Elev	99%	97.5%	1%	2.5%	5%	Range	1%	2.5%	5%
Casper AP	42 5	106 3	5319	-11	-5	92/58	90/57	87/57	31	63	61	60
Cheyenne AP	41 1	104 5	6126	-9	-1	89/58	86/58	84/57	30	63	62	60
Cody AP	44 3	109 0	5090	-19	-13	89/60	86/60	83/59	32	64	63	61
Evanston	41 2	111 0	6860	-9	-3	86/55	84/55	82/54	32	59	58	57
Lander AP (S)	42 4	108 4	5563	-16	-11	91/61	88/61	85/60	32	64	63	61
Laramie AP (S)	41 1	105 3	7266	-14	-6	84/56	81/56	79/55	28	61	60	59
Newcastle	43 5	104 1	4480	-17	-12	91/64	87/63	84/63	30	69	68	66
Rawlins	41 5	107 1	6736	-12	-4	86/57	83/57	81/56	40	62	61	60
Rock Springs AP	41 4	109 0	6741	-9	-3	86/55	84/55	82/54	32	59	58	57
Sheridan AP	44 5	107 0	3942	-14	-8	94/62	91/62	88/61	32	65	65	63
Torrington	4 0	104 1	4098	-14	-8	94/62	91/62	88/61	30	66	65	63

SOURCE: *ASHRAE Handbook—1985 Fundamentals.* Reprinted by permission.

TABLE 13-2 Climatic Conditions for Canada[a]

Col. 1	Col. 2		Col. 3		Col. 4	WINTER[e], F		SUMMER[f], F						
						Col. 5		Col. 6			Col. 7	Col. 8		
Province and Station[b]	Lati-tude[c]		Longi-tude[c]		Eleva-tion[d]	Design Dry-Bulb		Design Dry Bulb and Mean Coincident Wet-Bulb			Mean Daily	Design Wet Bulb		
	°	'	°	'	Ft	99%	97.5%	1%	2.5%	5%	Range	1%	2.5%	5%
ALBERTA														
Calgary AP	51	6	114	1	3540	−27	−23	84/63	81/61	79/60	25	65	63	62
Edmonton AP	53	34	113	31	2219	−29	−25	85/66	82/65	79/63	23	68	66	65
Grande Prairie AP	55	11	118	53	2190	−39	−33	83/64	80/63	78/61	23	66	64	62
Jasper	52	53	118	4	3480	−31	−26	83/64	80/62	77/61	28	66	64	63
Lethbridge AP (S)	49	38	112	48	3018	−27	−22	90/65	87/64	34/63	28	68	66	65
McMurray AP	56	39	111	13	1216	−41	−38	86/67	82/65	79/64	26	69	67	65
Medicine Hat AP	50	1	110	43	2365	−29	−24	93/66	90/65	37/64	28	70	68	66
Red Deer AP	52	11	113	54	2965	−31	−26	84/65	81/64	78/62	25	67	66	64
BRITISH COLUMBIA														
Dawson Creek	55	44	120	11	2164	−37	−33	82/64	79/63	76/61	26	66	64	62
Fort Nelson AP (S)	58	50	122	35	1230	−43	−40	84/64	81/63	78/62	23	67	65	64
Kamloops CO	50	43	120	25	1133	−21	−15	94/66	91/65	88/64	29	68	66	65
Nanaimo (S)	49	11	123	58	230	16	20	83/67	80/65	77/64	21	68	66	65
New Westminster	49	13	122	54	50	14	18	84/68	81/67	78/66	19	69	68	66
Penticton AP	49	28	119	36	1121	0	4	92/68	89/67	87/66	31	70	68	67
Prince George AP(S)	53	53	122	41	2218	−33	−28	84/64	80/62	77/61	26	66	64	62
Prince Rupert CO	54	17	130	23	170	−2	2	64/59	63/57	61/56	12	60	58	57
Trail	49	8	117	44	1400	−5	0	92/66	89/65	86/64	33	68	67	65
Vancouver AP (S)	49	11	123	10	16	15	19	79/67	77/66	74/65	17	68	67	66
Victoria CO	48	25	123	19	228	20	23	77/64	73/62	70/60	16	64	62	60
MANITOBA														
Brandon	49	52	99	59	1200	−30	−27	89/72	86/70	83/68	25	74	72	70
Churchill AP (S)	58	45	94	4	155	−41	−39	81/66	77/64	74/62	18	67	65	63
Dauphin AP	51	6	100	3	999	−31	−28	87/71	84/70	81/68	23	74	72	70
Flin Flon	54	46	101	51	1098	−41	−37	84/68	81/66	79/65	19	70	58	67
Portage la Prairie AP	49	54	98	16	867	−28	−24	88/73	86/72	83/70	22	76	74	71
The Pas AP (S)	53	58	101	6	894	−37	−33	85/68	82/67	79/66	20	71	69	68
Winnipeg AP (S)	49	54	97	14	786	−30	−27	89/73	86/71	84/70	22	75	73	71

[a] This table was prepared by ASHRAE Technical Committee 4.2 using data compiled from official weather stations where hourly weather observations are made by trained observers.

[b] When airport temperature observations were used to develop design data, "AP" follows the station name. Data for stations followed by "CO" come from office locations within an urban area and generally reflect an influence of the surroundings area. Stations without designation can be considered semirural and can be directly compared to most airport data.

[c] Latitude, for use in calculating solar loads, and longitude are given to the nearest minute.

[d] Elevations are ground elevations for each station as of 1964. Temperature readings are generally made at an elevation of 5 ft above ground.

[e] The winter design data are based on the month of January only.

[f] The summer design data are based on the month of July only.

TABLE 13-2 Climatic Conditions for Canada (*Continued*)

Col. 1	Col. 2		Col. 3		Col. 4	Col. 5		Col. 6			Col. 7	Col. 8		
	Lati-tude[c]		Longi-tude[c]		Eleva-tion[d]	WINTER[e], F		SUMMER[f], F						
Province and Station[b]						Design Dry-Bulb		Design Dry Bulb and Mean Coincident Wet-Bulb			Mean Daily	Design Wet Bulb		
	°	'	°	'	Ft	99%	97.5%	1%	2.5%	5%	Range	1%	2.5%	5%
NEW BRUNSWICK														
Campbellton CO	48	0	66	40	25	−18	−14	85/68	82/67	79/66	21	72	70	68
Chatham AP	47	1	65	27	112	−15	−10	89/69	85/68	82/67	22	72	71	69
Edmundston CO	47	22	68	20	500	−21	−16	87/70	83/68	80/67	21	73	71	69
Fredericton AP (S)	45	52	66	32	74	−16	−11	89/71	85/69	82/68	23	73	71	70
Moncton AP (S)	46	7	64	41	248	−12	− 8	85/70	82/69	79/67	23	72	71	69
Saint John AP	45	19	65	53	352	−12	− 8	80/67	77/65	75/64	19	70	68	66
NEWFOUNDLAND														
Corner Brook	48	58	57	57	15	− 5	0	76/64	73/63	71/62	17	67	66	65
Gander AP	48	57	54	34	482	− 5	− 1	82/66	79/65	77/64	19	69	67	66
Goose Bay AP (S)	53	19	60	25	144	−27	−24	85/66	81/64	77/63	19	68	66	64
St. John's AP (S)	47	37	52	45	463	3	7	77/66	75/65	73/64	18	69	67	66
Stephenville AP	48	32	58	33	44	− 3	4	76/65	74/64	71/63	14	67	66	65
NORTHWEST TERR.														
Fort Smith AP (S)	60	1	111	58	665	−49	−45	85/66	81/64	78/63	24	68	66	65
Frobisher AP (S)	63	45	68	33	68	−43	−41	66/53	63/51	59/50	14	54	52	51
Inuvik (S)	68	18	133	29	200	−56	−53	79/62	77/60	75/59	21	64	62	61
Resolute AP (S)	74	43	94	59	209	−50	−47	57/48	54/46	51/45	10	50	48	46
Yellowknife AP	62	28	114	27	682	−49	−46	79/62	77/61	74/60	16	64	63	62
NOVA SCOTIA														
Amherst	45	49	64	13	65	−11	− 6	84/69	81/68	79/67	21	72	70	68
Halifax AP (S)	44	39	63	34	83	1	5	79/66	76/65	74/64	16	69	67	66
Kentville (S)	45	3	64	36	40	− 3	1	85/69	83/68	80/67	22	72	71	69
New Glasgow	45	37	62	37	317	− 9	− 5	81/69	79/68	77/67	20	72	70	69
Sydney AP	46	10	60	3	197	− 1	3	82/69	80/68	77/66	19	71	70	68
Truro CO	45	22	63	16	131	− 8	− 5	82/70	80/69	78/68	22	73	71	70
Yarmouth AP	43	50	66	5	136	5	9	74/65	72/64	70/63	15	68	66	65

ONTARIO

Belleville	44	9	77	24	250	−11	− 7	86/73	84/72	82/71	20	75	74	73
Chatham	42	24	82	12	600	0	3	89/74	87/73	85/72	19	76	75	74
Cornwall	45	1	74	45	210	−13	− 9	89/73	87/72	84/71	21	75	74	72
Hamilton	43	16	79	54	303	− 3	1	88/73	86/72	83/71	21	76	74	73
Kapuskasing AP (S)	49	25	82	28	752	−31	−28	86/70	83/69	8C/67	23	72	70	69
Kenora AP	49	48	94	22	1345	−32	−28	84/70	82/69	8C/68	19	73	71	70
Kingston	44	16	76	30	300	−11	− 7	87/73	84/72	8C/71	20	75	74	73
Kitchener	43	26	80	30	1125	− 6	− 2	88/73	85/72	85/71	23	76	74	72
London AP	43	2	81	9	912	− 4	0	87/74	85/73	85/72	21	71	70	73
North Bay AP	46	22	79	25	1210	−22	−18	84/68	81/67	79/66	20	75	74	68
Oshawa	43	54	78	52	370	− 6	− 3	88/73	86/72	84/71	20	71	70	73
Ottawa AP (S)	45	19	75	40	413	−17	−13	90/72	87/71	84/70	21	75	73	72
Owen Sound	44	34	80	55	597	− 6	− 2	84/71	82/70	83/69	21	73	72	70
Peterborough	44	17	78	19	635	−13	9	87/72	85/71	83/70	21	75	73	72
St. Catharines	43	11	79	14	325	− 1	3	87/73	85/72	83/71	20	76	74	73
Sarnia	42	58	82	22	625	0	3	88/73	86/72	84/71	19	76	71	73
Sault Ste. Marie AP	46	32	84	30	675	−17	−13	85/71	82/69	79/68	22	73	71	70
Sudbury AP	46	37	80	48	1121	−22	−19	86/69	83/67	81/66	22	72	70	68
Thunder Bay AP	48	22	89	19	644	−27	−24	85/70	83/68	80/67	24	72	70	68
Timmins AP	48	34	81	22	965	−33	−29	87/69	84/68	81/66	25	72	72	68
Toronto AP (S)	43	41	79	38	578	− 5	− 1	90/73	87/72	85/71	20	75	74	73
Windsor AP	42	16	82	58	637	0	4	90/74	88/73	86/72	20	77	75	74

TABLE 13-2 Climatic Conditions for Canada (Continued)

Col. 1	Col. 2		Col. 3		Col. 4	Col. 5		Col. 6			Col. 7	Col. 8		
	Latitude		Longitude		Elevation	WINTER, F		SUMMER, F						
						Design Dry-Bulb		Design Dry Bulb and Mean Coincident Wet-Bulb			Mean Daily	Design Wet Bulb		
Province and Station	°	′	°	′	Ft	99%	97.5%	1%	2.5%	5%	Range	1%	2.5%	5%
PRINCE EDWARD ISLAND														
Charlottetown AP (S)	46	17	63	8	186	−7	−4	80/69	78/68	76/67	16	71	70	68
Summerside AP	46	26	63	50	78	−8	−4	81/69	79/68	77/67	16	72	70	68
QUEBEC														
Bagotville AP	48	20	71	0	536	−28	−23	87/70	83/68	80/67	21	72	70	68
Chicoutimi	48	25	71	5	150	−26	−22	86/70	83/68	80/67	20	72	70	68
Drummondville	45	53	72	29	270	−18	−14	88/72	85/71	82/69	21	75	73	71
Granby	45	23	72	42	550	−19	−14	88/72	85/71	83/70	21	75	73	72
Hull	45	26	75	44	200	−18	−14	90/72	87/71	84/70	21	75	73	72
Megantic AP	45	35	70	52	1362	−20	−16	86/71	83/70	81/69	20	74	72	71
Montreal AP (S)	45	28	73	45	98	−16	−10	88/73	85/72	83/71	17	75	74	72
Quebec AP	46	48	71	23	245	−19	−14	87/72	84/70	81/68	20	74	72	70
Rimouski	48	27	68	32	117	−16	−12	83/68	79/66	76/65	18	71	69	67
St. Jean	45	18	73	16	129	−15	−11	88/73	86/72	84/71	20	75	74	72
St. Jerome	45	48	74	1	556	−17	−13	88/72	86/71	83/70	23	75	73	72
Sept. Iles AP (S)	50	13	66	16	190	−26	−21	76/63	73/61	70/60	17	67	65	63
Shawinigan	46	34	72	43	306	−18	−14	86/72	84/70	82/69	21	74	72	71
Sherbrooke CO	45	24	71	54	595	−25	−21	86/72	84/71	81/69	20	74	73	71
Thetford Mines	46	4	71	19	1020	−19	−14	87/71	84/70	81/69	21	74	72	71
Trois Rivières	46	21	72	35	50	−17	−13	88/72	85/70	82/69	23	74	72	71
Val d'Or AP	48	3	77	47	1108	−32	−27	85/70	83/68	80/67	22	72	70	68
Valleyfield	45	16	74	6	150	−14	−10	89/73	86/72	84/71	20	75	74	72
SASKATCHEWAN														
Estevan AP	49	4	103	0	1884	−30	−25	92/70	89/68	86/67	26	72	70	69
Moose Jaw AP	50	20	105	33	1857	−29	−25	93/69	89/67	86/66	27	71	69	68
North Battleford AP	52	46	108	15	1796	−33	−30	88/67	85/66	82/65	23	69	68	66
Prince Albert AP	53	13	105	41	1414	−42	−35	87/67	84/66	81/65	25	70	68	67
Regina AP	50	26	104	40	1884	−33	−29	91/69	88/68	84/67	26	72	70	68
Saskatoon AP (S)	50	10	106	41	1645	−35	−31	89/68	86/66	83/65	26	70	68	67
Swift Current AP (S)	50	17	107	41	2677	−28	−25	93/68	90/66	87/65	25	70	69	67
Yorkton AP	51	16	102	28	1653	−35	−30	87/69	84/68	80/66	23	72	70	68
YUKON TERRITORY														
Whitehorse AP (S)	60	43	135	4	2289	−46	−43	80/59	77/58	74/56	22	61	59	58

SOURCE: *ASHRAE Handbook—1985 Fundamentals.* Reprinted by permission.

TABLE 13-3 Climatic Conditions for Other Countries

Col. 1 Country and Station	Col. 2 Latitude and Longitude °	Col. 3 Elevation, Ft	Winter, F — Col. 4			Summer, F — Col. 5 Design Dry-Bulb			Col. 6 Outdoor Daily Range F deg	Col. 7 Design Wet-Bulb		
			Mean of Annual Extremes	99%	97½%	1%	2½%	5%		1%	2½%	5%
AFGHANISTAN												
Kabul	34 35N/ 69 12E	5955	2	6	9	98	96	93	32	66	65	64
ALGERIA												
Algiers	36 46N/ 3 03E	194	38	43	45	95	92	89	14	77	76	75
ARGENTINA												
Buenos Aires	34 35S/ 58 29W	89	27	32	34	91	89	86	22	77	76	75
Cordoba	31 22S/ 64 15W	1388	21	28	32	100	96	93	27	76	75	74
Tucuman	26 50S/ 65 10W	1401	24	32	36	102	99	96	23	76	75	74
AUSTRALIA												
Adelaide	34 56S/138 35E	140	36	38	40	98	94	91	25	72	70	68
Alice Springs	23 48S/133 53E	1795	28	34	37	104	102	100	27	75	74	72
Brisbane	27 28S/153 02E	137	39	44	47	91	88	86	18	77	76	75
Darwin	12 28S/130 51E	88	60	64	66	94	93	91	16	82	81	81
Melbourne	37 49S/144 58E	114	31	35	38	95	91	86	21	71	69	68
Perth	31 57S/115 51E	210	38	40	42	100	96	93	22	76	74	73
Sydney	33 52S/151 12E	138	38	40	42	89	84	80	13	74	73	72
AUSTRIA												
Vienna	48 15N/ 16 22E	644	− 2	6	11	88	86	83	16	71	69	67
AZORES												
Lajes (Terceira)	38 45N/ 27 05W	170	42	46	49	80	78	77	11	73	72	71
BAHAMAS												
Nassau	25 05N/ 77 21W	11	55	61	63	90	89	88	13	80	80	79
BANGLADESH												
Chittagong	22 21N/ 91 50E	87	48	52	54	93	91	89	20	82	81	81
BELGIUM												
Brussels	50 48N/ 4 21E	328	13	15	19	83	79	77	19	70	68	67
BERMUDA												
Kindley AFB	33 22N/ 64 41W	129	47	53	55	87	86	85	12	79	78	78
BOLIVIA												
La Paz	16 30S/ 68 09W	12001	28	31	33	71	69	68	24	58	57	56

TABLE 13-3 Climatic Conditions for Other Countries (Continued)

Col. 1 Country and Station	Col. 2 Latitude and Longitude	Col. 3 Elevation, Ft	Winter, F Col. 4 Mean of Annual Extremes	Col. 4 99%	Col. 4 97½%	Summer, F Col. 5 Design Dry-Bulb 1%	Col. 5 2½%	Col. 5 5%	Col. 6 Outdoor Daily Range F deg	Col. 7 Design Wet-Bulb 1%	Col. 7 2½%	Col. 7 5%
BRAZIL												
Belem	1 27S/ 48 29W	42	67	70	71	90	89	87	19	80	79	78
Belo Horizonte	19 56S/ 43 57W	3002	42	47	50	86	84	83	18	76	75	75
Brasilia	15 52S/ 47 55W	3442	46	49	51	86	88	86	17	75	74	74
Curitiba	25 25S/ 49 17W	3114	28	34	37	82	88	82	21	75	74	74
Fortaleza	3 46S/ 38 33W	89	66	69	70	89	90	89	17	79	78	78
Porto Alegre	30 02S/ 51 13W	33	32	37	40	91	92	89	20	76	76	75
Recife	8 04S/ 34 53W	97	67	69	70	95	87	86	10	78	77	77
Rio De Janeiro	22 55S/ 43 12W	201	56	58	60	94	92	90	11	80	79	78
Salvador	13 00S/ 38 30W	154	65	67	68	88	87	86	12	79	79	78
Sao Paulo	23 33S/ 46 38W	2608	36	42	46	86	84	82	18	75	74	74
BRITISH HONDURAS												
Belize	17 31N/ 88 11W	17	55	60	62	90	90	89	13	82	82	81
BULGARIA												
Sofia	42 42N/ 23 20E	1805	−2	3	8	89	86	84	26	71	70	69
BURMA												
Mandalay	21 59N/ 96 06E	252	50	54	56	104	102	101	30	81	80	80
Rangoon	16 47N/ 96 09E	18	59	62	63	100	98	95	25	83	82	82
CAMBODIA												
Phnom Penh	11 33N/104 51E	36	62	66	68	98	96	94	19	83	82	82
CHILE												
Punta Arenas	53 10S/ 70 54W	26	22	25	27	68	66	64	14	56	55	54
Santiago	33 27S/ 70 42W	1706	27	32	35	90	89	88	32	71	70	69
Valparaiso	33 01S/ 71 38W	135	39	43	46	81	79	77	16	67	66	65
CHINA												
Chungking	29 33N/106 33E	755	34	37	39	99	97	95	18	81	80	79
Shanghai	31 12N/121 26E	23	16	23	26	94	92	90	16	81	81	80
COLOMBIA												
Barranquilla	10 59N/ 74 48W	44	66	70	72	95	94	93	17	83	82	82
Bogota	4 36N/ 74 05W	8406	42	45	46	72	70	69	19	60	59	58
Cali	3 25N/ 76 30W	3189	53	57	58	84	82	79	15	70	69	68
Medellin	6 13N/ 75 36W	4650	48	53	55	87	85	84	25	73	72	72
CONGO												
Brazzaville	4 15S/ 15 15E	1043	54	60	62	93	92	91	21	81	81	81
CUBA												
Guantanamo Bay	19 54N/ 75 09W	21	60	64	66	94	93	92	16	82	82	82
Havana	23 08N/ 82 21W	80	54	59	62	92	91	89	14	81	81	81
CZECHOSLOVAKIA												
Prague	50 05N/ 14 25E	662	3	4	9	88	85	83	16	66	65	64
DENMARK												
Copenhagen	55 41N/ 12 33E	43	11	16	19	79	76	74	17	68	66	64

Location		57	61	63	65	92	90	88	16	81	80	80
DOMINICAN REPUBLIC												
Santo Domingo	18 29N/ 69 54W											
EQUADOR												
Guayaquil	2 10S/ 79 53W	20	61	64	65	92	91	89	20	80	80	79
Quito	0 13S/ 78 32W	9446	30	36	39	73	72	71	32	63	62	62
EGYPT												
Cairo	29 52N/ 31 20E	381	39	45	46	102	100	98	26	76	75	74
EL SALVADOR												
San Salvador	13 42N/ 89 13W	2238	51	54	56	98	96	95	32	77	76	75
ETHIOPIA												
Addis Ababa	9 02N/ 38 45E	7753	35	39	41	84	82	81	28	66	65	64
Asmara	15 17N/ 38 55E	7628	36	40	42	83	81	80	27	65	64	63
FINLAND												
Helsinki	60 10N/ 24 57E	30	-11	-7	-1	77	74	72	14	66	65	63
FRANCE												
Lyon	45 42N/ 4 47E	938	-1	10	14	91	89	86	23	71	70	69
Marseilles	43 18N/ 5 23E	246	23	25	28	90	87	84	22	72	71	69
Nantes	47 15N/ 1 34W	121	17	22	26	86	83	80	21	73	69	67
Nice	43 42N/ 7 16E	39	31	34	37	87	85	83	15	73	72	72
Paris	48 49N/ 2 29E	164	16	22	25	89	86	83	21	70	68	67
Strasbourg	48 35N/ 7 46E	465	9	11	16	86	83	80	20	70	69	67
FRENCH GUIANA												
Cayenne	4 56N/ 52 27W	20	69	71	72	92	91	90	17	83	83	82
FEDERAL REPUBLIC GERMANY												
Berlin (West)	52 27N/ 13 18E	187	6	7	12	84	81	78	19	68	67	66
Hamburg	53 33N/ 9 58E	66	10	12	16	80	76	73	13	68	66	65
Hannover	52 24N/ 9 40E	561	7	16	20	82	78	75	17	68	67	65
Mannheim	49 34N/ 8 28E	359	2	8	11	85	85	82	18	71	69	68
Munich	48 09N/ 11 34E	1729	-1	5	9	86	83	80	18	68	66	64

TABLE 13-3   Climatic Conditions for Other Countries (Continued)

Col. 1 Country and Station	Col. 2 Latitude and Longitude °	Col. 3 Elevation, Ft	Winter, F			Summer, F							
			Col. 4			Col. 5 Design Dry-Bulb			Col. 6 Outdoor Daily Range F deg	Col. 7 Design Wet-Bulb		
			Mean of Annual Extremes	99%	97½%	1%	2½%	5%		1%	2½%	5%
GHANA												
Accra	5 33N/ 0 12W	88	65	68	69	91	90	89	13	80	79	79
GIBRALTAR												
Gibraltar	36 09N/ 5 22W	11	38	42	45	92	89	86	14	76	75	74
GREECE												
Athens...........	37 58N/ 23 43E	351	29	33	36	96	93	91	18	72	71	71
Thessaloniki	40 37N/ 22 57E	78	23	28	32	95	93	91	20	77	76	75
GREENLAND												
Narsarssuaq	61 11N/ 45 25W	85	−23	−12	− 8	66	63	61	20	56	54	52
GUATEMALA												
Guatemala City ...	14 37N/ 90 31W	4855	45	48	51	83	82	81	24	69	68	67
GUYANA												
Georgetown.......	6 50N/ 58 12W	6	70	72	73	89	88	87	11	80	79	79
HAITI												
Port Au Prince ...	18 33N/ 72 20W	121	63	65	67	97	95	93	20	82	81	80
HONDURAS												
Tegucigalpa	14 06N/ 87 13W	3094	44	47	50	89	87	85	28	73	72	71
HONG KONG												
Hong Kong	22 18N/114 10E	109	43	48	50	92	91	90	10	81	80	80
HUNGARY												
Budapest	47 31N/ 19 02E	394	8	10	14	90	86	84	21	72	71	70
ICELAND												
Reykjavik	64 08N/ 21 56E	59	8	14	17	59	58	56	16	54	53	53
INDIA												
Ahmedabad	23 02N/ 72 35E	163	49	53	56	109	107	105	28	80	79	78
Bangalore	12 57N/ 77 37E	3021	53	56	58	96	94	93	26	75	74	74
Bombay	18 54N/ 72 49E	37	62	65	67	96	94	92	13	83	81	81
Calcutta	22 32N/ 88 20E	21	49	52	54	98	97	96	22	82	82	82
Madras	13 04N/ 80 15E	51	61	64	66	104	102	101	19	84	83	83
Nagpur	21 09N/ 79 07E	1017	45	51	54	110	108	107	30	79	79	78
New Delhi	28 35N/ 77 12E	703	35	39	41	110	107	105	26	83	82	82
INDONESIA												
Djakarta	6 11S/106 50E	26	69	71	72	90	89	88	14	81	79	78
Kupang	10 10S/123 34E	148	63	66	68	94	93	92	20	80	80	80
Makassar	5 08S/119 28E	61	64	66	68	90	89	88	17	81	80	79
Medan...........	3 35N/ 98 41E	77	66	69	71	92	91	90	17	81	80	80
Palembang	3 00S/104 46E	20	67	70	71	92	91	91	17	80	80	79
Surabaya	7 13S/112 43E	10	64	66	68	91	90	89	18	80	79	79
IRAN												
Abadan	30 21N/ 48 16E	7	32	39	41	116	113	110	32	82	81	81
Meshed	36 17N/ 59 36E	3104	3	10	14	99	96	93	29	68	67	66
Tehran	35 41N/ 51 25E	4002	15	20	24	102	100	98	27	75	74	73

	Coordinates	Elev										
IRAQ												
Baghdad	33 20N/ 44 24E	111	27	32	35	113	111	108	34	73	72	72
Mosul	36 19N/ 43 09E	730	23	29	32	114	112	110	40	73	72	72
IRELAND												
Dublin	53 22N/ 6 21W	155	19	24	27	74	72	70	16	65	64	62
Shannon	52 41N/ 8 55W	8	19	25	28	76	73	71	14	65	64	63
IRIAN BARAT												
Manokwari	0 52S/134 05E	62	70	71	72	89	88	87	12	82	81	81
ISRAEL												
Jerusalem	31 47N/ 35 13E	2485	31	36	38	95	94	92	24	70	69	69
Tel Aviv	32 06N/ 34 47E	36	33	39	41	96	93	91	16	74	73	72
ITALY												
Milan	45 27N/ 9 17E	341	12	18	22	89	87	84	20	76	75	74
Naples	40 53N/ 14 18E	220	28	34	36	91	88	86	19	74	73	72
Rome	41 48N/ 12 36E	377	25	30	33	94	92	89	24	74	73	72
IVORY COAST												
Abidjan	5 19N/ 4 01W	65	64	67	69	91	90	88	15	83	82	81
JAPAN												
Fukuoka	33 35N/130 27E	22	26	29	31	92	90	89	20	82	80	79
Sapporo	43 04N/141 21E	56	-7	1	5	86	83	80	20	76	74	72
Tokyo	35 41N/139 46E	19	21	26	28	91	89	87	14	81	80	79
JORDAN												
Amman	31 57N/ 35 57E	2548	29	33	36	97	94	92	25	70	69	68
KENYA												
Nairobi	1 16S/ 36 48E	5971	45	48	50	81	80	78	24	66	65	65
KOREA												
Pyongyang	39 02N/125 41E	186	-10	-2	3	89	87	85	21	77	76	76
Seoul	37 34N/126 58E	285	-1	7	9	91	89	89	16	81	79	78
LEBANON												
Beirut	33 54N/ 35 28E	111	40	42	45	93	91	90	15	78	77	76

TABLE 13-3 Climatic Conditions for Other Countries *(Continued)*

Col. 1 Country and Station	Col. 2 Latitude and Longitude	Col. 3 Elevation, Ft	Winter, F			Summer, F						
			Col. 4			Col. 5 Design Dry-Bulb			Col. 6 Outdoor Daily Range F deg	Col. 7 Design Wet-Bulb		
			Mean of Annual Extremes	99%	97½%	1%	2½%	5%		1%	2½%	5%
LIBERIA												
Monrovia	6 18N/ 10 48W	75	64	68	69	90	89	88	19	82	82	81
LIBYA												
Benghazi	32 06N/ 20 04E	82	41	46	48	97	94	91	13	77	76	75
MADAGASCAR												
Tananarive	18 55S/ 47 33E	4531	39	43	46	86	84	83	23	73	72	71
MALAYSIA												
Kuala Lumpur	3 07N/101 42E	127	67	70	71	94	93	92	20	82	82	81
Penang	5 25N/100 19E	17	69	72	73	93	93	92	18	82	81	80
MARTINIQUE												
Fort De France	14 37N/ 61 05W	13	62	64	66	90	89	88	14	81	81	80
MEXICO												
Guadalajara	20 41N/103 20W	5105	35	39	42	93	91	89	29	68	67	66
Merida	20 58N/ 89 38W	72	56	59	61	97	95	94	21	80	79	77
Mexico City	19 24N/ 99 12W	7575	33	37	39	83	81	79	25	61	60	59
Monterrey	25 40N/100 18W	1732	31	38	41	98	95	93	20	79	78	77
Vera Cruz	19 12N/ 96 08W	184	55	60	62	91	89	88	12	83	83	82
MOROCCO												
Casablanca	33 35N/ 7 39W	164	36	40	42	94	90	86	50	73	72	70
NEPAL												
Katmandu	27 42N/ 85 12E	4388	30	33	35	89	87	86	25	78	77	76
NETHERLANDS												
Amsterdam	52 23N/ 4 55E	5	17	20	23	79	76	73	10	65	64	63
NEW ZEALAND												
Auckland	36 51S/174 46E	140	37	40	42	78	77	76	14	67	66	65
Christ Church	43 32S/172 37E	32	25	28	31	82	79	76	17	68	67	66
Wellington	41 17S/174 46E	394	32	35	37	76	74	72	14	66	65	64
NICARAGUA												
Managua	12 10N/ 86 15W	135	62	65	67	94	93	92	21	81	80	79
NIGERIA												
Lagos	6 27N/ 3 24E	10	67	70	71	92	91	90	12	82	82	81
NORWAY												
Bergen	60 24N/ 5 19E	141	14	17	20	75	74	73	21	67	66	65
Oslo	59 56N/ 10 44E	308	− 2	0	4	79	77	74	17	67	66	64
PAKISTAN												
Karachi	24 48N/ 66 59E	13	45	49	51	100	98	95	14	82	82	81
Lahore	31 35N/ 74 20E	702	32	35	37	109	107	105	27	83	83	81
Peshawar	34 01N/ 71 35E	1164	31	35	37	109	106	103	29	81	80	79
PANAMA AND CANAL ZONE												
Panama City	8 58N/ 79 33W	21	69	72	73	93	92	91	18	81	81	80

Location	Lat/Long	Elev										
PAPUA NEW GUINEA												
Port Moresby	9 29S/147 09E	126	62	67	69	92	91	90	14	80	80	79
PARAGUAY												
Asuncion	25 17S/ 57 30W	456	35	43	46	100	98	96	24	81	81	80
PERU												
Lima	12 05S/ 77 03W	394	51	53	55	85	85	84	17	76	75	74
PHILIPPINES												
Manila	14 35N/120 59E	47	69	73	74	94	92	91	20	82	81	81
POLAND												
Krakow	50 04N/ 19 57E	723	−2	2	6	84	81	78	19	68	67	66
Warsaw	52 13N/ 21 02E	394	−3	3	8	84	81	78	19	71	70	68
PORTUGAL												
Lisbon	38 43N/ 9 08W	313	32	37	39	89	86	83	16	69	68	67
PUERTO RICO												
San Juan	18 29N/ 66 07W	82	65	67	68	89	88	87	11	81	80	79
RUMANIA												
Bucharest	44 25N/ 26 06E	269	−2	3	8	93	91	89	26	72	71	70
SAUDI ARABIA												
Dhahran	26 17N/ 50 09E	80	39	45	48	111	110	108	32	86	85	84
Jedda	21 28N/ 39 10E	20	52	57	60	106	103	100	22	85	84	83
Riyadh	24 39N/ 46 42E	1938	29	37	40	110	108	106	32	78	77	76
SENEGAL												
Dakar	14 42N/ 17 29W	131	58	61	62	55	93	91	13	81	80	80
SINGAPORE												
Singapore	1 18N/103 50E	33	69	71	72	92	91	90	14	82	81	80
SOMALIA												
Mogadiscio	2 02N/ 49 19E	39	67	69	70	91	90	89	12	82	82	81
SOUTH AFRICA												
Capetown	33 56S/ 18 29E	55	36	40	42	93	90	86	20	72	71	70
Johannesburg	26 11S/ 28 03E	5463	26	31	34	85	83	81	24	70	69	69
Pretoria	25 45S/ 28 14E	4491	27	32	35	90	87	85	23	70	69	68

TABLE 13-3 Climatic Conditions for Other Countries (Continued)

Col. 1 Country and Station	Col. 2 Latitude and Longitude		Col. 3 Elevation, Ft	Winter, F Col. 4			Summer, F Col. 5 Design Dry-Bulb			Col. 6 Outdoor Daily Range F deg	Col. 7 Design Wet-Bulb		
				Mean of Annual Extremes	99%	97½%	1%	2½%	5%		1%	2½%	5%
SOUTH YEMEN													
Aden	12 50N/	45 02E	10	63	68	70	102	100	98	11	83	82	82
SOVIET UNION													
Alma Ata	43 14N/	76 53E	2543	−18	−10	−6	88	86	83	21	69	68	67
Archangel	64 33N/	40 32E	22	−29	−23	−18	75	71	68	13	60	58	57
Kaliningrad	54 43N/	20 30E	23	−3	−1	−6	83	80	77	17	67	66	65
Krasnoyarsk	56 01N/	92 57E	498	−41	−23	−27	84	80	76	12	64	62	60
Kiev	50 27N/	30 30E	600	−12	−5	−1	87	84	81	22	69	68	67
Kharkov	50 00N/	36 14E	472	−19	−10	3	87	84	82	23	69	68	67
Kuibyshev	53 11N/	50 06E	190	−23	−19	−13	89	85	81	20	65	64	63
Leningrad	59 56N/	30 16E	16	−14	−9	−5	78	75	72	15	67	66	65
Minsk	53 54N/	27 33E	738	−19	−11	1	80	77	74	16	69	67	65
Moscow	55 46N/	37 40E	505	−19	−11	1	84	81	78	21	70	69	68
Odessa	46 29N/	30 44E	214	−1	4	8	87	84	82	14	70	69	68
Petropavlovsk	52 53N/	158 42E	286	−9	−3	0	70	68	65	13	58	57	56
Rostov on Don	47 13N/	39 43E	159	−9	−2	4	90	87	84	20	70	69	68
Sverdlovsk	56 49N/	60 38E	894	−34	−25	−20	80	76	72	16	63	62	60
Tashkent	41 20N/	69 18E	1569	−4	−3	8	95	93	90	29	71	70	69
Tbilisi	41 43N/	44 48E	1325	12	18	22	87	85	83	18	68	67	66
Vladivostok	43 07N/	131 55E	94	−15	−10	−7	80	77	74	11	70	69	68
Volgograd	48 42N/	44 31E	136	−21	−13	−7	93	89	86	19	71	70	69
SPAIN													
Barcelona	41 24N/	2 09E	312	31	33	36	88	86	84	13	75	74	73
Madrid	40 25N/	3 41W	2188	22	25	28	93	91	89	25	71	69	67
Valencia	39 28N/	0 23W	79	31	33	37	92	90	88	14	75	74	73
SRI LANKA													
Colombo	6 54N/	79 52E	24	65	69	70	90	89	88	15	81	80	80
SUDAN													
Khartoum	15 37N/	32 33E	1279	47	53	56	109	107	104	30	77	76	75
SURINAM													
Paramaribo	5 49N/	55 09W	12	66	68	70	93	92	90	18	82	82	81
SWEDEN													
Stockholm	59 21N/	18 04E	146	3	5	8	78	74	72	15	64	62	60
SWITZERLAND													
Zurich	47 23N/	8 33E	1617	4	9	14	84	81	78	21	68	67	66
SYRIA													
Damascus	33 30N/	36 20E	2362	25	29	32	102	100	98	35	72	71	70
TAIWAN													
Tainan	22 57N/	120 12E	70	40	46	49	92	91	90	14	84	83	82
Taipei	25 02N/	121 31E	30	41	44	47	94	92	90	16	83	82	81

Location	Lat / Long	Elev										
TANZANIA												
Dar es Salaam	6 50S/ 39 18E	47	62	64	65	90	89	88	13	82	81	81
THAILAND												
Bangkok	13 44N/100 30E	39	57	61	63	9?	95	93	18	82	82	81
TRINIDAD												
Port of Spain	10 40N/ 61 31W	67	61	64	66	9?	90	89	16	80	80	79
TUNISIA												
Tunis	36 47N/ 10 12E	217	35	39	41	102	99	96	22	77	76	74
TURKEY												
Adana	36 59N/ 35 18E	82	25	33	35	100	97	95	22	79	78	77
Ankara	39 57N/ 32 53E	2825	2	9	12	94	92	89	28	68	67	66
Istanbul	40 58N/ 28 50E	59	23	28	30	9?	88	86	16	75	74	73
Izmir	38 26N/ 27 10E	16	24	27	29	98	96	94	23	75	74	73
UNITED KINGDOM												
Belfast	54 36N/ 5 55W	24	19	23	26	74	72	69	16	65	64	62
Birmingham	52 29N/ 1 56W	535	21	24	27	79	76	73	15	66	64	63
Cardiff	51 28N/ 3 10W	203	21	24	27	79	76	73	14	64	63	62
Edinburgh	55 55N/ 3 11W	441	22	25	28	73	70	68	13	64	62	61
Glasgow	55 52N/ 4 17W	85	17	21	24	74	71	68	13	63	63	61
London	51 29N/ 0 00	149	20	24	26	82	79	76	16	68	66	65
URUGUAY												
Montevideo	34 51S/ 56 13W	72	34	37	39	90	88	85	21	73	72	71
VENEZUELA												
Caracas	10 30N/ 66 56W	3418	49	52	54	84	83	81	21	70	69	69
Maracaibo	10 39N/ 71 36W	20	69	72	73	97	96	95	17	84	83	83
VIETNAM												
Da Nang	16 04N/108 13E	23	56	60	62	97	95	93	14	86	86	85
Hanoi	21 02N/105 52E	53	46	50	53	93	97	95	16	85	86	84
Ho Chi Minh City (Saigon)	10 47N/106 42E	30	62	65	67	93	91	89	16	85	84	83
YUGOSLAVIA												
Belgrade	44 48N/ 20 28E	453	4	9	13	92	89	86	23	74	73	72
ZAIRE												
Kinshasa (Leopoldville)	4 20S/ 15 18E	1066	54	60	62	92	91	90	19	81	80	80
Kisangani (Stanleyville)	0 26S/ 15 14E	1370	65	67	68	92	91	90	19	81	80	80

SOURCE: *ASHRAE Handbook—1985 Fundamentals*. Reprinted by permission.

13-1   Weather Extremes in Canada and the United States*

1. U.S. lowest annual mean temperature [−13°C (9°F)] and coolest summer average [2°C (36°F)]: Barrow, AK

2. U.S. lowest temperature [−62°C (−80°F)]: Prospect Creek, Endicott Mts., AK

3. Alaska's greatest snowfall in 24 h [157.5 cm (62 in)], in one storm [445.5 cm (175 in)], and in one season [2475 cm (974.5 in)]: all at Thompson Pass, AK

4. Canada's greatest snowfall in a climatological day [118 cm (46 in)]: Lakelse Lake, British Columbia

5. North America's greatest average yearly precipitation [650 cm (256 in)]: Henderson Lake, British Columbia, Canada

6. Canada's greatest 24-h rainfall [49 cm (19 in)]: Ucluelet Brynnor Mines, British Columbia

7. U.S. West Coast's foggiest place (average 2552 h/yr): Cape Disappointment, WA

8. North America's greatest snowfall in one season [2850 cm (1122 in)]: Rainier Paradise Ranger Station, WA

9. North America's greatest snowfall in one storm [480 cm (189 in)]: Mt. Shasta Ski Bowl, CA

10. North America's greatest depth of snow on the ground [1145.5 cm (451 in)]: Tamarack, CA

11. Western Hemisphere's highest temperature [57°C (134°F)] and highest summer average [37°C (98°F)]; U.S. highest annual mean temperature [26°C (78°F)] and lowest average yearly precipitation [4.1 cm (1.63 in)]: all at Death Valley, CA

12. U.S. longest dry period (767 days): Bagdad, CA

* Adapted from material appearing in Pauline Riordan and Paul G. Bourget, *World Weather Extremes,* compiled by the Geographic Sciences Laboratory, U.S. Army Engineer Topographic Laboratories, Ft. Belvoir, VA, December 1985. Used with permission.

13. U.S. coldest winter average temperature [–26.5°C (–16°F)]: Barter Island, AK

14. North America's lowest temperature, excluding Greenland [–63°C (–81°F)]: Snag, Yukon Territory, Canada

15. North America's highest sea-level air pressure [106.76 kPa (31.53 in)]: Mayo, Yukon Territory, Canada

16. Canada's greatest snowfall in one season [2446.5 cm (964 in)]: Revelstoke, Mt. Copeland, British Columbia

17. U.S. largest 24-h temperature fall [56°C (100°F)]: Browning, MT

18. U.S. lowest temperature, excluding Alaska [–56.5°C (–70°F)]: Rogers Pass, MT

19. North America's greatest 24-h snowfall [192.5 cm (76 in)]: Silver Lake, CO

20. Canada's heaviest hailstone [290 g (10.23 oz)]: Cedoux, Saskatchewan

21. Canada's highest temperature [45°C (113°F)]: Midale and Yellow Grass, Saskatchewan

22. U.S. largest 2-min temperature rise [27°C (49°F)]: Spearfish, SD

23. Three temperature rises and two falls of 22°C (40°F) or over during a 3-h, 10-min period: Rapid City, SD

24. World's greatest 42-min rainfall [30.5 cm (12 in)]: Holt, MO

25. U.S. largest hailstone circumference [44.5 cm (17.5 in)]: Coffeyville, KS

26. A 24-h rainfall of 109 cm (43 in): Alvin, TX

27. North America's lowest mean temperature for a month, excluding Greenland [–48°C (–54°C)], and Canada's lowest annual mean temperature [–19°C (–3°F)]: both at Eureka, Northwest Territories

28. Canada's highest average annual wind speed [36 km/h (22 mi/h)]: Cape Warwick, Resolution Island, Northwest Territories

29. Canada's highest maximum observed hourly wind speed [201 km/h (125 mi/h)]: Cape Hopes Advance, Quebec

30. U.S. East Coast's foggiest place (average 1580 h/yr): Moose Peak Lighthouse, Mistake Island, ME

31. World's highest surface wind peak gust [372 km/h (231 mi/h)] and 5-min wind speed [303 km/h (188 mi/h)]; U.S. highest average annual wind speed [56 km/h (35 mi/h)]: Mt. Washington, NH; Mt. Washington also had a mean wind speed for 24 h of 206 km/h (128 mi/h) and a mean wind speed for a month of 112 km/h (70 mi/h)

32. World's greatest 1-min rainfall [3.1 cm (1.23 in)]: Unionville, MD

33. North America's lowest sea-level air pressure [89.23 kPa (26.35 in)]: Matecumbe Key, FL

34. World's greatest average yearly precipitation [1168 cm (460 in)]: Mt. Waialeale, Kauai, HI

35. U.S. warmest winter average temperature [23°C (73°F)]: Honolulu, HI

36. U.S. greatest rainfall in 12 months [1878 cm (739 in)]: Kukui, Maui, HI

13-2   World Weather Extremes*

1. North America's lowest temperature (excluding Greenland) [−63°C (−81°F)]: Snag, Yukon Territory, Canada

2. North America's greatest snowfall in one season [2850 cm (1122 in)]: Rainier Paradise Ranger Station, WA

3. North America's greatest snowfall in one storm [480 cm (189 in)]: Mt. Shasta Ski Bowl, CA

---

* Adapted from material appearing in Pauline Riordan and Paul G. Bourget, *World Weather Extremes,* compiled by the Geographic Sciences Laboratory, U.S. Army Engineer Topographic Laboratories, Ft. Belvoir, VA, December 1985. Used with permission.

4. North America's greatest depth of snow on the ground [1145.5 cm (451 in)]: Tamarack, CA

5. Western Hemisphere's highest temperature [57°C (134°F)] and hottest summer average [37°C (98°F)]: both at Death Valley, CA

6. North America's lowest average yearly precipitation [3 cm (1.2 in)]: Bataques, Mexico

7. World's greatest average yearly precipitation [1168 cm (460 in)]: Mt. Waialeale, Kauai, HI

8. North America's lowest mean temperature for a month (excluding Greenland) [−48°C (−54°F)]: Eureka, Northwest Territories, Canada

9. North America's greatest average yearly precipitation [650 cm (256 in)]: Henderson Lake, British Columbia, Canada

10. U.S. largest 24-h temperature fall [56°C (100°F)]: Browning, MT

11. U.S. largest 2-min temperature rise [27°C (49°F)]: Spearfish, SD

12. North America's greatest 24-h snowfall [192.5 cm (76 in)]: Silver Lake, CO

13. World's highest surface wind peak gust [372 km/h (231 mi/h)] and 5-min wind speed [303 km/h (188 mi/h)]: Mt. Washington, NH; Mt. Washington also had a mean wind speed for 24 h of 206 km/h (128 mi/h) and a mean wind speed for a month of 112 km/h (70 mi/h)

14. World's greatest 1-min rainfall [3.1 cm (1.23 in)]: Unionville, MD

15. World's greatest 42-min rainfall [30.5 cm (12 in)]: Holt, MO

16. U.S. largest hailstone circumference [44.5 cm (17.5 in)]: Coffeyville, KS

17. 24-h rainfall of 109 cm (43 in) (possibly the world's greatest on flat terrain): Alvin, TX

18. Thule, Greenland, had a peak wind gust of 333 km/h (207 mi/h)

19. Greenland's lowest temperature [–66°C (–87°F)]: Northice

20. South America's greatest average yearly precipitation [899 cm (354 in)]: Quibdo, Colombia

21. World's lowest average yearly precipitation [0.08 cm (0.03 in)] and no rain for longer than 14 consecutive years: Arica, Chile

22. South America's highest temperature [49°C (120°F)]: Rivadavia, Argentina

23. South America's lowest temperature [–33°C (–27°F)]: Sarmiento, Argentina

24. Bahia Felix, Chile, averages 325 days/yr with rain

25. Bessans, France, had a snowfall of 172 cm (68 in) in 19 h

26. Europe's highest temperature [50°C (122°F)]: Seville, Spain

27. Europe's greatest average yearly precipitation [465 cm (183 in)]: Crkvice, Yugoslavia

28. World's greatest 20-min rainfall [20.5 cm (8.10 in)]: Curtea-de-Arges, Romania

29. Europe's lowest temperature [–55°C (–67°F)]: Ust'Shchugor, U.S.S.R.

30. Europe's lowest average yearly precipitation [16 cm (6.4 in)]: Astrakhan, U.S.S.R.

31. World's highest sea-level air pressure [108.38 kPa (32.01 in)]: Agata, U.S.S.R.

32. Africa's lowest temperature [–24°C (–11°F)]: Ifrane, Morocco

33. World's highest temperature [58°C (136°F)]: El Azizia, Libya

34. Africa's greatest average yearly precipitation [1029 cm (405 in)]: Debundscha, Cameroon; Debundscha, Cameroon, also has a 191-cm (75-in) average variability of annual precipitation

35. Africa's lowest average yearly precipitation [<0.25 cm (<0.1 in)]: Wadi Halfa, Sudan

36. Dallol, Ethiopia, has an annual mean temperature of 35°C (94°F) (possibly the world's highest)

37. Assab, Ethiopia, has 29°C (84°F) average afternoon dewpoint in June

38. Kampala, Uganda, averages 242 days/yr with thunderstorms

39. World's greatest 12-h rainfall [135 cm (53 in)]: Belouve, La Réunion Island

40. World's greatest 24-h rainfall [188 cm (74 in)] and 5-day rainfall [386 cm (152 in)]: Cilaos, La Réunion Island

41. Themed, Israel, has a 94 percent variability of annual precipitation

42. Asia's highest temperature [54°C (129°F)]: Tirat Tsvi, Israel

43. Persian Gulf had a 36°C (96°F) sea-surface temperature

44. Asia's lowest average yearly precipitation [4.6 cm) (1.8 in)]: Aden, South Yemen

45. Verkhoyansk, U.S.S.R., has a difference of 102°C (183°F) between lowest and highest recorded temperatures

46. Northern hemisphere's lowest temperature [–68°C (–90°F)]: Verkhoyansk and Oimekon, U.S.S.R.

47. Eastern Sayan Region of U.S.S.R. has 81°C (146°F) mean annual temperature range

48. Lhasa, Tibet, had a 100 percent relative variability of annual precipitation

49. World's greatest 12-month rainfall [2647 cm (1042 in)] and 1-month rainfall [930 cm (366 in)]; Asia's greatest average yearly precipitation [1143 cm (450 in)]: Cherrapunji, India

50. Miyakojima Island, Ryukyu Islands, had a peak gust of 306 km/h (190 mi/h)

51. Northern hemisphere's greatest 24-h rainfall [125 cm (49 in)]: Paishih, Taiwan

52. World's lowest sea-level air pressure [87.00 kPa (25.69 in)] (estimated by dropsonde in eye of typhoon tip): in area of 17°N 138°E

53. Bogor, Indonesia, averaged 322 days/yr with thunderstorms

54. Marble Bar, Western Australia, had temperatures of 38°C (100°F) or above on 162 consecutive days

55. Australia's highest temperature [53°C (128°F)]: Cloncurry, Queensland

56. Australia's lowest average yearly precipitation [10 cm (4.05 in)]: Mulka, South Australia

57. Australia's greatest average yearly precipitation [455 cm (179 in)]: Tully, Queensland

58. Australia's greatest 24-h rainfall [91 cm (36 in)]: Crohamhurst, Queensland

59. Australia's lowest temperature [–22°C (–8°F)]: Charlotte Pass, New South Wales

60. South Pole's highest temperature: –14°C (7.5°F)

61. South Pole has 463 W/m^2 (955 ly) average daily insolation in December

62. Plateau Station, Antarctica, had a mean temperature for a month of –73°C (–100°F) and an annual mean temperature of –57°C (–70°F)

63. World's lowest temperature [–89°C (–129°F)]: Vostok, Antarctica

64. Port Martin, Antarctica, had a mean wind speed for 24 h of 174 km/h (108 mi/h) and a mean wind speed for a month of 105 km/h (65 mi/h)

65. Vanda Station, Antarctica, had a maximum temperature of 15°C (59°F) (possibly Antarctica's highest)

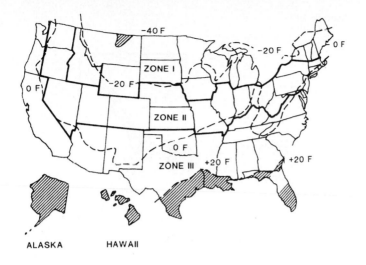

ALASKA   HAWAII

Zones include areas with design temperatures about as follows: Zone I, −20°F and lower; Zone II, 0 to −20°F; and Zone III, above 0°F. Cross-hatched areas are outside Zones I and III.

FIGURE 13-1 Condensation zones in the United States. (*Source:* ASHRAE Handbook—1989 Fundamentals. *Reprinted by permission.*)

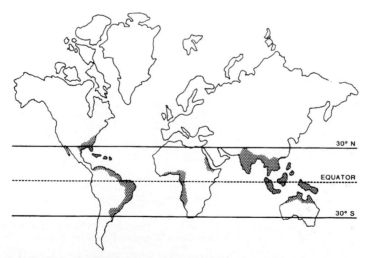

FIGURE 13-2 Humid climates generally occur in latitudes 30°S to 30°N in low coastal regions. (*Source:* ASHRAE Handbook—1989 Fundamentals. *Reprinted by permission.*)

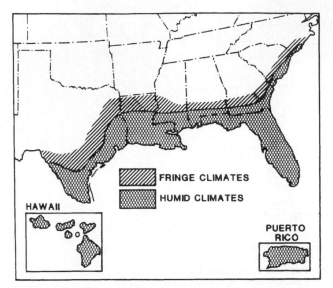

FIGURE 13-3 Humid climates in the continental United States.
(*Source:* ASHRAE Handbook—1989 Fundamentals. *Reprinted
by permission.*)

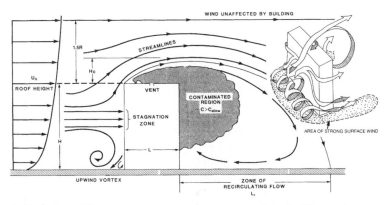

FIGURE 13-4 Flow patterns around a rectangular building. (*Source:*
ASHRAE Handbook—1989 Fundamentals. *Reprinted by permission.*)

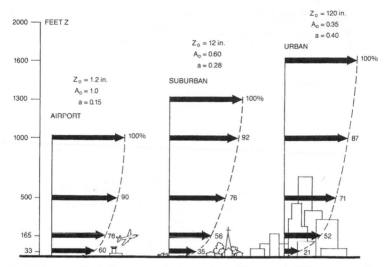

FIGURE 13-5 Typical mean wind speed profiles over different terrain roughness. (*Source:* ASHRAE Handbook—1989 Fundamentals. *Reprinted by permission.*)

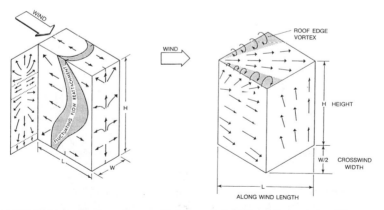

FIGURE 13-6 Surface flow patterns and building dimensions. (*Source:* ASHRAE Handbook—1989 Fundamentals. *Reprinted by permission.*)

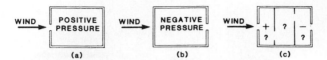

Pressures in Buildings Resulting from Wind.

(a) With upstream opening only, pressure is positive.
(b) With downstream opening only, pressure is Negative.
(c) Pressures are as shown if openings are equal in shape and area. With unequal openings, pressures can be either negative or positive in each space, depending on relative areas of openings.

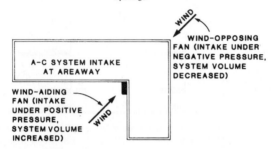

FIGURE 13-7 Sensitivity of system volume to locations of building openings, intakes, and exhausts. (*Source:* ASHRAE Handbook—1989 Fundamentals. *Reprinted by permission.*)

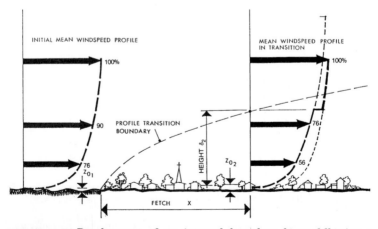

FIGURE 13-8 Development of an internal boundary layer following a change in terrain roughness. (*Source:* ASHRAE Handbook—1989 Fundamentals. *Reprinted by permission.*)

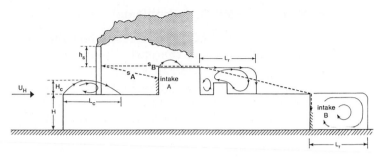

FIGURE 13-9 Flow recirculation regions and exhaust to intake stretched string distances. (*Source:* ASHRAE Handbook—1989 Fundamentals. *Reprinted by permission.*)

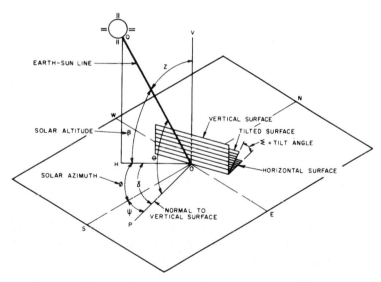

FIGURE 13-10 Solar angles for vertical and horizontal surfaces. (*Source:* ASHRAE Handbook—1989 Fundamentals. *Reprinted by permission.*)

13-3 Roof-Snow and Ice-Dam Problems*

The intent of this publication is to awaken public interest in the behavior of roof-snow and the part it plays in ice-dam formation. Each year, ice-dams cause millions of dollars in damage to homes located in the "snow-belt." Yet there is little understanding among homeowners, builders, and suppliers as to the cause of these ice-dams and the remedial measures that can be undertaken to prevent further damages to residential units.

Many areas and facets of roof-snow could not be included in this publication. Further in-depth investigation by better equipped technicians is recommended. The interpretations presented are generally based upon first-hand observations; perhaps higher levels of technical sophistication will be needed to convince reluctant building interests that they are clearly "involved" in the problems of roof-snow behavior. Satisfactory servicing of ice-dammed buildings requires the enthusiastic support of responsible and knowledgeable suppliers of both materials and services.

In general, ice-dams are formed when attic heat moves upward to warm the roof and melt roof-snow at or near ridge areas. Melting of snow occurs at the snow-shingle interface and runs downward (under the snow) as snow-water. At or near the edge of the roof, colder conditions exist that usually result in the freezing of the snow-water, thus forming the ice-dam. Subsequent melting of roof-snow usually accumulates as a pocket of snow-water that eventually backs up under the shingles to cause major damage in the plateline area. This damage can appear in the form of soaked (inefficient) insulation; stained, cracked, and spalled plaster or sheetrock; damp, odorous, and rotting wall cavities; and stained, blistered, and peeling wall paint,

---

* The text material and illustrations in this and the following subsections were originally prepared by Howard L. Grange and Lewis T. Hendricks and published under the title *Roof-Snow Behavior and Ice-Dam Prevention in Residential Housing,* by the Agriculture Extension Service, University of Minnesota (Extension Bulletin 399-1976). Permission has been granted by Minnesota Extension Services—University of Minnesota for its use in this sourcebook.

both inside and outside the house. Insulation alone will not solve the ice-dam problem nor will contemporary ventilation techniques. Other commonly used techniques such as heating cables and the removal of snow at roof edges are of little value in combating ice-dam problems. Rather, it is the proper use of insulation and ventilation in conjunction with correct house design that offers the best solutions to the problem.

As a final note, the authors do not intend this report to be used as a "book of instructions" or as a maintenance guide. Improvisations by do-it-yourselfers or untrained installers will most likely result in blunders, disappointments, and disillusionment in the bright promises of the observations, comments, and products mentioned in this report.

The explanations, interpretations, and suggestions of this report are founded on a 10-year study of actual roofs and visible roof-snow behavior in several thousand observations as well as color slide documentations in 36 states and 5 Canadian provinces.

Visible evidence everywhere has indicated the frustrations of building owners and the inability, or unwillingness, of building professionals to adequately cope with extensive damages caused by eave ice-dams and blocked snow-water penetrations.

These observations can be verified by streetside observation during almost every winter in the Twin Cities in late February, three winters out of four in Chicago, and at least occasional winters throughout latitudes as far south as the Ohio River.

Although snowloading of structures has been well recognized and documented in architectural texts and structural design data, few references (13, 3, 6, 7) to roof-snow melting and ice-dam behavior have been found in technical literature. Hopefully, this report will bridge a sadly neglected informational gap and also stimulate homeowner response and action to eliminate this needless, but widespread and devastating, fault of building design and performance.

13-4   Recognition of Ice-Dams

Our investigations have revealed that many extensively involved and troubled homeowners are reluctant to admit (and discuss) *their individual* problems with ice-dams. The most common response to our question has been, *"What are ice-dams?"*

Figure 13-11 shows late winter snow on a roof, a bulge of ice attached to the eave, with numerous icicles that may drip from it on warmer days, and stained siding that may be optimistically written off as *"a bit of rust stain from the overloaded rain gutters."* Similarly, a wetted interior ceiling and wall are hopefully nothing more than a *"bad place in the shingles that we will fix next spring."* Perhaps this is true, but the attacks are more likely to be as indicated in Fig. 13-12 revealing the "actions" inside the cornices and attic spaces.

The ice-loaded eave, the puffed snow-blanket, and the dangling icicles of Fig. 13-11 may be all that is apparent; but inside the cornices, within the attic spaces, and within the wall cavity, the destructive actions—seldom detected by

FIGURE 13-11   Typical snow-country eave in late winter.

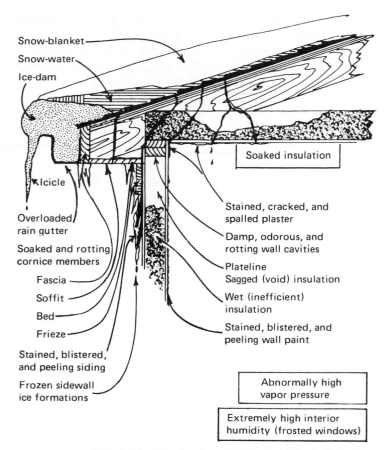

Snow-blanket

Snow-water

Ice-dam

Icicle

Overloaded rain gutter

Soaked and rotting cornice members

Fascia

Soffit

Bed

Frieze

Stained, blistered, and peeling siding

Frozen sidewall ice formations

Soaked insulation

Stained, cracked, and spalled plaster

Damp, odorous, and rotting wall cavities

Plateline
Sagged (void) insulation

Wet (inefficient) insulation

Stained, blistered, and peeling wall paint

Abnormally high vapor pressure

Extremely high interior humidity (frosted windows)

FIGURE 13-12 This sketch of the ice-dam problem identifies both the ice-dam and its damages. Of course, all the damages illustrated may (or may not) occur at any one instance. The damages illustrated here are far more common and costly than is generally acknowledged.

direct observation—are illustrated for easy recognition in Fig. 13-12. The existence of these hidden actions is occasionally betrayed on the exterior by wetted walls, by massive wall ice formations, by wall stains, and subsequently—usually later in the spring—by blistered and peeling paint.

Inside the house, the distraught housewife may not identify ice-blocked snow-water as the source of wetted ceiling and walls, of stained and peeling paint or blistered, cracked, and spalling plaster, and perhaps the offensive, yet difficult to locate, odor of mildewed and rotting wood.

Unseen, *and detectable only indirectly,* are the damages in the wall, ceiling, and cornice cavities, such as soaked, sagging inefficient insulation; mildewed and rotting wood structural members; rust-marked, stain-oozing degraded sidewalls; and corrosion of nails, flashings, and fasteners. Hidden cavities are also slow to dry out after the roof-snow season. Such severely soaked insulation and wall cavities contribute substantially to house moisture and to damages associated with excessive vapor pressures in residential interiors.

13-5   Occurrence of Roof-Snow and Ice-Dam Problems

Ice-dams develop on buildings whenever and wherever roof-snow accumulates on roofs to depths of an inch or two and when the weather turns or remains generally below freezing for 3 or 4 days.

No precise limits of snow-depth and temperature conditions have been determined, nor would such specific limits be significant; wide variations in the range of roof-snow depth and temperature depend on the whims of nature, the direction and velocity of winds, and the shape of the roof.

In general, the deeper the snow, either by single incident snowfall or by accumulations, and the lower (and more persistent) the subfreezing temperatures, the more inevitable and sizeable are the ice formations for typical houses.

A 5-year study (1) of winter conditions in 25 locations sought to develop an index of snow-cold severity for comparisons of relative ice-dam susceptibility in such diverse locations as *Sault Ste. Marie,* Michigan (Ontario); *Minneapolis,* Minn.; *Madison,* Wis.; *Chicago,* Ill.; *Pittsburgh,* Penn.; and *Lexington,* Ky. Snowfalls and roof-damages, as anticipated, proved generally more severe, frequent, and enduring with increasing north latitude locations. Other climatic factors may supersede latitude.

Wind direction and velocity may free some roofs of snow or selectively clear certain planes and drift deeply on another plane or in a susceptible valley.

Heavy snowstorms characteristically move as low pressure frontal systems across the midwest in approximately southwesterly to northeasterly directions. Such blizzard-type winds blow moisture in from the southeast, shifting to the northeast and to heavy snow as the storm front moves in. Snowfalls diminish as winds shift to the north and northwest as the front passes and cold-temperature air follows. Thus, deepest snow and drifts occur normally on northerly and westerly roof planes and valleys. Such selective orientation of deepest snow helps explain why snow depths and snow-water damages are generally greater on northerly and westerly roof planes and valleys. Wind and complex roof-structures may confuse observers of roof-snow actions unless wind factors are carefully recognized and considered.

Farm homes and buildings with wind-exposed roofs are less prone to roof-snow problems. Clearly, roof planes freed of snow will avoid ice-dams and snow-water difficulties.

Affluent suburban homes where trees have been carefully preserved or where trees have been densely re-established have a much higher incidence and severity of eave-ice damming than do rural or windswept (treeless) new developments. High winds, or the lack of winds, that may or may not blow snow from roofs are key factors in the relative depths and damages of roof-snow.

Within metropolitan areas, there may be startling variations in ice-dam susceptibility from dense residential areas to the windswept, thinly built-up, outlying areas.

There are striking local variations, such as Chicago's lakefront in relation to its western suburbs. The moderating effect of large bodies of water distorts the norms of latitude, and the greater incidence of snow in South Bend's Chicago-polluted air indicates need for further study of the snow/cold variables that lead to roof-snow and snow-water damages. Snow density, as it affects both roof loads and snow-water quantity, has received little attention and consideration.

The snow/cold conditions for severe roof-snow difficulties vary little from year to year for the Soo and the Twin Cities, where coping with snow and cold is accepted as part of life. In Madison and Chicago latitudes, severe snow/cold may occur in only 2 years out of 5, and, in more southerly latitudes, damaging roof-snow may occur only once in several years. Curiously, homeowner attitudes appear to be optimistically myopic in locations where chances of ice-dammed snow-water attacks are less than 50 percent; this is sadly shortsighted, because one ice-dammed snow-water attack in a 10-year period can be devastating.

13-6   Identification of Snow-Water Damages

Icicles

Icicles, sparkling in the sun of midwinter days, are only indicators of the interactions of roof-snow, the efficiency of ceiling insulation, and the effects of warmed attic spaces. Decorative icicles hanging beyond the eaves are solidified snow-water that is out and over the dams and are hazardous only to those who carelessly knock them down.

Roof-ice glacier actions

The visible troublemaker on snow-decked buildings is the massive ice formed from successive flows of roof snow-water freezing as it emerges from the protective insulating blanket of roof-snow. The disruptive forces of expanding ice-actions may damage roof shingles and the roof-deck. The glacierlike forces on roofs and in valleys deserve better consideration by building designers.

Rain gutters

Rain gutters (eaves troughs) are often condemned as responsible for ice-dammed eaves. Ice-filled rain gutters may contribute to concentrated overloads at the eaves. Massive ice in (and around) eaves troughs inevitably slows melting and interferes with desirable roof drainage and runoffs; however, rain gutters, like icicles, are simply symptomatic of snow-water actions, rather than causative.

Snow-water penetrations

The deceptive offender on winter roofs is snow-water. Its many actions are disguised in successive transformations. It develops from snow-crystals of varying densities into snow-water, and it may transform on cooling into massive ice-dams that block further runoff at the eave. Roof-snow can sporadically alter its water-ice character with thermal variations in atmosphere and roof-deck thermal relationships. Snow-water deceptions have been inadequately recognized and confronted.

Snow-water blocked by ice-dams may develop to depths sufficient to penetrate cracks and flaws in a deck that was not built to resist water attack from below and beneath it. Alternate freezing and thawing increases both the incidence and the magnitude of roof-deck flaws. Snow-water penetration into the structure may follow paths and channels difficult to analyze as the source of sidewall soaking. Penetrating snow-water often seeps down rafters, along plates, into wall cavities, and saturates sidewall insulation to reduce its efficiency, passes into and through sheathing-siding as either water or vapor, and may stain or blister the thin skins of paint to effect a thoroughly undesirable appearance and a lowered protective performance. The ultimate damage through rotting of structure by successive waves of snow-water attacks may destroy the building, even when cosmetically covered with vapor and water-impervious materials.

The amount of sidewall moisture originating from snow-water is impossible to determine because most of the evidence is hidden initially by the snow-cover and, after penetration, by the roof and sidewall coverings. Actions within the sidewall cavities can be deduced from the sometimes visible wetting and from the occasional frozen sidewall-icicle, the developing stains, and the subsequent paint blisters. Winter snow-water seepages may be difficult to identify, but the musty odor of wet and rotting wall and cornice members may permeate the house during summer. Massive sidewall soaking may saturate cavity insulations and retain moist and rotting conditions for months.

Frozen sidewall leaks

Visible evidence of the penetration of ice-dammed snow-water into attic-warmed cornices is not rare in the upper

Midwest in December through March. Intensely cold weather may freeze the cornice leaks in a stop-action sidewall ice-formation, as displayed in Fig. 13-13. The visible ice may be apparent for only a few days. With even slight warming, the frozen action may melt to an almost invisible flow. Similarly, ice-blocked snow-water undoubtedly also flows on the interior of the wall cavity where its detection is difficult; evidence of its presence may subsequently develop as stain, blistering, and peeling of the painted siding.

Sidewall degradation

Snow-water penetrations through roof-decks, into cornices, and seeping generally within and without the side wall cause not only paint failures, but general degradation of the structure. Insulation wetted by water seepages loses most of its efficiency with consequent increases in snow-melting thermals in the attics and cornices.

FIGURE 13-13   Frozen sidewall leaks emanating from ice-blocked eave.

Plateline thermals

Perhaps the most neglected heat losses of well-insulated modern houses occur at the exterior platelines. Sagging side-wall insulation—due to settling, wetting, or careless installation—adds to the overlooked heat losses in the critical plateline region. Such heat losses further aggravate snow-water penetrations by supporting the liquid ice-dammed pool in a most vulnerable position for penetration damage on the roof immediately above the eave platelines. The conditions promote snow-water seepage during extended periods of cold/snowy winters to soak insulation and wall cavities. Complete drying is extremely sluggish, and the rot-promoting conditions may persist for several months.

Other damages

A reexamination of the damaging effects shown in Fig. 13-12 will identify and explain the damages of ice-dammed snow-water far better than words. All damages illustrated have been observed and recorded in a series of private audio-visual reports (2). These audio-visual reports form much of the data substantiating these comments. The dank and rotting odor of some of the soaked structures is the one feature that cannot be described or sensed with the audio-visual data.

13-7 Homeowner Confrontations with Roof-Snow Problems

Expert advice and the state of the art

With periodic regularity, matched only by falling leaves and snow, newspaper and magazine articles offer warnings, tips, and counsel for homeowners faced with the annual siege of ice-dammed snow-water on roofs. With good intentions, but with a deplorable revelation of the inadequacies of the art, "expert advisors" are prone to contribute to the problem rather than to its solution.

Roof-snow removals

The most common, and apparently reasonable, advice is to shovel or rake the snow from the roof, since with no snow, there will be no snow-water, or ice-dam, and, ostensibly, no problems. However, homeowner reactions to such news article suggestions often develop even more frustrating side effects than the predictable hazards of falling bodies and broken bones following each snowfall.

Partial roof-snow removals

Most homeowners are seldom equipped with sufficient incentive and enthusiasm to achieve more than partial snow removal. Equipment, energy, dedication, and the many required roof-snow removals appear to limit the effort to a short reach uproof. Such partial snow removals leave substantial snow-blankets on the upper part of the roof slope.

With only partial removals, uproof snow will melt and run downslope to emerge from the protective snow-blanket, freeze, and build into an ice-dam that blocks snow-water uproof in what may be a more vulnerable leak position than at the eaves (Figs. 13-14 and 13-15). Such secondary ice-dams—oftentimes multiple uproof ice-dams—are visible to observers who take the time and winter-cold discomfort to examine the reaction of homeowners to the "shovel-it-off" advice of newspaper authorities.

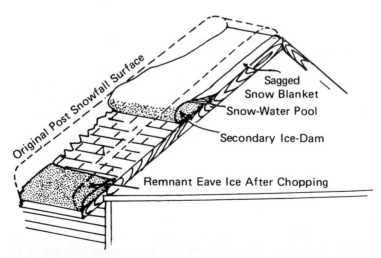

FIGURE 13-14   Partial roof-snow removals develop secondary ice-dams.

FIGURE 13-15   Secondary dam developed at snow-blanket terminus at the limit of partial snow-removal reach.

Complete removal of roof-snow, of course, prevents ice formations and snow-water damages. In many cities that are annually plagued with roof-snow the yellow pages of the phone directory list firms pleased to be paid to remove each successive snowfall. The recurrent expense discourages many homeowners. Many budget-strapped and ill-advised homeowners attack the roof-snow and eave-ice in a desperate do-it-yourself assault of shoveling, chipping, hosing, picking, chiseling, and (believe it or not) blow-torching to rid their eaves of ice-dammed snow-water.

Neither professionals, who are seldom reported falling from the treacherous roof planes, nor the amateurs, who frequently drop into the news as killed or injured, should risk cavorting around on slippery roofs. Nor should prudent caretakers permit such stomping around on winter-embrittled roofs that are not designed to resist such riproaring gyrations. Roofing companies are pleased to shovel off roof-snow as off-season employment; they may score a "double-shot" in returning later to repair or replace the shattered shingles and leaking roof.

Eave heat tapes (electric cables)

Building owners, under the duress of ice-dam inundations, often install electric cables along the eaves and in the valleys. Unfortunately, the melting effectiveness of such heat tapes is limited to only a few inches from the cable; the typical zigzag pattern of installation betrays this localization of melting effectiveness.

Heat-tape secondary ice-dams

The characteristic sawtooth melting of snow within a few inches of the heat tape develops a limited and selective removal of snow and ice; partial removals, whether by raking, shoveling, or localized heat cables, often develop uproof secondary ice-dams. Heat-tape secondary dams are observable along the terminal lines of the protective snow-blanket in a zigzag pattern prescribed by the limited melting zone of the heat tape.

Figure 13-16 shows the characteristic pattern of limited zone melting of an electric heat-taped eave. Figure 13-17 verifies the formation of the heat-taped secondary dam immediately above the highly localized melt zone of the heat tape.

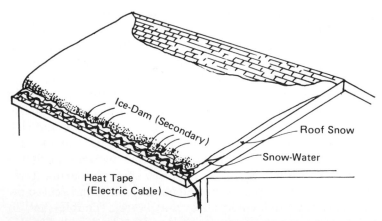

FIGURE 13-16   Characteristic zigzag pattern of limited melting zone of eave heat tape.

FIGURE 13-17   Secondary ice-dam developing uproof above heat zone of electric cable.

Valley heat tapes

Electric cables have been frequently observed in their zigzag pattern in valleys. Limited melt zones near the cable are often flanked with secondary ice-dams that obstruct normal drainage into the valleys. It is difficult to justify the expenses of electric cables, their installation, their recurrent power demands, and their adverse side effects in light of their limited effectiveness. Employing heat to melt ice into water on a troubled roof where heat has already produced the potentially penetrating snow-water suggests such approaches are not only wrong solutions, but are clearly complications of the eave-ice problem. Surprisingly extensive use of eave and valley heat tapes may be attributed to the futility and desperation of frustrated homeowners seeking "pushbutton gadgets" to alleviate their problems. Heat tapes are more numerous on roofs in latitudes where winters are not severe than where eave-ice is an annual assault and the futility and wasted expenses of localized melting are better recognized.

Counteracting roof thermals

Confronting roof-snow with controlled thermal gradients to melt the snow (as with electric heat cables) has been studied and developed for structures in the specific climate of the California High Sierras at Lake Tahoe (13). The energy costs and the adaptation to a specific climate of freeze-at-night and thaw-by-day limit the usefulness of houses designed to establish thermal gradient patterns from eave to ridge to melt troublesome roof-snow. Such counter-gradient installations in the upper Midwest would effect a return to the old *heat waster* (Fig. 13-22) performances of roof-snow melting.

Hosing with tap water

Another method of thermal attack on roof-snow has been employed in emergency situations to remove roof-snow and ice. Simply hosing down the roof (on mild days) from ridge to eave with tap water will remove roof-snow with reasonable efficiency; the method might be acceptable *if* the owner has no concern for the ice-formations that coat and overload the shrubbery around the house. Melting snow with flows of water is a method that has most of the hazards to life, limbs, and shingles of the ill-conceived shoveling techniques.

Eave flashings

In the snow-country near the Canadian border from Maine to British Columbia, many houses can be observed with eave flashings of metal (Fig. 13-18). The 2- to 3-foot widths of eave flashing seldom improve the appearance of the house, but distressed people living in deep snow-country seem willing to forgo the esthetics of appearance for some measure of protection from snow-water penetrations.

Flashings of smooth metal (or roll roofing) serve as leak protection devices along the leak-vulnerable eaves. The smooth (usually aluminum) edge-band eases the slough-off of eave snow and ice along the eave projections where attic heat is insufficient to sustain further down-roof flows of snow-water. With the lowest edge of the snow-blanket a few feet above the eave, *secondary ice-dams will form and build above the metallic flashed eave.*

FIGURE 13-18 Eave flashings of aluminum.

Heat tapes fastened to eave flashings have been observed in several states and provinces. Such improvisations suggest a high degree of owner frustration and compounded desperation in such "systems" of ice-dam prevention. The practical usefulness of such metallic eave flashings appears to be in the leak protection of a continuous (monolithic) sheet of material.

Metal roofs

Sealed-seam metallic roof coverings are common on homes of deep-snow ski areas from Stowe, Vt., to Aspen, Colo. These interlocked (monolithic) coverings indicate an awareness of the damage threats of roof-snow and ice. Such "total roof flashings" are directed toward leak protection and appear to defy or ignore ice-dams. Metal roofs apparently serve well (especially on steeply pitched roofs) by simply sloughing off the troublesome roof-snow.

Double roofs

Double-surface roofs constructed with cool-air venting spaces between them are reportedly used in European

snow-countries to prevent ice formations (13). The report
also suggests this device has found little acceptance in this
country because of the excessive costs of an insulated inner
roof with a second weather-roof spaced atop by the furring
supports. In several thousand observations, only one building (at Lake Louise, Alberta) was detected as newly constructed with an inner roof to retain the attic heat and an
offset top surface (of corrugated steel) to protect against
snow and rain. Furring strips running from eave to ridge
supported the "weather-roof" and provided space between
the two roofs for a gravitational flow of air from eave to
vented ridge. Presumably, the inner roof was well insulated
to retain and conserve attic and room heat. What little heat
does escape the inner roof should air-wash upward and out
the ridge *without melting the weather-roof snow-blanket
from its underside.*

When such vent-spaced double roofs are functioning properly, roof-snow is not melted by room and attic heat, and
snow-water does not flow down the roof-slope beneath the
snow-cover to freeze into ice formations at the eaves. The
double roof is, in effect, a "cold roof" and is a natural method
of avoiding destructive ice-dams.

Explanations of roof-snow melting behavior considerations should precede further considerations of cold-roof performances. An understanding of roof-snow behavior (and
misbehavior) on familiar roof designs and shapes leads to
logical explanations of the errors and follies of our present
approaches to ice-dam problems and results in an appreciation of the natural efficiency of a simple adaptation of the
"cold-roof concept" of eave-ice prevention.

A following section of this publication shows how the concept of a "cold-roof" can be achieved by practical methods
without the excessive expense of the furred-out double roof.

13-8   Why the Neglect of Ice-Dams?

Anyone recognizing the extent, frequency, and severity of
roof-snow damages may be puzzled by homeowners' strangely
apathetic and passive tolerance toward houses plagued with
such a visible deficiency.

Respected technical professionals have generally neglected, ignored, or denied the impacts and importance of roof-snow behavior. It may appear incredible—as it was for us early in this research—that technically competent and responsible professionals in building design and supply have so strangely failed to attend and resolve the destructive impacts of eave ice-dams and roof-snow.

Some possible explanations of this neglect are offered here, not to fuel an argument, but to encourage evaluation of these observations that may initially appear to affront established custom and conflict with the (limited) writings of respected technical professionals.

With ice-dammed snow-water visibly soaking the ceilings and sidewalls of their homes, distraught owners have had no obvious target for their wrath. Homeowners cannot identify any specific material or product manufacturer to blame. Materials suppliers and installers neither recognize nor acknowledge any involvement with (or responsibility for) the ice-dammed destruction. Technically responsible authorities have generally ignored the extent, severity, and increasing intensity of roof-snow damages; such experts must recognize the inadequate state of the art and endorse research and development programs if this correctable fault of building performance in the snow-country is to be eliminated.

The lack (or avoidance) of involvement in snow-water actions by building industry leaders is vividly indicated in an upper Midwest newspaper quote (7) of a building association chief executive in recommending heat tapes and shoveling the snow from the roof: "It's the homeowner's responsibility, just as much as shutting the windows when it rains or removing the snow from the sidewalk." The director of a highly respected government research laboratory confirmed those recommendations as the state of the art in coping with ice-dams.

Such building industry attitudes are in puzzling conflict with private communications with federal housing officials and another metropolitan builders' association executive who said such ice-dammed snow-water "is the greatest single complaint we receive from homeowners."

Homeowners, with little competent technical support to guide them, turn to various makeshift methods, gadgets, attitudes, and escapes from the attacks of snow-water and eave-ice. A few methods are partially helpful, some are monstrously hazardous, some are illusionary, and some are little more than deceptions. Distraught homeowners are reluctant to discuss degradations of their property. They have developed, perhaps with the support of compensatory insurance settlements, an amazingly apathetic, stoical, or calloused acceptance of this "unmentionable" of north country homes.

A fundamental reason for the apparent indifference of technical professionals to the damages of ice-dammed snow-water may be found in the assignment of vapor condensate as the almost sole source of sidewall moisture problems. Sidewall inundations from eave-blocked snow-water penetrations have been ignored or assigned an insignificant role by most technologists in residential housing.

Our observations and recorded color slide data indicate a substantial, if not a major, role of snow-water in sidewall moisture accumulations and damages.

It will be productive to assign competent technical attention and consideration to long-neglected ice-dammed snow-water as a significant source of sidewall moisture and damage. As promised in this report, *eave-ice can be prevented, and with the prevention of eave-ice, homeowners can be assured there will be no sidewall moisture originating from blocked eaves.*

It is not productive to argue the relative importance (or destructions) of vapor condensate versus ice-dammed snow-water as sources of unwanted sidewall moisture. It is important to note that, even with the best recommendations for blocking vapor movement into sidewalls, moisture damages continue to occur with embarrassing frequency in spite of scrupulously specified (and installed) vapor barrier installations.

Vapor condensate theory

A classic study and report (5) of the causes of sidewall moisture in the 30s—when houses were initially tightened with insulation and asphalt shingles—identified and emphasized sidewall moisture sources as vapor movement from the tightened interior. Unfortunately, eave-blocked snow-water was apparently neither considered nor researched.

Acceptance of vapor movements as the major (if not the only) troublesome source of sidewall soaking has permeated all segments of the responsible building industry. This is evident in pamphlets from paint, vapor barrier, insulation, design, louvering, and associated manufacturers and associations (see Refs. 4, 5, 6, 8, 10, 11, 12, 14, 16, 17, 22, 23, 24, 25). Four decades of neglect of ice-dammed snow-water as a second (and possibly major) cause of sidewall soaking may be explained, but to continue to ignore snow-water attacks is unconscionable.

Definition of sidewall damage region

The classic definition of geographic limits of sidewall moisture troubles as the "January Mean Temperatures of 35°F," is illustrated in Fig. 13-19. The origin of this geographic definition of sidewall troubles can best be understood from the following quote from page 15 of FPL Report No. 1710 (11), originally issued in September 1947:

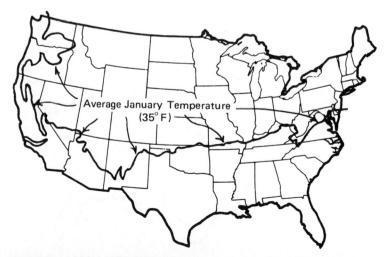

FIGURE 13-19 Map of the United States associating sidewall moisture damage susceptibility with temperature (Ref. 6).

The Forest Products Laboratory has been receiving reports of condensation in houses from various parts of the country for many years. In normal or mild winters most of these are from areas north of the Ohio River, but after a severe winter, such as occurs every 4 or 5 years, the reports are more numerous and include many from areas farther south. On a basis of these reports, it has been established that condensation problems may be expected in houses in those parts of the country where the average January temperature, according to Weather Bureau reports, is 35°F. or lower.

It is interesting, if not amazingly revealing, to find that the January 35°F mean temperature geographical limit of sidewall moisture problems coincides almost precisely with the U.S. Dept. of Commerce Weather Bureau's map of the 6- to 8-inch mean annual total snowfall. [See Fig. 13-20 reproduced from U.S. Department of Commerce Weather Bureau Map and Data (18).] Had those geographical limits of trouble been identified by *snowfall criteria* instead of *temperature criteria,* ice-dams (as well as vapor) might have received some, if not a major, consideration in sidewall moisture problems.

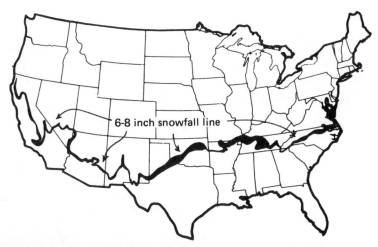

FIGURE 13-20   U.S. Weather Bureau map and data (Ref. 18) of mean annual total snowfall with 6- to 8-in line delineated.

Ice-dams overlooked

Ice-dams and damages have been almost ignored in reports of sidewall moisture problems throughout several revisions and building industry derivations of the classic vapor study (5); a 1972 revision (6) of the original research has added sketches and brief explanations of ice-dams caused by insufficient insulation and ventilation.

In discussing what appears to have been an honest, but possibly embarrassing, oversight of the important factor of roof-snow effects on sidewall degradation, we do not want to be argumentative. We believe it is in the public interest to re-examine all sources of sidewall moisture.

13-9   Typical Roof-Snow Melting Actions

Contrary to popular belief, the radiant energy of the sun does not provide the major source of heat that melts roof-snow on snow-country buildings. During cold winter days, attic air warms the roof-deck, which then melts roof-snow from its underside.

Light and fluffy snow crystals serve as unusually efficient insulation that slows atmospheric melting, supports roof-contact melting, and maintains fluid flows of roof-deck snow-water during subfreezing weather.

In Fig. 13-21, "heat" arrows are used to indicate thermal movement, direction, and concentration. Room heat is shown passing sluggishly (slowed by insulation) into the attic where it warms the attic air. As attic air rises in temperature, it expands, becomes lighter, and rises to accumulate and increase air temperatures in the higher attic spaces at or near the ridge. Attic air warms the roof-deck, making it warmest near the ridge and less warm toward and along the eaves. These thermal gradients of the roof-deck cause the roof-snow cover to melt initially and more intensely at the warmer ridge zone.

Roof-snow, melted from its underside by the heat of the attic-warmed roof-deck, flows downroof as snow-water under the insulative snow-blanket. When typical snowstorms are followed by subfreezing cold, melted snow under the snow-blanket flows slowly downroof to emerge from the snow-cover and freeze into accumulating ice at the edge of

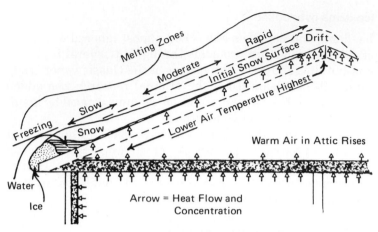

FIGURE 13-21   Schematic representation of roof-snow melting.

the snow-blanket. Normally, ice-dams form and develop along the eavelines where the snow-blanket ceases to protect the fluid flow and where the roof-extension beyond the plates has less (or no) attic heat to keep it liquid.

As the snow-water emerges to freeze into ice masses along the edges of the snow-blanket, subsequent water beneath the protective blanket may be blocked from runoff by the "ice-dams"; it may freeze or be held as pools under the insulating blanket of snow. Blocked by massive ice-formations, snow-water pools develop and may rise on roofs to heights sufficient to penetrate between and under the shingles and roof-deck. Snow-water penetrations through spaces, cracks, and flaws of the roof-deck lead to soaking and extensive damages to the roof, cornices, and sidewalls. Such typical roof-snow behavior is illustrated in Fig. 13-21, and damages are shown in Fig. 13-12.

The major destructive agent on snow-covered roofs is the ice-blocked snow-water, which is kept liquid by attic heat and protected from atmospheric freezing cold by the insulative snow-blanket. The roof-snow covers and hides the deceptive liquid-ice-liquid (and vapor) transformations energized by attic-heated roof-decks.

Roof-melted snow-water accumulates and develops ice-formations at any roof location where there ceases to be sufficient attic heat to counteract freezing or where the insulating snow-blanket is terminated. If the snow-blanket is undisturbed, the ice-dam will form at the eave; if snow is removed part way uproof from the eave, the ice-dam will form at that newly established snow/freezing air interface.

Intermittent and sporadic melting and freezing of roof-snow and snow-water is difficult to follow because most of the actions are invisible beneath the covering of roof-snow.

Simple corroboration of these actions of roof-snow can be observed from streetside. Thin coatings of frost or light snow showers reveal many highly localized sources of attic heat, such as: chimney; bathroom; dryer; vent-stacks; voids of carelessly applied ceiling insulation; platelines, because of their higher thermal contribution; and the pattern or ridge area heat concentrations.

Such visible indicators of roof-snow behavior develop more slowly with heavy snow-blankets and during severely cold weather. Attic-heat actions are visibly evident in terms of the diminishing depths of roof-snow near the ridge where the warm air of the attic rises to produce thermal concentration. Roof-snow patterns provide proof that roof-snow melting proceeds primarily from roof-decks warmed by the internally developed attic or building heat.

With very short (or no) cornice projections, massive ice and blocked snow-water pools develop at the platelines where attic heat and sidewall heat leakage are greatest; observations suggest that such short cornice houses have the greater incidence of *visible* sidewall soaking because pools are more likely to develop directly over the wall lines. Cornice projections, increased insulation, steeply pitched roofs, "cathedral" ceiling structures, knee-walled half stories, split-levels, lean-tos, and improved louvering practices have varying influences on the formation of snow-water and eave-ice.

13-10 Roof-Snow Performances on Pre-30's Houses

A review of the progress in house construction since the 1930's reveals some of the reasons for increasingly troublesome roof-snow problems. These considerations also disclose inadequacies of present construction methods, materials, codes, and standards in coping with roof-snow.

Typical American houses built before 1930 were constructed with relatively steeply pitched roofs, with "open" (spaced) sheathing, and mostly with wood shingles; these (cedar) shingles had spaces between them that closed when wetted. Such gaps between shingles served to ventilate attics and to cool the roof-decks.

Most "pre-30's" houses had no insulation in either ceilings or walls; indeed, there was little demand for heat conservation when rooms were selectively *warmed* by stoves rather than heated by a central furnace. Some homes had central heating, of course, but only a few had insulation in ceilings and sidewalls. Wood and coal were plentiful; there were no outraged public demands for energy conservation. Nor were paint staining, blistering, and peeling of the traditional wood siding embarrassing and expensive problems. *The houses built in those days were not plagued with paint problems until they were modernized;* "updating" with asphalt shingles, insulation, and other installations designed to "tighten the old heat waster" and conserve heat led to sidewall moisture and paint failures of alarming severity.

Numerous pre-1930 houses in our country have been "modernized" with tight roofing, variable (but almost always inadequate) insulation, ventilation, and vapor barriers. Because of the highly heated air of the attic, the warmed roof melts roof-snow rapidly; in only a few days, roof-snow is transformed to snow-water and/or ice.

Figure 13-22 is a schematic representation of the roof-snow melting of such houses. Heat concentrations and heat escape movements are indicated in the quantity and directions of the arrows.

Notice the hot air concentration (as heat arrows) in the high spaces of the attic where the warmest air rises and is trapped. Such ridge buildup of attic heat initiates and supports the most intense melt actions on the warmer roof-deck nearest the ridge. Roof-snow converted to snow-water flows downroof under the protection of the snow cover. On the old heat wasters, the excessive attic heat supports rapid flows of

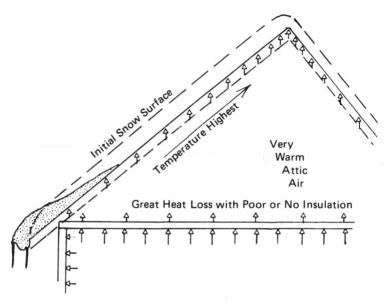

FIGURE 13-22 Schematic representation of thermal influences on roof-snow melting of pre-1930's houses ("the old heat wasters").

snow-water downroof to the cooler eave projection and beyond. When air temperatures are less than extremely cold and attic heat is in plentiful supply, rapid melting of roof-snow quickly clears the roof of snow—with only a gutterful or eave-line of ice and icicles.

Such roofs flow fast and furiously; roof-snow disappears initially at the ridge, next from the midrun, and last from the eave. "Old heat wasters" flow comparatively rapidly and copiously with snow-water; their roofs clear of snow more rapidly than do neighboring moderns; snow-water penetrations are of comparatively short duration; and massive eave-ice may be all that remains of the roof-snow until the next snowfall.

13-11   Roof-Snow Performances on Modern Houses

Roof-snow blanketing on modern houses, in contrast with "old heat wasters," melts slowly; beneath the cover, water pools may threaten penetration into the house for periods of weeks or months in the snow/cold of the upper Midwest.

Modern houses, as a type apart from the old heat wasters, are generally considered adequately insulated and ventilated since they meet contemporary standards, codes, and recommendations.

With codes usually stated as minimums in thickness (or R-factor), builders tend to insulate (and ventilate) close to the minimums to keep building costs competitive. Few houses, other than those specified for electric heat, have more than the minimums. Generally, there is an unquestioning public acceptance of the minimum standards as "adequate."

Residential areas are dominated by "old heat wasters" in older communities and by many "adequately" insulated and ventilated moderns in the newer developments. Roof-snow melting may be observed as fast and furious for "old heat wasters"; in contrast, modern houses melt roof-snow with comparative slowness, greater deception, and often more insidious damages. Simply increasing insulation will not stop the flow of waste heat into attics; insulation will only slow the movement of thermal energy to economic and ecologic levels of acceptance. Warm, if not hot, attics are an undesirable characteristic of millions of "modern" American homes.

Thermal movements and concentrations for moderns are represented schematically by the arrows in Fig. 13-23. Compared with the "old heat wasters," there are fewer arrows of heat passing through the ceiling into the attic. Because insulation only slows heat escape, some waste heat passes into the attic air in contact with the roof-deck. With less heat entering the attic, there will be less intense heating of the roof; indeed, if heat escape is very slight because of generously insulated ceilings, only slight warming of the roof-deck may be expected. However, with present attic venting standards, the roof is often warmer than the freezing atmospheric air; imperceptibly slow and intermittent melting of the underside of the snow-blanket will (and does) develop roof snow-water that persists and threatens roof penetrations for periods of several weeks or months.

Roof-snow melting of modern houses will usually initiate near the ridge where the thermal concentrations of the attic

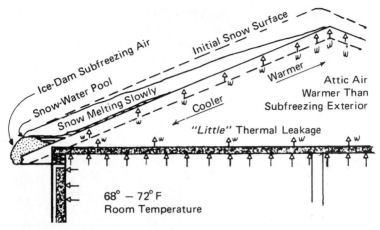

FIGURE 13-23 Schematic representation of thermal influences on roof-snow melting of modern (contemporary) houses.

create the warmest gradients. Streetside observations reveal these thermal gradients in the sagging patterns of roof snow-blankets. Ridge-snow, melted by the warm roof-deck, will flow downroof under the insulative snow-blanket to freeze into ice—either at the edge of the roof snow-blanket with the subfreezing air or where the roof thermals will not maintain the liquid flow. Roof snow-water may sporadically freeze and thaw as it progresses downroof to eventually emerge from the snow-cover where it meets the subfreezing cold air, solidifies, and grows into massive ice-dams with blocked snow-water pools immediately uproof. With long periods of formation, snow-water pools, and potential penetrations, modern houses are more prone to severe damages from snow-water than are the icicle-trimmed old-timers.

The sagging roof-snow pattern of Fig. 13-24 demonstrates the effects and locations of thermal gradients on this 1962 house. Differences in melting actions as delicate as the shingle pattern are evident. The deeper snow along the rake and eave cornices is a result of cool roof projections beyond the attic heat supply. The eave ice-dam, with its probable snow-water pool immediately uproof, is topped and protected with an insulative snow-cover along the 24-inch eave projection.

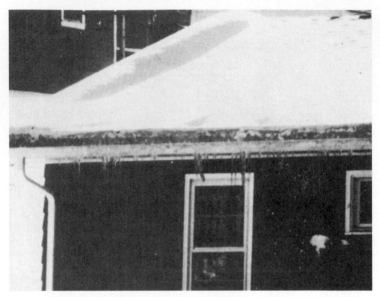

FIGURE 13-24 Telescopic lens photo of shrinking and sagging roof-snow on a modern (1962) roof.

Thermal patterns on this roof are somewhat more pronounced than on many contemporaries, but the roof-heat pattern is characteristic for houses built under minimum standards.

Observations attest to this slow, deceptive, and insidiously penetrating attack by roof-snow on "adequately" insulated, and even on electrically heated houses. The problems of roof-snow melting and its damages can be seen on almost all buildings: the older poorly or noninsulated houses and on moderns complying with the latest insulation and ventilation building codes and standards. Contemporary specifications may provide only illusions of adequate protection.

Ice-dams and snow-water damages occur on almost all homes, ranging from "old heat wasters" through modernized old timers and contemporary moderns to the thermally efficient electric heated homes: *all develop ice-dams,* either fast and furiously flowing with snow-water or slow insidious seepage for as much as 3 months' duration. The roof performances can be visibly verified almost every year during

February and March in the Twin Cities and in those recurring "hard" winters in Chicago's western suburbs.

The present emphasis (and sales promotions) of "thicker" insulation to offset the inflating costs of residential heating should be considered in relation to ice-dam formation. If energy conservation is limited to only greater insulation depths (and R-factor improvements) without recognizing and considering roof-snow behavior, *sidewall soaking and related damages can be expected to intensify* to epidemic proportions, possibly more confusing and devastating than the scourge of baffling sidewall degradations of the 1930's and 1940's.

These explanations of roof-snow behavior ranging from old heat wasters to the "adequate" moderns are intended to stimulate thought and consideration of roof-snow actions; they are also essential for a fair appreciation of the extent of this wintertime scourge of the snow-country where *almost no contemporary dwellings are immune to roof-snow-water damages.*

Observations and recognition of roof-snow actions on contemporary housing point to an appalling inadequacy in present standards and lack of building industry response to this neglected misperformance of snow-country buildings.

13-12   Roof-Snow Behavior on a Cold-Roof

Elimination (or control) of undesirable attic heat suggests a fundamental approach to eave-ice prevention; in essence, the object is to develop inexpensive cold-roof houses.

The effectiveness of cold-surface snow behavior to avoid massive ice formations is exhibited everywhere in winterized snow country. There are no massive ice formations on felled timber, rock outcrops, woodpiles, picnic tables, sheds, patios, marquees, and *unheated buildings!*

Cold-surface reaction to covers of snow is the way nature confronts and directs orderly melting and dissipation of snow. Ice may accumulate as glaciers from snow-packs or snow-water accumulations; however, generally, *nature is not troubled with ice-dams.* The human race, with its heated constructions, has created the monster, and as we errant humans are so often reminded, "It is not nice to fool Nature!"

Examination and interpretation of an idealized cold-roof performance should prove helpful, encouraging thought and action for practical application and adaptation of the cold-roof principle of eave ice-prevention. Such an ideally functioning house is illustrated in Fig. 13-25.

Continuous removal of attic air to maintain the roof-deck at atmospheric air temperatures requires minimizing the amount of heat entering the attic, together with a scavenging free movement of air from eave inlets to uproof (ridge) outlets. The small thin arrows in Fig. 13-25 illustrate escaping room heat passing sluggishly through the insulated ceiling. Arrows are few because heat loss of this very efficiently insulated house is minimal compared to the conventional modern house of Fig. 13-23.

In this idealized sketch, none of the heat arrows contact and warm the roof. When comparatively insignificant amounts of heat are lost through the well-insulated ceiling, the expansion of the warmed attic air causes it to rise slowly and be swept out the ridge vents by the natural flow of air from eave to ridge. Incoming atmospheric air flowing under the roof-deck keeps the roof cool.

For an ideally functioning cold-roof house, the snow-blanketed roof-deck will be maintained near the atmospheric temperature: the roofside of the snow-blanket will remain frozen during cold weather; on warm (or sunny) days, roof-

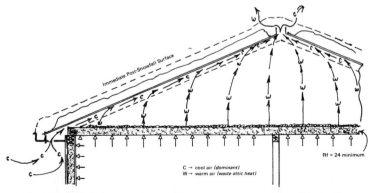

FIGURE 13-25 Schematic representation of roof-snow melting behavior on a cold-roof house.

snow melting will progress from the exposed atmospheric side of the roof-snow blanket.

13-13 Cold-Roof Performance in Warm Climates

The performance and desirability of cold-roof designs for warmer climates and for *better summer characteristics* should be recognized; cold-roof technology should be made available for all homeowners.

Because of this publication's concern for and emphasis on roof-snow, the full performances and benefits of cold-roof house construction may appear beneficial and limited to only deep snow-country buildings.

Attic heat buildups are possibly more undesirable and uneconomical in summer than they are in winter. The cold-roof design, with its attic air-sweep of air, merits its adaptation to all climates.

Attic air temperatures build to the 140–150°F range under the hot summer sun. With a room cooled to 75°F and the attic at 150°F, the heat flow impact is equivalent to the typical winter conditions of the room at 72°F and the attic at −3°F. The need for generous insulation to slow the heat flow is easily recognized. For both summer comfort and air-conditioning (cooling) energy conservation, sales promoters urge homeowners to purchase attic exhaust fans.

Electrically operated exhaust fans consume energy in their operation to conserve energy. Designed cold (atmospheric) roofs flush out the rising (expanded and lightened) air in the free air-flow movements from the lower eave inlets to higher ridge outlets.

No energy-consuming exhaust fans are needed in the cold-roof attics in the summer, either in the north country or in the hot deep South.

13-14 Unique Characteristics of Cold-Roof Snow Patterns

Roof-snow disappearance patterns display sharply contrasting differences between cold-roofs and others; these contrasts provide visible evidence of a cold-roof performance; on cold roofs, the edges are the first to melt, and the center of the blanket is the last to disappear. Roof-snow disappearance patterns provide a convenient means to evaluate thermal gradients on subject buildings and for monitoring the effectiveness of cold-roof conversions.

The snow on the roof of the cold-roof house of Fig. 13-26 is in a midroof pattern that is in sharp contrast with most (occupied) residential buildings. Notice the rake areas have cleared, and the eave edge is also free of both snow and ice; snow-water drainage has been as free-flowing and orderly as the run-off rains. This performance can be better appreciated in its contrast with the roof-snow pattern of Fig. 13-24, where attic-warmed ridge and midslope areas melt first, the cooler rake edges follow, and the eave-ice persists in its drainage-blocking action until it, at last, melts.

FIGURE 13-26 Residual snow pattern on a cold-roof house specifically designed for ice-dam immunity.

13-15 Detection of Cold-Roof Efficiencies in Roof-Snow Patterns

The distinctive patterns that develop as roof-snow dissipates provide a convenient way to recognize and evaluate a subject roof's characteristics in eave-ice resistance.

The sketches illustrating three types of melt patterns developed from the snowfall of Fig. 13-27 are shown in Figs. 13-28, 13-29, and 13-30. These three house types relate to an "old heat waster," a modern, and a cold-roof. These roof-snow patterns may be correlated to the melting performances earlier described for these three characteristic types of residential housing.

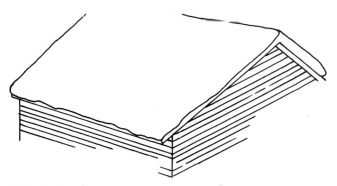

FIGURE 13-27 Typical post-snowfall pattern.

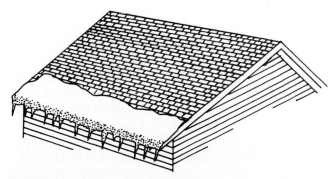

FIGURE 13-28 "Old heat waster" after a few days.

FIGURE 13-29   Modern house roof after several days or weeks.

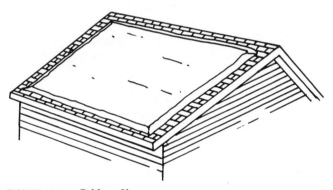

FIGURE 13-30   Cold-roof house.

A properly functioning cold-roof can often be visibly identified in the unique ruffling and eroding of the roof-snow edges into the peculiarly concave shapes indicated in Fig. 13-30. Cold-roof snow edges at ridge, rakes, and eaves contrast revealingly with the convex eave and rake edges of roof-snow on almost all contemporary buildings.

When there is no attic (or underroof) heat to melt the snow-cover from its underside, the snow-blanket on the cold-roof will dissipate initially and progressively from the exposed edges at ridge, rakes, and eaves.

On a cold-roof, snow melting and compacting are accompanied by wind actions that activate blow-offs, sublimation, evaporation, melting, and a generally orderly free-flowing dissipation and drainage at the greater surface exposure of the edges.

Deep roof-snow has long been recognized (and proclaimed in insulation promotions) as an indicator of well-insulated houses. Unfortunately, the depth of snow accumulations does not indicate the combined, or balanced, efficiencies of insulation and ventilation.

How roof-snow "shapes up" after a period of exposure to winds, sun, and interior heating energy will provide a simple means to evaluate the insulation-ventilation characteristics of any subject house or building.

As you look for the characteristic edge dissipation patterns of cold roofs, you will find almost none on contemporary housing, *unless* you chance upon a residence where the occupant has vacated and the house-heat is cut back or off.

Simple visible verification of most of the observations and comments of this publication may be monitored in roof-snow patterns, especially in *late* winter.

With reasonable recognition of the erratics and complexities of wind-drifted snows, interpretations of roof-snow behavior should prove stimulating and productive.

13-16   Eave-Ice Prevention Technology

Basic guidelines

The fundamental concept of a cold-roof is quite simple: adaptation of the basic guidelines to new construction, and to existing buildings, may be accomplished with *apparently* little more than these two modifications of conventional residential construction:

1. *Insulate* room ceilings far more than customary to minimize heat losses and attic temperatures.

2. *Ventilate* profusely at all eaves and ridge for a natural flow of air to sweep out the warmed attic air.

Limitations, exceptions, difficulties, and solutions

Earlier in this publication, better eave and ridge venting was suggested, as well as improved insulation for homes. Such broad suggestions, like our basic guidelines, may be technically profound, but of dubious practical value without further explanations of the details and problems of their applications.

The *apparently* simple application of the basic guidelines for eave-ice prevention may often be difficult, sometimes less than fully satisfactory, and downright impractical for a few building types.

The cold-roof concept is simple; its applications are complex.

The following discussions on the complexities of eave-ice prevention technology are not intended to discourage corrections. The difficulties are described and the limitations are interpreted *to prevent misapplications* and possible disillusionment with the effectiveness and widespread adaptability of *competently installed eave-ice constructions and corrections.* Cautious homeowners should be aware of these difficulties and limitations so they may judge applicator capability as well as the quality of the completed application.

The trouble with insulation

"If a little is good, more is better." Certainly this phrase has its limits of judicious application, but in present standards (and existing installations) of ceiling insulation, much, much, *much more* will be better for 99+ percent of residential housing; this is especially important if attics are to be properly vented and eave-ice is to be prevented.

Apologetic marketing of ceiling insulation still persists since the early introductory sales of the 1930's when the additional expense of an unfamiliar item met "dubious-buyer" resistance. Justification and redemption of insulation costs through amortization of "additional outlays" has, until the recent energy conservational promotions, limited installations of ceiling insulation to little, or no more than, the Federal Housing Authority (FHA) minimum standard specifications.

No stipulations of the amount of insulation (or louvering) can be recommended *as sufficient,* partly because of the innumerable variables in residential designs and requirements, but mostly because there should be no interpretation of such a stipulated "sufficiency" that might inhibit an owner's urge to "really insulate (and ventilate) the house!"

A homeowner might be satisfied that he or she has accomplished all that is feasible in ceiling insulation when room to attic heat losses are reduced to the level required for efficient electric heating. Such quality of room heat retention is in harmony with the needs and promises of contemporary promotions for energy conservation.

Budget-strapped homeowners may be somewhat reluctant to purchase these "extra" insulation improvements, but the savings in energy outlays alone should not be difficult to appreciate in this era of runaway fuel costs. The savings available through prevention of snow-water damages may be more indirect and difficult to foresee; however, coupled with the fuel savings, *"maximum" insulation can be doubly rewarding!* This may mean that as much as 10 to 12 inches of insulation should be applied in attic areas; *AND it must be intelligently and carefully installed.*

Building code specifications usually specify the insulation requirements in terms of *"minimum* standards." Cost-conscious builders, under the pressure of price competition, tend to supply the bare minimum. Of course, additional insulation may be purchased, and there are many high-grade installations by quality builders; proof of these "quality jobs" can be observed in deep and winter-enduring roof-snow that melts and converts very, very slowly into ice-blocked eaves. Unwittingly the quality conscious builder or owner *rarely matches the adequate insulation with a comparable maximum of ventilation.*

Much contemporary residential construction barely meets minimum building codes; in many cases, houses are deficient not only in "thickness" or R-factor (that is, resistance to heat loss), but in addition, ceilings are prone to numerous voids and discontinuities in the protective insulation.

Builders often pay too little attention to the care with which insulation is installed. Continuity of the protective ceiling blanket is frequently interrupted by bridging, wires, recessed ceiling fixtures, scuttles, stair-entrances, and a variety of occupant actions such as careless attic storage, TV, Hi-Fi, and associated attic uses and misuses.

Extremely conscientious ceiling insulation application and care must not be considered trivial; cold-roofs for eave-ice prevention cannot tolerate the flaws, voids, thin spots, and discontinuities that are frequently associated with present construction practices.

Seldom considered and rarely examined attic spaces are too easily disregarded, ignored, forgotten, and neglected! *At least biennial inspections of ceiling insulation should be a maintenance routine;* sagged, compressed, wetted, blown-out, disturbed, or other diminished insulation should be located and corrected. Energy conservation programs may suggest your concern for insulation efficiencies; cold-roof ice-prevention *demands* it!

Disruptions in the ceiling insulation are costly not only in wasted room heat, but also because such flaws lead to roof hot spots that develop intensified localized roof-snow melting. Such hot spots are sometimes identified by spot shrinkages in the snow and related ice-dams downroof at the eave. An attic door or scuttle inadvertently left open may be revealed in the sagging snow-pattern on the roof.

Uninsulated chimneys (furnace flues), gas vents, bathroom, kitchen, clothes dryers, and any other warm exhaust equipment may contribute to a troublesome supply of attic heat; some pipes may have to be re-routed, and all of such attic heat contributors should be carefully wrapped with effective insulation. Eave-ice prevention technology requires *minimization of all attic heat sources!*

"Reading" roof-snow patterns to reveal insulation defenses of houses is as easy for the trained ice-prevention technologist as "reading a pass defense" is for an all-pro quarterback. The insulation picture blurs for ice-prevention analysts only when extremes of attic ventilation are encountered. Because almost all contemporary houses are vent-specified for *no more than minimums for vapor removal,* streetside observers will find properly vented houses extremely rare.

Ventilation problems

Some building owners may find it necessary to dispel the notion that attic heat is an asset to be retained. Perhaps such misconceived reverence for attic heat was reasonable in early dwellings before insulation was available to slow escaping room heat to insignificant heat losses.

Anyone who has endured the discomforts of an uninsulated summer resort or motel cabin where sleep was delayed until early morning (when the overly abundant attic heat had dissipated) can appreciate the importance of ventilation in preventing attic heat buildup in summer; prevention of eave-ice with cold-roofs and cold attics demands a similar ventilation removal of attic heat in winter.

Builders and homeowners reluctant to accept ventilator-to-ceiling-area ratios of 1/900 to "as much as" 1/250 may be shocked with cold-roof designs demanding infinitely greater louvering—both in size, number, and distribution.

Standards in venting specifications, such as those quoted for attic vapor control, are both inadequate and meaningless in eave-ice prevention. Roof shapes, cornice styles, and appearance esthetics restrict many installations; available positions for louver distribution may limit ideal venting. Some cold-roof installations may have to settle for less than maximum ventilation; ice-prevention performances may, therefore, be proportionately less than fully effective.

Louver products of various types and shapes are presently available for venting both eaves and ridges.

Triangular gable louvers may be satisfactory for vapor dissipation if they are large enough and if they are strategically placed to utilize wind or air currents; seldom are gable louvers suitable for development of the eave-inlet ridge-outlet patterns of air-wash required for a properly functioning cold-roof.

Pitched bonnet-type roof vents, either powered or gravity exhausting, are seldom used in sufficient quantity and distribution to achieve the blanketing air-wash needed for complete cooling of the roof-surface.

Rectangular soffit louvers, circle-spot louvers, and alternated perforated soffit panels usually limit the eave inlets to less than desired inlet free-air venting and distribution. Generously perforated soffit materials may promote embarrassing wind-blasting of insulation. On house designs of little or no cornice projection, there is little or no room for soffit louvers; a proprietary dual-walled fascia (19) has been developed and effectively tested for such soffitless house designs, but the product is not commercially available.

Strip louvers on ridge and eaves are the simplest products for securing a good distribution of eave inlets and ridge outlets. Most ridge strip louvers are apparently designed to be unobtrusively small to avoid affronts to residential esthetics in styling. Such size limitations may leave much to be desired in deep-snow country where such ridge venting may become submerged and plugged by deep snow.

Because the naturally rising action of warming air generates eave-to-ridge movements for cooling roofs, the stack effect of cupolas and chimneys may find greater adaptations in future ice-prevention designs; a simple corbelling of the brick chimney as it passes through the attic can supply additional attic-venting flues and also contribute to the balanced appearance of a massive chimney emerging from the roof-ridge. Such stack actions, integrated with free distribution of eave inlets in soffit or fascias, effectively assure cool attics for summer comfort and cold-roofs for winter ice-prevention.

There appears to be a challenge for creative designers in the development of innovative devices for venting roofs for ice-prevention in harmony with the esthetics of acceptable appearance. Perhaps the multiple roof attractions of Chinese and Japanese structures may find a justifiable place in American home designs.

The watchword for venting attics, as with ceiling insulation, is simply: *"You can only install too little, never too much!"*

And yet, under certain *careless insulation practices,* it may, at first, appear that there is too much ventilation!

Too much venting

Where insulation is not carefully fastened or protected near the eaves, strong wind-blasts through freely vented eaves may blow insulation away to expose room ceiling areas to the attic's winterized temperatures. When the bared ceiling areas drop below the dew point or freezing point, those ceiling surfaces will develop wetted, stained, and frosted ceiling spots where room moisture has condensed on the cold ceiling areas.

When insulation has been blown backward from the plateline area, you commonly encounter a very infuriated homeowner, a bewildered builder, and a worker who has been ordered by the builder to "stuff the stuff along the platelines" to block recurrent wind-blasting of the plateline insulation. The builder blames the unhappy situation on the designer who "uses too much ventilation" and attempts to assure the hostile homeowner that "nothing like this will happen again!" No, it won't happen again when the venting is all but completely choked off, but it won't be long before the "ice-dam cometh" on the eave again, *again,* and *again!*

Is the blunder of insulation blow-outs due to "excessive" venting, or is it due to careless insulation application? If eave-ice prevention is an objective, there is no doubt that insulation must be more carefully positioned, and perhaps fastened, than has been customarily practiced by the building trades.

Restricted plateline gaps

The plateline region has been misunderstood, unrecognized, or ignored by many builders as a critically vulnerable region for heat-loss as well as a restricted passage that almost always limits and often inhibits air-flow from eave inlets to uproof outlets.

Exterior wall-to-ceiling corner joints are characterized by unavoidable cracks of: doubled plates; exterior sheathing edges; lath or wallboard joints at the interior corners; and usually *ceiling insulation that has been terminated at the platelines.* Wall insulation tends to sag beneath the plates and loses both its effectiveness as an insulator (and possibly its positioning) when moisture from condensate *or blocked snow-water* wets and soaks the sidewall insulation. Thermographic analyses disclose exterior platelines as major heat loss areas. The state of the art of residential design, construction, and research indicates gross neglect of this critical heat-loss region. Excessive heat losses at the exterior platelines also contribute to snow-water penetrations by supporting the liquid pool immediately over the plates. Quality construction calls for consideration, reduction, and preferably elimination of the plateline heat concentrations.

The exterior platelines require more and carefully positioned insulation; they must also retain spaces for a free flow of air past the usually restricted gaps above the plates. Such a dual requirement appears to produce an impasse for the majority of homes where the roof-pitch, rafter-heel-cut, and sheathing allow less than 6 inches of clearance—often as little as 3½ inches.

The plateline gap difficulty is sketched in Fig. 13-31 as it appeared in a recent publication (of the Forest Products Laboratory, Madison, Wis.). Notice that free air-flows are indicated, but at an unfortunate sacrifice of insulation that was probably not intended. That sketch not only helps illustrate the restrictive nature of such exterior platelines, but it also demonstrates that the state of the art recognizes the *need* for adequate air flow into attics, but *has failed to*

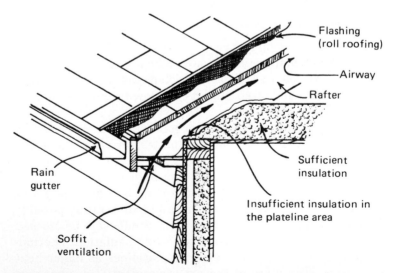

FIGURE 13-31   The plate-gap (as the FPL report in Ref. 6 shows it).

explain HOW such essential airflow and insulation can be accomplished!

The sketch of Fig. 13-32 illustrates a solution for compromising the restricted passages between *air-flow demands* and *insulation optimums*. The proprietary product (20), called "Air Passage Protector," is not commercially available.

New construction applications of this air passage protector add little to labor costs. Placement requires only positioning. The subsequently fastened sheathing holds the roll on to the rafters with the fins upward; the resilient insulation inserted beneath it supports and forces it upward against the sheathing, but is limited by the fins to assure air channels of fin depth between the insulation and the roof sheathing. Any fin on a rafter is simply crushed as the sheathing is nailed.

Positioned tightly between the sheathing and insulation, the spaced-fin roll not only creates and assures air channels at the restricted plateline gaps, but also serves as an inte-

Bridging the plateline gaps

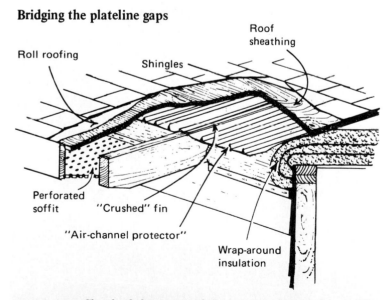

FIGURE 13-32 Sketch of plate-gap with "air passage protector" in position between sheathing and wall plates.

rior flashing to help protect this leak-prone eave-zone from inadvertent roof-leaks above it.

With little additional labor cost, the roll-out-on-rafters procedures may be used on existing houses about to be re-shingled. Often, especially where there has been a history of ice-dammed snow-water on the eaves, the weak and rotted roof sheathing should be replaced; the roll-out-on-rafters installation of the plateline air-passage protector (and flash-ing) material can be accomplished as simply as in new con-struction (see Fig. 13-33).

Where reshingling is not needed on an existing dwelling, the insulation baffles may be prepared as finned *panels* sized for simple insertion *between* the rafters to bridge the

FIGURE 13-33   Worker applying air passage protector along eaves.

plateline insulation. Such insertions may be made from the attic where that approach is feasible, or from ladder or scaffold along the exterior with the fascia removed for direct access to the plateline gaps.

13-17   Proven Cold-Roof Performers

Existing houses which have been corrected for ice-dams offer convincing proof of the effectiveness and practicality of the materials and methods developed to convert roofs decked with thermal gradients into cold-roof performers. Proof of performances of these developments can be observed on a few houses in southwestern Wisconsin that were planned and constructed to perform as cold roofs. Most of the owners are unaware of the unique eave-ice protection built into their homes.

Figure 13-34 is a photo of a Colonial-type house constructed in 1933; it was generously insulated (even by today's standards). It has no vapor barriers; it had a 39-year history of blistered and peeling paint; it has also been decorated with eave ice-dams almost every winter until 1972.

FIGURE 13-34 Well-insulated house built in 1933 with its almost annual eave ice-dam.

In the summer of 1971, the house in Fig. 13-34 was converted (with the materials and methods described in this publication) into a cold-roof house. Since the correction was made, the house has had no eave ice-dams.

Converting the house of Fig. 13-34 into a cold-roof performer not only eliminated the eave ice-dams, but altered the roof-snow melting pattern, as may be seen in Fig. 13-35. Notice the roof-snow edges at both rake and eave exhibit a ruffled concavity where wind actions have initiated roof-snow dissipation and removal. This house now sheds both rain *and snow-water* in an orderly drainage.

One of the most interesting "conversions" has been performed on a contemporary house distressingly troubled with ice-dammed snow-water. After corrections were made in late December, the snow-water penetrations stopped, *ice-dam growth ceased,* and by late February, the ruffled concave eave pattern (characteristic of a cold-roof) was evident and reassuring of the effectiveness of this rare, and not normally recommended, midwinter correction.

Detailed observations and comments of these and other "case history" performances in these developments have been logged in several audio-visual reports. The audio-

FIGURE 13-35 The house in Fig. 13-34 after its modifications into a cold-deck performer.

FIGURE 13-36   A house planned and constructed for cold-roof performance, but without the expense of the double-roof structure.

visual presentations have been prepared for training installation personnel in the whys, wheres, whens, hows, and how-nots of the practical application of ice-prevention technology (see Fig. 13-36, for example).

References

1. "An Introduction To The IDP Rating." A survey and analysis of "Tough Winters" at twenty-five stations over five years—1967–1971. By Howard L. Grange, R.P.E. An unpublished report of *Ice-Dam-Potential.*
2. "Audio-Visual Reports For Tape-Color Slide Review and Presentation" by Howard L. Grange, R.P.E. Personal papers and audio visuals prepared for consultation and teaching the methods and materials of Ice-Prevention Technology.
3. "Comparative Study Of The Effectiveness Of Fixed Ventilating Louvers" by H.S. Henrichs. A report for HC Products Co., Princeville, Ill. 61559.
4. "Condensation In Farm Buildings" by L.V. Teesdale, Engineer, Forest Product Laboratory. Information Reviewed and Reaffirmed January 1956 FPL No. 1186.
5. "Condensation Problems In Modern Buildings" by L.V. Teesdale, Engineer, Forest Products Laboratory Report for presentation before Conference on Air Conditioning, University of Illinois, March 8–9, 1939.

6. "Condensation Problems: Their Prevention And Solution" by L.O. Anderson, Engineer, Forest Products Laboratory, Forest Service, U.S. Department of Agriculture. Research Paper FPL 132, 1972.

7. "Home Builder Says Ice In Eaves May Cause Roof To Leak" by John Newhouse in Wisconsin State Journal, newspaper article of January 3, 1970.

8. "Insulation: Where And How Much" by Laurence Shuman. Technical Reprint Series No. 4 from HHFA Technical Bulletin No. 3 (March 1948) Housing and Home Finance Agency. From Supt. of Documents.

9. "Local Climatological Data." United States Department of Commerce, National Oceanic And Atmospheric Administration, Environmental Data Service, National Climatic Center, Federal Building, Asheville, N.C. 28801.

10. "Moisture Condensation" by Frank B. Rowley, Director, University of Minnesota Engineering Experiment Station; University of Illinois Bulletin, Vol. 44 No. 34, Jan. 27, 1947. Issued by The Small Homes Council Circular Series Index Number F6.2.

11. "Remedial Measures For Building Condensation Difficulties." Original reported dated September 1947, written by L.V. Teesdale, Engineer, Information reviewed and reaffirmed August 1962, Forest Products Laboratory Report N. 1710.

12. "Save Your Home From the MENACE OF MOISTURE." A pamphlet issued by the National Paint, Varnish, and Lacquer Association, Inc.

13. "Snow Country Design" by Ian Mckinlay, A.I.A. and W.F. Willis. R.P.E. procured from Mackinlay/Winnacker AIA & Associates, 5238 Claremont Ave., Oakland, Ca. 94618 for nominal fee.

14. "Some Construction Defects Which Permit Water To Menace Homes." Circular 763, issued October 1953 by the Scientific Section, National Paint, Varnish, and Lacquer Association, Inc.

15. "The Wald-Way System of Component House Construction" by Howard L. Grange, R.P.E. The book is a licensed private communication.

16. "Thermal Insulation Made Of Wood-Base Materials, Its Application And Use In Houses" by L.V. Teesdale, Engineer, Forest Product Laboratory, Revised October 1958, FPL No. 1740.

17. "Understanding The Mechanisms Of Deterioration Of House Paint" by F.L. Browne, Forest Products Laboratory. Reprint from November 1959, Forest Products Journal (Vol IX, No. 11).

18. United States Department of Commerce Weather Bureau Map and Data "Mean Monthly, and Annual Total Snowfall (inches." U.S. Dept. of Commerce, National Oceanic and Atmospheric Administration, Environmental Data Service, National Climatic Center, Federal Building, Asheville, NC 28801 *"Taken from the Climatic Atlas of the U.S."*

19. "U.S. Patent #2,797,180 'Ventilated Roof Construction'" Howard L. Grange, Inventor. Canadian Patent #949718.

20. "U.S. Patent #3,683,785 'Roof Construction Providing Air Flow From Eave to Ridge'" Howard L. Grange, Inventor. Canadian #995,869.

21. "Weather And The Hand Of Man." Environmental Science Services Administration Pamphlet of the Department of Commerce.

22. "What To Do About Condensation" by E.R. Queer and E.R. McLaughlin, Engineering Professors of Penn State College, a technical reprint from HHFA Technical Bulletin No. 4 (May 1948) of Housing And Home Finance Agency, Division Of Housing Research. From Supt. of Documents.

23. "Wood Decay In Houses and How To Control It." Home and Garden Bulletin No. 73, U.S. Department of Agriculture. From Supt. of Documents.
24. "Wood Handbook." Handbook No. 72. Forest Products Laboratory, Forest Service, United States Department of Agriculture. From Supt. of Documents.
25. "Wood-Frame Construction" by L.O. Anderson and O.C. Heyer, Engineers, Forest Products Laboratory, Forest Service, U.S. Department of Agriculture. Handbook No. 73; from Supt. of Documents.

NOTES

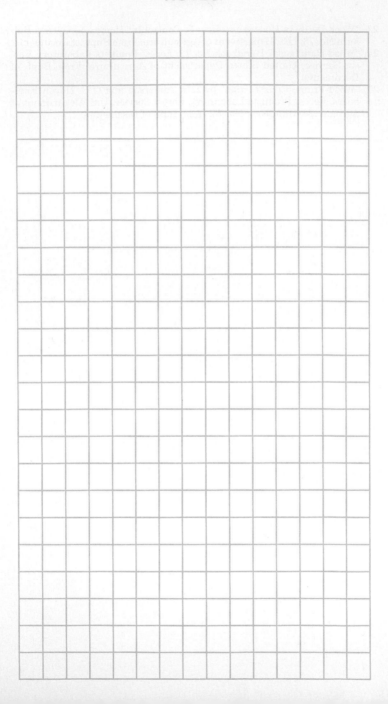

Graphic Standards
and Symbols

SYMBOL MEANING	SYMBOL	SYMBOL MEANING	SYMBOL
POINT OF CHANGE IN DUCT CONSTRUCTION (BY STATIC PRESSURE CLASS)		SUPPLY GRILLE (SG)	20 · 12 SG / 700 CFM
DUCT (1ST FIGURE, SIDE SHOWN 2ND FIGURE, SIDE NOT SHOWN)	20 · 12	RETURN (RG) OR EXHAUST (EG) GRILLE (NOTE AT FLR OR GLG)	20 · 12 RG / 700 CFM
ACOUSTICAL LINING DUCT DIMENSIONS FOR NET FREE AREA		SUPPLY REGISTER (SR) (A GRILLE · INTEGRAL VOL CONTROL)	20 · 12 SR / 700 CFM
DIRECTION OF FLOW		EXHAUST OR RETURN AIR INLET CEILING (INDICATE TYPE)	20 · 12 GR / 700 CFM
DUCT SECTION (SUPPLY)	S 30 · 12	SUPPLY OUTLET, CEILING, ROUND (TYPE AS SPECIFIED) INDICATE FLOW DIRECTION	20 / 700 CFM
DUCT SECTION (EXHAUST OR RETURN)	E OR R / 20 · 12	SUPPLY OUTLET, CEILING, SQUARE (TYPE AS SPECIFIED) INDICATE FLOW DIRECTION	12 · 12 / 700 CFM
INCLINED RISE (R) OR DROP (D) ARROW IN DIRECTION OF AIR FLOW	R	TERMINAL UNIT (GIVE TYPE AND OR SCHEDULE)	T U
TRANSITIONS GIVE SIZES NOTE F O T FLAT ON TOP OR F O B FLAT ON BOTTOM IF APPLICABLE		COMBINATION DIFFUSER AND LIGHT FIXTURE	
STANDARD BRANCH FOR SUPPLY & RETURN (NO SPLITTER)	S R	DOOR GRILLE	DG / 12 · 6
SPLITTER DAMPER		SOUND TRAP	ST
VOLUME DAMPER MANUAL OPERATION	VD	FAN & MOTOR WITH BELT GUARD & FLEXIBLE CONNECTIONS	
AUTOMATIC DAMPERS MOTOR OPERATED	SEC MOD	VENTILATING UNIT (TYPE AS SPECIFIED)	
ACCESS DOOR (AD) ACCESS PANEL (AP)	OR AD	UNIT HEATER (DOWNBLAST)	
FIRE DAMPER SHOW ◄VERTICAL POS. SHOW ◆HORIZ. POS	FD AD	UNIT HEATER (HORIZONTAL)	
SMOKE DAMPER	SD / AD	UNIT HEATER (CENTRIFUGAL FAN) PLAN	
CEILING DAMPER OR ALTERNATE PROTECTION FOR FIRE RATED CLG		THERMOSTAT	T
TURNING VANES		POWER OR GRAVITY ROOF VENTILATOR-EXHAUST (ERV)	
FLEXIBLE DUCT FLEXIBLE CONNECTION		POWER OR GRAVITY ROOF VENTILATOR-INTAKE (SRV)	
GOOSENECK HOOD (COWL)		POWER OR GRAVITY ROOF VENTILATOR-LOUVERED	
BACK DRAFT DAMPER	BDD	LOUVERS & SCREEN	36 · 24L

FIGURE 14-1   Symbols for ventilation and air conditioning. (*From* HVAC Duct Construction Standards—Metal and Flexible, *1st ed., © 1985 by the Sheet-Metal and Air Conditioning Contractor's National Association. Used with permission of the copyright holder.*)

Symbol	Description	Symbol	Description
———	Sanitary drain	—LPS—	Steam (low pressure)
––SS––	Sub soil drain	—MPS—	Steam (med. pressure)
––S––	Storm drain	—HPS—	Steam (high pressure)
—AW—	Acid waste	—LPR—	Condensate (low pressure)
––AV––	Acid vent	—MPR—	Condensate (med. pressure)
––––––	Vent	—HPR—	Condensate (high pressure)
—CW—	Cold water	—HWS—	Heating water supply
—DS—	Distilled water	—HWR—	Heating water return
—DW—	Drinking water	—EGS—	Ethylene glycol supply
—DWC—	Drinking water circulating	—EGR—	Ethylene glycol return
—HW—	Hot water	—CWS—	Chilled water supply
—HWC—	Hot water circulating	—CWR—	Chilled water return
–HW 180°–	Hot water 180°	—CPD—	Condensate or vacuum pump discharge
–HWC 180°–	Hot water 180° circulating	—CD—	Condensate drain
—TW—	Tempered water	—FOS—	Fuel oil supply
—TWC—	Tempered water circulating	—FOR—	Fuel oil return
—F—	Fire protection water service	––FOV––	Fuel oil vent
—SM—	Sprinkler main	—RL—	Refrigerant liquid line
—SP—	Stand pipe	—RS—	Refrigerant suction line
—G—	Natural gas	—RD—	Refrigerant hot gas discharge line
—A—	Compressed air	—CS—	Condenser water supply
—PN—	Pneumatic tube	—CR—	Condenser water return
—V—	Vacuum line	—BBO—	Boiler blow off
—OX—	Oxygen line	—ES—	Exhaust steam
—LPG—	L.P. gas	—HCWS—	Heating & cooling water supply
▬▬▬▬	Sanitary drain underground	—HCWR—	Heating & cooling water return
——◀——	Direction of flow	—①—	Balancing valve
—▨▨▨—	Flexible connector	→⊂—	Running trap
—▽—	Strainer	͞CO OR ͞CO	Clean out
—⊣∣⊢—	Union	—↺—	Vacuum breaker
—⊏▭⊐—	Expansion joint	—▷◁—	Gate valve
—⦵—	Automatic control valve	—⊗—	Globe valve
—⦵—	Pressure regulating valve	—↙—	Check valve
—⦵—	Safety relief valve		
—⦵—	Blow off valve	—Ⓡ—	Pressure relief valve
—▨#/h—	F&T trap (cap #/hr)	—▷◁—	Stop & waste valve
—⊗—	Thermostatic trap	—ↄ—	Stop & waste valve (in riser)
—Ō—	Radiator valve	—✳—	Automatic 3-way valve
—∅—	Flow control valve	—●—	Gas cock

FIGURE 14-2 Standard mechanical symbols. (*Source: A. M. Khashab,* Heating, Ventilating and Air Conditioning Systems Estimating Manual, *McGraw-Hill, New York, 1977. Used with permission.*)

Symbol	Description	Symbol	Description
	Air bleeder valve (radiant panel)		Fire alarm valve (sprinkler)
	Air vent		
	Gauge cock		Cold weather valve (sprinkler)
	Solenoid valve		Alarm gong (sprinkler)
	Thermostatic expansion valve (Refrigerant)	HB	Hose bibs
	Back pressure valve (refrigerant)	WH	Wall hydrant
	Sight glass	FH	Fire hydrant
	Pipe or round duct riser	MH	Manhole
GCO	Grade cleanout	SW	Street washer
WC	Water closet (tank type)	BT	Bath tub
WC	Water closet	SHD	Shower drain
UR	Urinal	FD	Floor drain
LAV	Lavatory	RD	Roof drain
KS	Kitchen sink	DS	Down spout
SS	Service sink	UH-1	Unit heater-propeller type
CSS	Clinic service sink	CUH	Cabinet unit heater
CRS	Classroom sink	9'0"-1-MBH	Fin tube (Figures = elements length, mark & MBH)
DF	Drinking fountain	C-1	Convector
EWC	Electric water cooler	UV-1	Unit ventilator
SM	Shower head	ID	Indirect waste drain
CS	Counter sink		
12/6	Rectangular duct—first figure is side show	12 X 6 G-1 250 6'-0"	Return or exh. grill (reg. similar) Figures = size, mark, CFM & distance above floor
12"	Round duct (figure=size)		
	Canvas connection		Round ceiling diffuser
	Volume damper		Square ceiling diffuser (supply)
	Turning vanes		Round ceiling diffuser (supply & return)
	Extractor		Square ceiling diffuser (supply & return)
	Supply duct		Square ceiling diffuser 3 way throw
	Return or exhaust duct		Square ceiling diffuser 2 way throw
ST	Sound trap		Motorized damper
12 X 6 R-1 250 6'-0"	Supply register (grill simular) figures = size, mark, cfm & distance above floor		Gravity damper
		FD-1	Fire damper
	Fire hose cabinet	1	Heating riser No.
F-1	Special fixture	1	Exhaust fan riser No.
	Column number	1	Detail number
11	Plumbing riser No.	M-2	Drawing number
101	Room number		

FIGURE 14-2 (*Continued*)

TABLE 14-1 Mechanical Drawing Symbols

Symbol	Description	Abbr.	Meaning
—— S ——	STEAM PIPE	MBH	THOUSAND BTU PER HOUR
– – – C – – –	CONDENSATE RETURN PIPE	GPM	GALLONS PER MINUTE
—— HWS ——	HOT WATER SUPPLY PIPE	CFM	CUBIC FEET PER MINUTE
– – HWR – –	HOT WATER RETURN PIPE	◯	ROUND
—— CWS ——	CHILLED WATER SUPPLY PIPE	⬚	SQUARE
– – CWR – –	CHILLED WATER RETURN PIPE	SA	SUPPLY AIR
—— HCS ——	COMB HOT - CHILLED WATER SUPPLY	RA	RETURN AIR
– – HCR – –	COMB HOT - CHILLED WATER RETURN	OA	OUTSIDE AIR
—— CS ——	CONDENSER WATER SUPPLY PIPE	EA	EXHAUST AIR
– – CR – –	CONDENSER WATER RETURN PIPE	HSWR	HIGH SIDEWALL REGISTER
—— D ——	DRAIN PIPE FROM COOLING COIL	HSWG	HIGH SIDEWALL GRILLE
—— FOS ——	FUEL OIL SUPPLY PIPE	LSWR	LOW SIDEWALL REGISTER
—— FOR ——	FUEL OIL RETURN PIPE	LSWG	LOW SIDEWALL GRILLE
—— R ——	REFRIGERANT PIPE	CSR	CEILING SUPPLY REGISTER
PIPE RISING	PIPE RISING	CR	CEILING REGISTER
PIPE TURNING DOWN	PIPE TURNING DOWN	CG	CEILING GRILLE
UNION	UNION	FR	FLOOR REGISTER
REDUCER - CONCENTRIC	REDUCER - CONCENTRIC	FG	FLOOR GRILLE
REDUCER - ECCENTRIC	REDUCER - ECCENTRIC	CD	CEILING DIFFUSER
STRAINER	STRAINER	TV	TURNING VANES
GATE VALVE	GATE VALVE	AE	AIR EXTRACTOR
GLOBE VALVE	GLOBE VALVE	SD	SPLITTER DAMPER
VALVE IN RISER	VALVE IN RISER	MD	MANUAL DAMPER
CHECK VALVE	CHECK VALVE	FD	FIRE DAMPER
PRESSURE REDUCING VALVE	PRESSURE REDUCING VALVE	DL	DUCT LINER IN DUCT
PRESSURE RELIEF VALVE	PRESSURE RELIEF VALVE	AHU	AIR HANDLING UNIT
SQUARE HEAD COCK	SQUARE HEAD COCK	BU	BLOWER UNIT
BALANCING VALVE	BALANCING VALVE	FCU	FAN COIL UNIT
3-WAY CONTROL VALVE	3-WAY CONTROL VALVE	HWC	HOT WATER CONVECTOR
2-WAY CONTROL VALVE	2-WAY CONTROL VALVE	UV	UNIT VENTILATOR
PITCH PIPE MINIMUM 1"/40'	PITCH PIPE MINIMUM 1"/40'	WH	WALL HEATER
ANCHOR LOCATION	ANCHOR LOCATION	UH	UNIT HEATER
FLEXIBLE PIPE CONNECTION	FLEXIBLE PIPE CONNECTION	WF	WALL FIN RADIATION
IN-LINE PUMP	IN-LINE PUMP	PRV	POWER ROOF VENTILATOR
BOTTOM TAKE-OFF	BOTTOM TAKE-OFF	UVS	UTILITY VENT SET
TOP TAKE-OFF	TOP TAKE-OFF	PF	PROPELLER FAN
		ⓉT	THERMOSTAT
PRESSURE GAUGE	PRESSURE GAUGE	Ⓣ N	NIGHT THERMOSTAT
THERMOMETER	THERMOMETER	Ⓣ M	THERMOSTAT - HEATING ONLY
HOT WATER RISER	HOT WATER RISER	Ⓣ C	THERMOSTAT - COOLING ONLY
CHILLED WATER RISER	CHILLED WATER RISER	Ⓣ—○	THERMOSTAT - REMOTE BULB
FAN COIL UNIT	FAN COIL UNIT	6'-8"	MOUNTING HEIGHT ABOVE FINISHED FLOOR
EQUIPMENT AS INDICATED	EQUIPMENT AS INDICATED	NIC	NOT IN CONTRACT
AIR INTO REGISTER	AIR INTO REGISTER		
AIR OUT OF REGISTER	AIR OUT OF REGISTER	⬙	SUPPLY AIR DUCT SECTION
AIR FLOW THRU UNDERCUT OR LOUVERED DOOR	AIR FLOW THRU UNDERCUT OR LOUVERED DOOR		
TURNING VANES	TURNING VANES	◁	RETURN OR EXHAUST DUCT SECTION
AIR EXTRACTOR	AIR EXTRACTOR	⬚	FLEXIBLE DUCT CONNECTION

SOURCE: John E. Traister, *Practical Drafting for the HVAC Trades,* Prentice-Hall, Englewood Cliffs, NJ. Used with permission of the author.

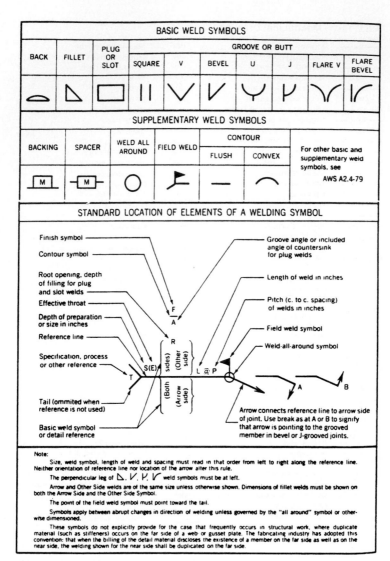

FIGURE 14-3 Welded joints standard symbols. (*Reproduced courtesy of American Institute of Steel Construction.*)

TABLE 14-2   Abbreviations of Organizations and Agencies

The names of technical societies, trade organizations, and governmental agencies are abbreviated as follows in HVAC documents:

AABC	Associated Air Balance Association
ABMA	American Boiler Manufacturers Association
AGA	American Gas Association
ADC	Air Diffusion Council
AMCA	Air Moving and Conditioning Association
ANSI	American National Standards Institute
ASHRAE	American Society of Heating, Refrigerating, and Air-Conditioning Engineers
ASME	American Society of Mechanical Engineers
ASTM	American Society for Testing and Materials
AWWA	American Water Works Association
FIA	Factory Insurance Association
FM	Factory Mutual Insurance Company
IBR	Institute of Boiler and Radiator Manufacturers
MCAA	Mechanical Contractors Association of America
NFPA	National Fire Protection Association
NBS	National Bureau of Standards
NEC	National Electrical Code NFPA pamphlet No. 70
NEMA	National Electrical Manufacturers Association
SBI	Steel Boiler Institute
SMACNA	Sheet Metal and Air Conditioning Contractors National Association
UL	Underwriters' Laboratories Incorporated

SOURCE: A. M. Khashab, *Heating, Ventilating, and Air Conditioning Systems Estimating Manual,* McGraw-Hill, New York, 1977. Used with permission.

TABLE 14-3   General Abbreviations

HVAC terms and units are abbreviated as follows. SI and ANSI abbreviations are shown in parentheses when they differ from the common HVAC abbreviations

ac	Alternating current
ACU	Air-conditioning unit
AHU	Air-handling unit
amp (A)	Ampere
ATC	Automatic temperature control
atm	Atmospheric
auto	Automatic
avg	Average
AWG	American wire gauge
BE	Beveled end
BF	Board foot
bhp	Brake horsepower or boiler horsepower
bldg	Building
bsmt	Basement
Btu	British thermal unit
Btuh (Btu/h)	British thermal unit per hour
°C	Degree centigrade (celsius)
CC	Cooling coil
C-C	Center to center
cc (cm^1)	Cubic centimeter
cf (ft^1)	Cubic foot
cfm (ft^1/min)	Cubic feet per minute
CH	Chiller
CI	Cast iron
cm	Centimeter
const	Construction
CS	Commercial standard
CT	Cooling tower
cu ft (ft^3)	Cubic foot

TABLE 14-3 General Abbreviations (*Continued*)

cu in (in³)	Cubic inch
cu m (m³)	Cubic meter
cu yd (yd³)	Cubic yard
CV	Constant volume
CY (yd³)	Cubic yard
db (dB)	Decibel
DBT	Dry-bulb temperature
dc	Direct current
DD	Dual duct
deg (°C or °F)	Degree
dia	Diameter
dia-in	Diameter inch
dia-mm	Diameter millimeter
disch	Discharge
dist	Distribution
dn	Down
DPT	Dew-point temperature
DWV	Drainage, waste, vent
D-X	Direct-expansion
EDR	Equivalent direct radiation
eff	Efficiency
cxh	Exhaust
exp	Expansion
°F	Degree Fahrenheit
F&T	Float and thermostatic
FBM	Foot board measure
FCU	Fan-coil unit
FtoF	Face to face
FIG	Figure
flex	Flexible
flr	Floor
FOR	Fuel oil return

TABLE 14-3   General Abbreviations (*Continued*)

FOS	Fuel oil supply
fpm (ft/min)	Feet per minute
FS	Fuel supply
ft	Foot
ga	Gauge or gage
gal	Gallon
gph (gal/h)	Gallons per hour
gpm (gal/min)	Gallons per minute
hex	Hexagonal
horz	Horizontal
H.P.	High pressure
hp	Horsepower
hr (h)	Hour
ht	Height
HV	Heating and ventilating
HVAC	Heating, ventilating, and air conditioning
IB	Iron body
ID	Inside diameter
in	Inches
IPS	Iron-pipe size
IU	Induction unit
jnt	Joint
K	Degree Kelvin
kg	Kilogram
kv (kV)	Kilovolt
kw (kW)	Kilowatt
lb	Pound
ld	Load
LF	Linear feet
LH	Latent heat
L.P.	Low pressure
liq	Liquid

TABLE 14-3   General Abbreviations (*Continued*)

m	Meter
M (k)	Thousand
max	Maximum
MB	Mixing box
MBH	Thousand British thermal units per hour
M/C	Machine
mech	Mechanical
M&F	Male and female
MIN	Minimum
min	Minute
mm	Millimeter
mph (mi/h)	Miles per hour
MPS (m/s)	Meters per second
mtd	Mounted
MZ	Multizone
No.	Number
NP	Nominal pipe size
OAI	Outside-air intake
OD	Outside diameter
O&P	Overhead and profit
OSA	Outdoor supply air
OS&Y	Outside screw and yoke
oz	Ounce
P	Pump
pct	Percentage
PE	Plain end
PG	Pressure gage
pneu	Pneumatic
pr	Pair
PRV	Pressure-reducing valve
PSF (lb/ft^2)	Pounds per square foot
psi (lb/in^2)	Pounds per square inch

TABLE 14-3 General Abbreviations (*Continued*)

PVC	Polyvinyl chloride
R	Degree Rankine
RA	Return air
Rej.	Reject
ret	Return
RF	Refrigeration machine
RH	Relative humidity
rm	Room
rnd	Round
rpm (r/min)	Revolutions per minute
SA	Supply air
sch	Schedule
scr	Screw
sec (s)	Second
SH	Sensible heat
spec	Specification
sp gr	Specific gravity
sq	Square
SR	Steam return
sq in (in^2)	Square inch
sq ft (ft^2)	Square foot
sq yd (yd^2)	Square yard
SP	Static pressure
SS	Stainless steel
std	Standard
stm	Steam
supp	Supply
SZ	Single-zone
T (ton)	Ton
T&C	Thread and coupling
temp	Temperature
thermo	Thermostat or thermometer

TABLE 14-3 General Abbreviations (*Continued*)

TR	Ton of refrigeration
USG	United States gauge
USS	United States standard
V	Valve
V	Volt
VAV	Variable air volume
vel	Velocity
vert	Vertical
vol	Volume
VV	Variable volume
WBT	Wet-bulb temperature
wt	Weight
xhvy	Extra heavy
yd	Yard
yr	Year
%	Percentage
#	Number or pound

SOURCE: A. M. Khashab, *Heating, Ventilating and Air Conditioning Systems Estimating Manual,* McGraw-Hill, New York, 1977. Used with permission.

TABLE 14-4   ASHRAE SI for HVAC and R Conversions

Multiply	By	To Obtain	Multiply	By	To Obtain
acre	0.405	ha	Btu/(lb · °F) (specific heat, c)	4.19	kJ/(kg · K)
bar	100	kPa			kJ/(kg · °C)
barrel (42 US gal, petroleum)	159	L	bushel	0.0352	m^3
			calorie, gram	4.19	J
Btu, IT	0.159	m^3	calorie, kilogram; kilocalorie	4.19	kJ
Btu/ft³	1.055	kJ	centipoise, viscosity, μ		
Btu/ft³	37.3	kJ/m³; J/L	(absolute, dynamic)	*1.00	mPa · s
Btu/gal	0.279	kJ/L	centistokes, kinematic viscosity, ν	*1.00	mm²/s
Btu · ft/h · ft² · °F	1.731	W/(m · K)	cost, $ per square (100 sq ft)	0.108	$/m²
Btu · in/(h · ft² · °F)			cost, $ per square foot	10.8	$/m²
(thermal conductivity, k)	0.144	W/(m · K)	cost, $ per pound	2.20	$/kg
		W/(m · °C)	cost, $ per ton (refrigeration)	*0.100	$/kW
Btu/h	0.293	W	dyne/cm²	11.4	Pa
Btu/ft²	11.4	kJ/m²	EDR hot water (150 Btu/h)	44.0	W
Btu/(y · ft²)	0.000293	kWh/(y · m²)	EDR steam (240 Btu/h)	70.3	W
		(not SI)	EER	0.293	COP
Btu/(y · ft²)	0.0000114	GJ/(y · m²)	fuel cost comparison @ 100% efficiency		
Btu/(h · ft²)	3.15	W/m²	cents per gallon	0.264	¢/L
Btu/(h · ft² · °F)			cents per gallon (no. 2 fuel oil)	0.0677	$/GJ
(overa heat trans coeff, U)	5.68	W/(m² · K)	cents per gallon (no. 6 fuel oil)	0.0632	$/GJ
(thermal conductance, C)		W/(m² · °C)	cents per gallon (propane)	0.113	$/GJ
Btu/lb	2.33	kJ/kg			

*Conversion factor is exact.

Units are US values unless otherwise noted.

cents per kWh	2.78	$/GJ
cents per therm	0.0948	$/GJ
ft	*0.3048	m
ft	*304.8	mm
ft/min, fpm	0.00508	m/s
ft/s, fps	*0.3048	m/s
ft of water	2.99	kPa
ft of water per 100 ft pipe	0.0981	kPa/m
ft²	0.0929	m²
ft²·h·°F/Btu (thermal resistance, R)	0.176	m²·K/W m²·°C/W
ft²/s, kinematic viscosity, ν	92 900	mm²/s
ft³	28.3	L
ft³	0.0283	m³
ft³/h, cfh	7.87	mL/s
ft³/min, cfm	0.472	L/s
ft³/s, cfs	28.3	L/s
ft·lb$_f$ (torque or moment)	1.36	N·m
ft·lb$_f$ (work)	1.36	J
ft·lb$_f$/lb (specific energy)	2.99	J/kg
ft·lb$_f$/min (power)	0.0226	W
gallon (US, 231 in³)	3.79	L
gallon	0.00379	m³
gph	1.05	mL/s
gpm	0.0631	L/s
gpm/ton refrigeration	0.0179	mL/J
grain (1/7000 lb)	0.0648	g
gr/gal	17.1	mg/L
gr/lb	0.143	g/kg
horsepower (boiler)	9.81	kW
horsepower (550 ft·lb$_f$/s)	0.746	kW
inch	*25.4	mm
in. of mercury (60°F)	3.38	kPa
in. of water (60°F)	249	Pa
mile	1.61	km
mile, nautical	1.85	km
mph	1.61	km/h
mph	0.44	m/s
millibar	*0.100	kPa
mm of mercury (60°F)	0.133	kPa
mm of water (60°F)	9.80	Pa
metre of water	9.80	kPa
ounce (mass, avoirdupois)	28.3	g
ounce (force or thrust)	0.278	N
ounce (liquid, US)	29.6	mL
ounce inch (torque, moment)	7.06	mN·m
ounce (avoirdupois) per gallon	7.49	g/L
perm (permeance)	57.4	ng/(s·m²·Pa)
perm inch (permeability)	1.46	ng/(s·m·Pa)
pint (liquid, US)	473	mL
pound		
lb (mass)	0.454	kg
lb (mass)	454	g
lb$_f$(force or thrust)	4.45	N
lb$_f$/ft (uniform load)	1.49	kg/m
lb$_m$/(ft·h) viscosity (absolute, dynamic, μ)	0.413	mPa·s
lb$_f$/(ft·s) viscosity (absolute, dynamic, μ)	1 490	mPa·s
lb/h	0.126	g/s
lb/min	0.00756	kg/s
lb of steam per hour @ 212°F (100°C)	0.284	kW
lb$_f$/ft²	47.9	Pa
lb$_f$·s/ft²viscosity (absolute dynamic μ)	47 900	mPa·s
lb/ft²	4.88	kg/m²
lb/ft³ (density, ϱ)	16.0	kg/m³
lb/gallon	120	kg/m³

TABLE 14-4 ASHRAE SI for HVAC and R Conversions (Continued)

Multiply	By	To Obtain	Multiply	By	To Obtain
in/100 ft, thermal expansion	0.833	mm/m	ppm (by mass)	*1.00	mg/kg
in·lb_f (torque or moment)	113	mN·m	psi	6.89	kPa
in²	645	mm²	quad	1.055	EJ
in³ (volume)	16.4	mL	quart (liquid U.S.)	0.946	L
in³/min (SCIM)	0.273	mL/s	square (100 sq ft)	9.29	m²
in³ (section modulus)	16 400	mm³	tablespoon (approximately)	15	mL
in⁴ (section moment)	416 000	mm⁴	teaspoon (approximately)	5	mL
km/h	0.278	m/s	therm (US)	105.5	MJ
kWh	*3.60	MJ	ton, long (2 240 lb)	1.016	t (tonne); Mg
kWh/(y·ft²)	0.0388	GJ/(y·m²)	ton, short (2 000 lb)	0.907	t (tonne); Mg
kWh/1000 cfm	2.12	J/L	ton, refrigeration (12 000 Btu/h)	3.52	kW
kilopond (kg force)	9.81	N	torr (1 mm Hg@0°C)	133	Pa
kip (1000 lb)	4.45	kN	watt per square foot	10.8	W/m²
kip/in² (ksi)	6.89	MPa	yd	*0.9144	m
litre	*0.001	m³	yd²	0.836	m²
micron of mercury (60°F)	133	mPa	yd³	0.765	m³
To Obtain	**By**	**Divide**	**To Obtain**	**By**	**Divide**

*Conversion factor is exact.

Units are US values unless otherwise noted.

SOURCE: *ASHRAE Handbook—1989 Fundamentals.* Reprinted by permission.

TABLE 14-5 Nomenclature

ACFM = actual cubic feet per minute (i.e. at process conditions)

A_p = cross sectional area of piston, sq in.

A_r = cross sectional area of piston rod, sq in.

Bhp = brake or shaft horsepower

C = cylinder clearance as a per cent of cylinder volume

C_p = specific heat at constant pressure, Btu/(lb · °F)

C_v = specific heat at constant volume, Btu/(lb · °F)

D = cylinder inside diameter, in.

d = piston rod diameter, in.

F = an allowance for interstage pressure drop, Eq 10.4

Ghp = gas horsepower, actual compression horsepower, excluding mechanical losses, bhp

H = head, ft · lb/lb

h = enthalpy, Btu/lb

ICFM = inlet cubic feet per minute, usually at suction conditions

k = isentropic exponent, C_p/C_v

MC_p = molal specific heat at constant pressure, Btu/(lb mol · °F)

MC_v = molar specific heat at constant volume, Btu/(lb mol · °F)

MW = molecular weight

mm cfd = million cubic ft/day

N = speed, rpm

N_m = molal flow, mols/min

n = polytropic exponent or number of mols

P = pressure, psia

P_c = critical pressure, psia

PD = piston displacement, ft³/min

P_t = pressure base used in the contract or regulation, psia

pP_c = pseudo critical pressure, psia

P_R = reduced pressure, P/P_c

pT_c = pseudo critical temperature, °R

P_v = partial pressure of contained moisture, psia

p = pressure, lb/ft²

Q = inlet capacity (ICFM)

R = universal gas constant

  = $10.73 \dfrac{\text{psia} \cdot \text{ft}^3}{\text{lb mol} \cdot °R}$

  = $1545 \dfrac{(\text{lb/ft}^2) \cdot \text{ft}^3}{\text{lb mol} \cdot °R}$ or $\dfrac{\text{ft} \cdot \text{lb}}{\text{lb mol} \cdot °R}$

  = $1.986 \dfrac{\text{Btu}}{\text{lb mol} \cdot °R}$

r = compression ratio, P_2/P_1

s = entropy, Btu/(lb · °R) or number of wheels

SCFM = standard cubic feet per minute measured at 14.7 psia and 60°F

stroke = length of piston movement, in.

T = absolute temperature, °R

T_c = critical temperature, °R

T_R = reduced temperature, T/T_c

t = temperature, °F

V = specific volume, ft³/lb

VE = volumetric efficiency, per cent

W = work, ft · lb

w = weight flow, lb/min

X = temperature rise factor

y = mol fraction

Z = compressibility factor*

Z_{avg} = average compressibility factor = $\dfrac{Z_1 + Z_d}{2}$*

η = efficiency, expressed as a decimal

Subscripts

d = discharge

is = isentropic process

p = polytropic process

S = standard conditions, usually 14.7 psia, 60°F

s = suction

t = total or overall

1 = inlet conditions

2 = outlet conditions

L = standard conditions used for calculation or contract

SOURCE: *Compressed Air and Gas Handbook,* 5th ed., 1988. Used with permission from CAGI, 1300 Sumner Ave., Cleveland, Ohio 44115.

NOTES

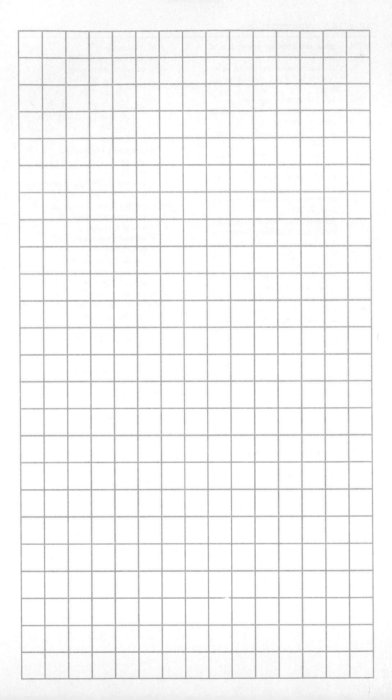

Metric Data
and Conversion*

* This section adapted with permission from R. O. Parmley, *Mechanical Components Handbook,* McGraw-Hill, New York, 1985.

15-1 Conversion Factors

Units derived from SI base units are shown in Table 15-1.

Table 15-2 gives the definitions of various units of measure that are exact numerical multiples of coherent SI units, and provides multiplication factors for converting numbers and miscellaneous units to corresponding new numbers and SI units.

The first two digits of each numerical entry represent a power of 10. An asterisk follows each number that expresses an exact definition. For example, the entry "–02 2.54*" expresses the fact that 1 inch = 2.54 × 10^{-2} meter, exactly, by definition. Most of the definitions are extracted from National Bureau of Standards documents. Numbers not followed by an asterisk are only approximate or are the results of physical measurements. The conversion factors are listed alphabetically and by physical quantity.

The listing by physical quantity includes only relationships which are frequently encountered and deliberately omits the many combinations of units which are used for more specialized purposes. Conversion factors for combinations of units are easily generated from numbers given in the alphabetical listing by the technique of direct substitution or by other well-known rules for manipulating units. These units are adequately discussed in many science and engineering textbooks and are not repeated here.

TABLE 15-1 Derived Units of the International System

Quantity	Name of unit	Unit symbol or abbreviation, where differing from base form	Unit expressed in terms of base or supplementary units†
Area	square meter		m²
Volume	cubic meter		m³
Frequency	hertz, cycle per second‡	Hz	s⁻¹
Density	kilogram per cubic meter		kg/m³
Velocity	meter per second		m/s
Angular velocity	radian per second	،	rad/s
Acceleration	meter per second squared		m/s²
Angular acceleration	radian per second squared		rad/s²
Volumetric flow rate	cubic meter per second		m³/s
Force	newton	N	kg·m/s²
Surface tension	newton per meter, joule per square meter	N/m, J/m²	kg/s²
Pressure	newton per square meter, pascal	N/m², Pa	kg/m·s²
Viscosity, dynamic	newton-second per square meter, pascal-second	N·s/m², Pa·s	kg/m·s
Viscosity, kinematic	meter squared per second		m²/s
Work, torque, energy, quantity of heat	joule, newton-meter, watt-second	J, N·m, W·s	kg·m²/s²
Power, heat flux	watt, joule per second	W, J/s	kg·m²/s³
Heat flux density	watt per square meter	W/m²	kg/s³
Volumetric heat release rate	watt per cubic meter	W/m³	kg/m·s³
Heat transfer coefficient	watt per square meter-kelvin	W/m²·K	kg/s³·K
Heat capacity (specific)	joule per kilogram-kelvin	J/kg·K	m²/s²·K
Capacity rate	watt per kelvin	W/K	kg·m²/s³·K
Thermal conductivity	watt per meter-kelvin	W/m·deg, J·m/s·m²·K	kg·m/s³·K
Quantity of electricity	coulomb	C	A·s
Electromotive force	volt	V, W/A	kg·m²/A·s³
Electric field strength	volt per meter		V/m
Electric resistance	ohm	Ω, V/A	kg·m²/A²·s³
Electric conductivity	ampere per volt-meter	A/V·m	A²s³/kg·m³
Electric capacitance	farad	F, A·s/V	A²s⁴/kg·m²
Magnetic flux	weber	Wb, V·s	kg·m²/A·s²
Inductance	henry	H, V·s/A	kg·m²/A²s²
Magnetic permeability	henry per meter	H/m	kg·m/A²s²
Magnetic flux density	tesla, weber per square meter	T, Wb/m²	kg/A·s²
Magnetic field strength	ampere per meter		A/m
Magnetomotive force	ampere		A
Luminous flux	lumen	lm	cd·sr
Luminance	candela per square meter		cd/m²
Illumination	lux, lumen per square meter	lx, lm/m²	cd·sr/m²

† Supplementary units are plane angle, radian (rad), solid angle, steradian (sr).
SOURCE: Adapted from T. G. Hicks, *Metrication Manual,* McGraw-Hill, New York, 1972. Used with permission of the publisher.

TABLE 15-2   Conversion Factors as Extracted Multiples of SI Units

To convert from	To	Multiply by
abampere	ampere	+01 1.00*
abcoulomb	coulomb	+01 1.00*
abfarad	farad	+09 1.00*
abhenry	henry	−09 1.00*
abmho	siemens	+09 1.00*
abohm	ohm	−09 1.00*
abvolt	volt	−08 1.00*
acre	meter^2	+03 4.046 873
ampere (international of 1948)	ampere	−01 9.998 35
angstrom	meter	−10 1.00*
are	meter^2	+02 1.00*
astronomical unit	meter	+11 1.495 979
atmosphere (standard)	pascal (newton/meter^2)	+05 1.013 250*
bar	pascal (newton/meter^2)	+05 1.00*
barn	meter^2	−28 1.00*
barrel (petroleum, 42 gallons)	meter^3	−01 1.589 873
barye	newton/meter^2	−01 1.00*
British thermal unit (ISO/TC 12)	joule	+03 1.055 06
British thermal unit (International Steam Table)	joule	+03 1.055 04
British thermal unit (mean)	joule	+03 1.055 87
British thermal unit (thermochemical)	joule	+03 1.054 350 264 488
British thermal unit (39°F)	joule	+03 1.059 67
British thermal unit (60°F)	joule	+03 1.054 68
bushel (U.S.)	meter^3	−02 3.523 907 016 688*
cable	meter	+02 2.194 56*
caliber	meter	−04 2.54*
calorie (International Steam Table)	joule	+00 4.1868
calorie (mean)	joule	+00 4.190 02
calorie (thermochemical)	joule	+00 4.184*
calorie (15°C)	joule	+00 4.185 80
calorie (20°C)	joule	+00 4.181 90
calorie (kilogram, International Steam Table)	joule	+03 4.1868
calorie (kilogram, mean)	joule	+03 4.190 02
calorie (kilogram, thermochemical)	joule	+03 4.184*
carat (metric)	kilogram	−04 2.00*
Celsius (temperature)	kelvin	$t_K = t_C = 273.15$
centimeter of mercury (0°C)	newton/meter^2	+03 1.333 22
centimeter of water (4°C)	newton/meter^2	+01 9.806 38
chain (engineer or ramden)	meter	+01 3.048*
chain (surveyor or gunter)	meter	+01 2.011 68*
circular mil	meter^2	−10 5.067 074 8
cord	meter^3	+00 3.624 556 3
coulomb (international of 1948)	coulomb	−01 9.998 35
cubit	meter	−01 4.572*

TABLE 15-2 Conversion Factors as Extracted Multiples of SI Units
(*Continued*)

To convert from	To	Multiply by
cup	meter^3	-04 2.365 882 365*
curie	disintegration/second	$+10$ 3.70*
day (mean solar)	second (mean solar)	$+04$ 8.64*
day (sidereal)	second (mean solar)	$+04$ 8.616 409 0
degree (angle)	radian	-02 1.745 329 251 994 3
denier (international)	kilogram/meter	-07 1.00*
dram (avoirdupois)	kilogram	-03 1.771 845 195 312 5*
dram (troy or apothecary)	kilogram	-03 3.887 934 6*
dram (U.S. fluid)	meter^3	-06 3.696 691 195 312 5*
dyne	newton	-05 1.00*
electron-volt	joule	-19 1.602 10
erg	joule	-07 1.00*
Fahrenheit (temperature)	kelvin	$t_K = (5/9) (t_F + 459.67)$
Fahrenheit (temperature)	Celsius	$t_C = (5/9) (t_F - 32)$
farad (international of 1948)	farad	-01 9.995 05
faraday (based on carbon 12)	coulomb	$+04$ 9.648 70
faraday (chemical)	coulomb	$+04$ 9.649 57
faraday (physical)	coulomb	$+04$ 9.652 19
fathom	meter	$+00$ 1.828 8*
fermi (femtometer)	meter	-15 1.00*
fluid ounce (U.S.)	meter^3	-05 2.957 352 956 25*
foot	meter	-01 3.048*
foot (U.S. survey)	meter	$+00$ 1200/3937*
foot (U.S. survey)	meter	-01 3.048 006 096
foot of water (39.2°F)	newton/meter^2	$+03$ 2.988 98
foot-candle	lumen/meter^2	$+01$ 1.076 391 0
foot-lambert	candela/meter^2	$+00$ 3.426 259
furlong	meter	$+02$ 2.011 68*
gal (galileo)	meter/second^2	-02 1.00*
gallon (U.K. liquid)	meter^3	-03 4.546 087
gallon (U.S. dry)	meter^3	-03 4.404 883 770 86*
gallon (U.S. liquid)	meter^3	-03 3.785 411 784*
gamma	tesla	-09 1.00*
gauss	tesla	-04 1.00*
gilbert	ampere-turn	-01 7.957 747 2
gill (U.K.)	meter^3	-04 1.420 652
gill (U.S.)	meter^3	-04 1.182 941 2
grad	degree (angular)	-01 9.00*
grad	radian	-02 1.570 796 3
grain	kilogram	-05 6.479 891*
gram	kilogram	-03 1.00*
hand	meter	-01 1.016*
hectare	meter^2	$+04$ 1.00*
henry (international of 1948)	henry	$+00$ 1.000 495
hogshead (U.S.)	meter^3	-01 2.384 809 423 92*
horsepower (550 foot lbf/second)	watt	$+02$ 7.456 998 7

TABLE 15-2   Conversion Factors as Extracted Multiples of SI Units (*Continued*)

To convert from	To	Multiply by
horsepower (boiler)	watt	+03 9.809 50
horsepower (electric)	watt	+02 7.46*
horsepower (metric)	watt	+02 7.354 99
horsepower (U.K.)	watt	+02 7.457
horsepower (water)	watt	+02 7.460 43
hour (mean solar)	second (mean solar)	+03 3.60*
hour (sidereal)	second (mean solar)	+03 3.590 170 4
hundredweight (long)	kilogram	+01 5.080 234 544*
hundredweight (short)	kilogram	+01 4.535 923 7*
inch	meter	−02 2.54*
inch of mercury (32°F)	pascal (newton/meter²)	+03 3.386 38
inch of mercury (60°F)	pascal (newton/meter²)	+03 3.376 85
inch of water (39.2°F)	pascal (newton/meter²)	+02 2.490 82
inch of water (60°F)	pascal (newton/meter²)	+02 2.488 4
joule (international of 1948)	joule	+00 1.000 165
kayser	1/meter	+02 1.00*
kilocalorie (International Steam Table)	joule	+03 4.186 74
kilocalorie (mean)	joule	+03 4.190 02
kilocalorie (thermochemical)	joule	+03 4.184*
kilogram mass	kilogram	+00 1.00*
kilogram force (kgf)	newton	+00 9.806 65*
kilopond force	newton	+00 9.806 65*
kip	newton	+03 4.448 221 615 260 5*
knot (international)	meter/second	−01 5.144 444 444
lambert	candela/meter²	+04 1/π*
lambert	candela/meter²	+03 3.183 098 8
langley	joule/meter²	+04 4.184*
lbf (pound force, avoirdupois)	newton	+00 4.448 221 615 260 5*
lbm (pound mass, avoirdupois)	kilogram	−01 4.535 923 7*
league (British nautical)	meter	+03 5.559 552*
league (international nautical)	meter	+03 5.556*
league (statute)	meter	+03 4.828 032*
light year	meter	+15 9.460 55
link (engineer or ramden)	meter	−01 3.048*
link (surveyor or gunter)	meter	−01 2.011 68*
liter	meter³	−03 1.00*
lux	lumen/meter²	+00 1.00*
maxwell	weber	−08 1.00*
meter	wavelengths Kr 86	+06 1.650 763 73*
micron	meter	−06 1.00*
mil	meter	−05 2.54*
mile (U.S. statute)	meter	+03 1.609 344*

TABLE 15-2   Conversion Factors as Extracted Multiples of SI Units (Continued)

To convert from	To	Multiply by
mile (U.K. nautical)	meter	+03 1.853 184*
mile (international nautical)	meter	+03 1.852*
mile (U.S. nautical)	meter	+03 1.852*
millibar	newton/meter^2	+02 1.00*
millimeter of mercury (0°C)	newton/meter^2	+02 1.333 224
minute (angle)	radian	−04 2.908 882 086 66
minute (mean solar)	second (mean solar)	+01 6.00*
minute (sidereal)	second (mean solar)	+01 5.983 617 4
month (mean calendar)	second (mean solar)	+06 2.628*
nautical mile (international)	meter	+03 1.852*
nautical mile (U.S.)	meter	+03 1.852*
nautical mile (U.K.)	meter	+03 1.853 184*
oersted	ampere/meter	+01 7.957 747 2
ohm (international of 1948)	ohm	+00 1.000 495
ounce force (avoirdupois)	newton	−01 2.780 138 5
ounce mass (avoirdupois)	kilogram	−02 2.834 952 312 5*
ounce mass (troy or apothecary)	kilogram	−02 3.110 347 68*
ounce (U.S. fluid)	meter^3	−05 2.957 352 956 25*
pace	meter	−01 7.62*
parsec	meter	+16 3.083 74
pascal	newton/meter^2	+00 1.00*
peck (U.S.)	meter^3	−03 8.809 767 541 72*
pennyweight	kilogram	−03 1.555 173 84*
perch	meter	+00 5.0292*
phot	lumen/meter^2	+04 1.00
pica (printers)	meter	−03 4.217 517 6*
pint (U.S. dry)	meter^3	−04 5.506 104 713 575*
pint (U.S. liquid)	meter^3	−04 4.731 764 73*
point (printers)	meter	−04 3.514 598*
poise	newton-second/meter^2	−01 1.00*
pole	meter	+00 5.0292*
pound force (lbf avoirdupois)	newton	+00 4.448 221 615 260 5*
pound mass (lbm avoirdupois)	kilogram	−01 4.535 923 7*
pound mass (troy or apothecary)	kilogram	−01 3.732 417 216*
poundal	newton	−01 1.382 549 543 76*
quart (U.S. dry)	meter^3	−03 1.101 220 942 715*
quart (U.S. liquid)	meter^3	−04 9.463 529 5
rad (radiation dose absorbed)	joule/kilogram	−02 1.00*
Rankine (temperature)	kelvin	$t_K = (5/9)t_R$
rayleigh (rate of photon emission)	1/second-meter^2	+10 1.00*
rhe	meter^2/newton-second	+01 1.00*
rod	meter	+00 5.0292*
roentgen	coulomb/kilogram	−04 2.579 76*
rutherford	disintegration/second	+06 1.00*

TABLE 15-2   Conversion Factors as Extracted Multiples of SI Units (*Continued*)

To convert from	To	Multiply by
second (angle)	radian	−06 4.848 136 811
second (ephemeris)	second	+00 1.000 000 000
second (mean solar)	second (ephemeris)	Consult American Ephemeris and Nautical Almanac
second (sidereal)	second (mean solar)	−01 9.972 695 7
section	meter^2	+06 2.589 988 110 336*
scruple (apothecary)	kilogram	−03 1.295 978 2*
shake	second	−08 1.00
skein	meter	+02 1.097 28*
slug	kilogram	+01 1.459 390 29
span	meter	−01 2.286*
statampere	ampere	−10 3.335 640
statcoulomb	coulomb	−10 3.335 640
statfarad	farad	−12 1.112 650
stathenry	henry	+11 8.987 554
statmho	mho	−12 1.112 650
statohm	ohm	+11 8.987 554
statute mile (U.S.)	meter	+03 1.609 344*
statvolt	volt	+02 2.997 925
stere	meter^3	+00 1.00*
stilb	candela/meter^2	+04 1.00
stoke	meter^2/second	−04 1.00*
tablespoon	meter^3	−05 1.478 676 478 125*
teaspoon	meter^3	−06 4.928 921 593 75*
ton (assay)	kilogram	−02 2.916 666 6
ton (long)	kilogram	+03 1.016 046 908 8*
ton (metric)	kilogram	+03 1.00*
ton (explosive energy of one ton of TNT)	joule	+09 4.184
ton (register)	meter^3	+00 2.831 684 659 2*
ton (short, 2000 pound)	kilogram	+02 9.071 847 4*
tonne	kilogram	+03 1.00*
torr (0°C)	newton/meter^2	+02 1.333 22
township	meter^2	+07 9.323 957 2
unit pole	weber	−07 1.256 637
volt (international of 1948)	volt	+00 1.000 330
watt (international of 1948)	watt	+00 1.000 165
yard	meter	−01 9.144*
year (calendar)	second (mean solar)	+07 3.1536*
year (sidereal)	second (mean solar)	+07 3.155 815 0
year (tropical)	second (mean solar)	+07 3.155 692 6
year 1900, tropical, Jan., day 0, hour 12	second (ephemeris)	+07 3.155 692 597 47*
year 1900, tropical, Jan., day 0, hour 12	second	+07 3.155 692 597 47

TABLE 15-2   Conversion Factors as Extracted Multiples of SI Units
(*Continued*)

To convert from	To	Multiply by
	LISTING BY PHYSICAL QUANTITY	
	Acceleration	
foot/second^2	meter/second^2	−01 3.048*
free fall, standard	meter/second^2	+00 9.806 65*
gal (galileo)	meter/second^2	−02 1.00*
inch/second^2	meter/second^2	−02 2.54*
	Area	
acre	meter^2	+03 4.046 856 422 4*
are	meter^2	+02 1.00*
barn	meter^2	−28 1.00*
circular mil	meter^2	−10 5.067 074 8
foot^2	meter^2	−02 9.290 304*
hectare	meter^2	+04 1.00*
inch^2	meter^7	−04 6.4516*
mile^2 (U.S. statute)	meter^2	+06 2.589 988 110 336*
section	meter^2	+06 2.589 988 110 336*
township	meter^2	+07 9.323 957 2
yard^2	meter^2	−01 8.361 273 6*
	Density	
gram/centimeter^3	kilogram/meter^3	+03 1.00*
lbm/inch^3	kilogram/meter^3	+04 2.767 990 5
lbm/foot^3	kilogram/meter^3	+01 1.601 846 3
slug/foot^3	kilogram/meter^3	+02 5.153 79
	Energy	
British thermal unit (ISO/TC 12)	joule	+03 1.055 06
British thermal unit (International Steam Table)	joule	+03 1.055 04
British thermal unit (mean)	joule	+03 1.055 87
British thermal unit (thermochemical)	joule	+03 1.054 350 264 488
British thermal unit (39°F)	joule	+03 1.059 67
British thermal unit (60°F)	joule	+03 1.054 68
calorie (International Steam Table)	joule	+00 4.1868
calorie (mean)	joule	+00 4.190 02
calorie (thermochemical)	joule	+00 4.184*
calorie (15°C)	joule	+00 4.185 80
calorie (20°C)	joule	+00 4.181 90
calorie (kilogram, International Steam Table)	joule	+03 4.1868
calorie (kilogram, mean)	joule	+03 4.190 02
calorie (kilogram, thermochemical)	joule	+03 4.184*

TABLE 15-2   Conversion Factors as Extracted Multiples of SI Units (*Continued*)

To convert from	To	Multiply by
electron-volt	joule	−19 1.602 10
erg	joule	−07 1.00*
foot-lbf	joule	+00 1.355 817 9
foot-poundal	joule	−02 4.214 011 0
joule (international of 1948)	joule	+00 1.000 165
kilocalorie (International Steam Table)	joule	+03 4.1868
kilocalorie (mean)	joule	+03 4.190 02
kilocalorie (thermochemical)	joule	+03 4.184*
kilowatt-hour	joule	+06 3.60*
kilowatt-hour (international of 1948)	joule	+06 3.600 59
ton (nuclear equivalent of TNT)	joule	+09 4.20
watt-hour	joule	+03 3.60*

Energy/area time		
Btu (thermochemical)/foot^2-second	watt/meter^2	+04 1.134 893 1
Btu (thermochemical)/foot^2-minute	watt/meter^2	+02 1.891 488 5
Btu (thermochemical)/foot^2-hour	watt/meter^2	+00 3.152 480 8
Btu (thermochemical)/inch^2-second	watt/meter^2	+06 1.634 246 2
calorie (thermochemical)/ centimeter^2-minute	watt/meter^2	+02 6.973 333 3
erg/centimeter^2-second	watt/meter^2	−03 1.00*
watt/centimeter^2	watt/meter^2	+04 1.00*

Force		
dyne	newton	−05 1.00*
kilogram force (kgf)	newton	+00 9.806 65*
kilopond force	newton	+00 9.806 65*
kip	newton	+03 4.448 221 615 260 5*
lbf (pound force, avoirdupois)	newton	+00 4.448 221 615 260 5*
ounce force (avoirdupois)	newton	−01 2.780 138 5
pound force, lbf (avoirdupois)	newton	+00 4.448 221 615 260 5*
poundal	newton	−01 1.382 549 543 76*

Length		
angstrom	meter	−10 1.00*
astronomical unit	meter	+11 1.495 978 9
cable	meter	+02 2.194 56*
caliber	meter	−04 2.54*
chain (surveyor or gunter)	meter	+01 2.011 68*
chain (engineer or ramden)	meter	+01 3.048*
cubit	meter	−01 4.572*
fathom	meter	+00 1.8288*
fermi (femtometer)	meter	−15 1.00*

**TABLE 15-2 Conversion Factors as Extracted Multiples of SI Units
(Continued)**

To convert from	To	Multiply by
foot	meter	−01 3.048*
foot (U.S. survey)	meter	+00 1200/3937*
foot (U.S. survey)	meter	−01 3.048 006 096
furlong	meter	+02 2.011 68*
hand	meter	−01 1.016*
inch	meter	−02 2.54*
league (U.K. nautical)	meter	+03 5.559 552*
league (international nautical)	meter	+03 5.556*
league (statute)	meter	+03 4.828 032*
light year	meter	+15 9.460 55
link (engineer or ramden)	meter	−01 3.048*
link (surveyor or gunter)	meter	−01 2.011 68*
meter	wavelengths Kr 86	+06 1.650 763 73*
micron	meter	−06 1.00*
mil	meter	−05 2.54*
mile (U.S. statute)	meter	+03 1.609 344*
mile (U.K. nautical)	meter	+03 1.853 184*
mile (international nautical)	meter	+03 1.852*
mile (U.S. nautical)	meter	+03 1.852*
nautical mile (U.K.)	meter	+03 1.853 184*
nautical mile (international)	meter	+03 1.852*
nautical mile (U.S.)	meter	+03 1.852*
pace	meter	−01 7.62*
parsec	meter	+16 3.083 74
perch	meter	+00 5.0292*
pica (printers)	meter	−03 4.217 517 6*
point (printers)	meter	−04 3.514 598*
pole	meter	+00 5.0292*
rod	meter	+00 5.0292*
skein	meter	+02 1.097 28*
span	meter	−01 2.286*
statute mile (U.S.)	meter	+03 1.609 344*
yard	meter	−01 9.144*

Mass		
carat (metric)	kilogram	−04 2.00*
dram (avoirdupois)	kilogram	−03 1.771 845 195 312 5*
dram (troy or apothecary)	kilogram	−03 3.887 934 6*
grain	kilogram	−05 6.479 891*
gram	kilogram	−03 1.00*
hundredweight (long)	kilogram	+01 5.080 234 544*
hundredweight (short)	kilogram	+01 4.535 923 7*
kgf-second²-meter (mass)	kilogram	+00 9.806 65*
kilogram mass	kilogram	+00 1.00*
lbm (pound mass, avoirdupois)	kilogram	−01 4.535 923 7*
ounce mass (avoirdupois)	kilogram	−02 2.834 952 312 5*
ounce mass (troy or apothecary)	kilogram	−02 3.110 347 68*

TABLE 15-2   Conversion Factors as Extracted Multiples of SI Units (Continued)

To convert from	To	Multiply by
pennyweight	kilogram	−03 1.555 173 84*
pound mass, lbm (avoirdupois)	kilogram	−01 4.535 923 7*
pound mass (troy or apothecary)	kilogram	−01 3.732 417 216*
scruple (apothecary)	kilogram	−03 1.295 978 2*
slug	kilogram	+01 1.459 390 29
ton (assay)	kilogram	−02 2.916 666 6
ton (long)	kilogram	+03 1.016 046 908 8*
ton (metric)	kilogram	+03 1.00*
ton (short, 2000 pound)	kilogram	+02 9.071 847 4*
tonne	kilogram	+03 1.00*

Power

To convert from	To	Multiply by
Btu (thermochemical)/second	watt	+03 1.054 350 264 488
Btu (thermochemical)/minute	watt	+01 1.757 250 4
calorie (thermochemical)/second	watt	+00 4.184*
calorie (thermochemical)/minute	watt	−02 6.973 333 3
foot-lbf/hour	watt	−04 3.766 161 0
foot-lbf/minute	watt	−02 2.259 696 6
foot-lbf/second	watt	+00 1.355 817 9
horsepower (550 foot lbf/second)	watt	+02 7.456 998 7
horsepower (boiler)	watt	+03 9.809 50
horsepower (electric)	watt	+02 7.46*
horsepower (metric)	watt	+02 7.354 99
horsepower (U.K.)	watt	+02 7.457
horsepower (water)	watt	+02 7.460 43
kilocalorie (thermochemical)/ minute	watt	+01 6.973 333 3
kilocalorie (thermochemical)/ second	watt	+03 4.184*
watt (international of 1948)	watt	+00 1.000 165

Pressure

To convert from	To	Multiply by
atmosphere	newton/meter^2	+05 1.013 25*
bar	newton/meter^2	+05 1.00*
barye	newton/meter^2	−01 1.00*
centimeter of mercury (0°C)	newton/meter^2	+03 1.333 22
centimeter of water (4°C)	newton/meter^2	+01 9.806 38
dyne/centimeter^2	newton/meter^2	−01 1.00*
foot of water (39.2°F)	newton/meter^2	+03 2.988 98
inch of mercury (32°F)	newton/meter^2	+03 3.386 389
inch of mercury (60°F)	newton/meter^2	+03 3.376 85
inch of water (39.2°F)	newton/meter^2	+02 2.490 82
inch of water (60°F)	newton/meter^2	+02 2.4884
kgf centimeter^2	newton/meter^2	+04 9.806 65*
kgf/meter^2	newton/meter^2	+00 9.806 65*
lbf/foot^2	newton/meter^2	+01 4.788 025 8
lbf/inch^2(psi)	newton/meter^2	+03 6.894 757 2
millibar	newton/meter^2	+02 1.00*

TABLE 15-2   Conversion Factors as Extracted Multiples of SI Units
(*Continued*)

To convert from	To	Multiply by
millimeter of mercury (0°C)	newton/meter^2	+02 1.333 224
pascal	newton/meter^2	+00 1.00*
psi (lbf/inch^2)	newton/meter^2	+03 6.894 757 2
torr (0°C)	newton/meter^2	+02 1.333 22

Speed

foot/hour	meter/second	−05 8.466 666 6
foot/minute	meter/second	−03 5.08*
foot/second	meter/second	−01 3.048*
inch/second	meter/second	−02 2.54*
kilometer/hour	meter/second	−01 2.777 777 8
knot (international)	meter/second	−01 5.144 444 444
mile hour (U.S. statute)	meter/second	−01 4.4704*
mile/minute (U.S. statute)	meter/second	+01 2.682 24*
mile/second (U.S. statute)	meter/second	+03 1.609 344*

Temperature

Celsius	kelvin	$t_K = t_C + 273.15$
Fahrenheit	kelvin	$t_K = (5/9)(t_F + 459.67)$
Fahrenheit	Celsius	$t_C = (5/9)(t_F - 32)$
Rankine	kelvin	$t_K = (5/9)t_R$

Time

day (mean solar)	second (mean solar)	+04 8.64*
day (sidereal)	second (mean solar)	+04 8.616 409 0
hour (mean solar)	second (mean solar)	+03 3.60*
hour (sidereal)	second (mean solar)	+03 3.590 170 4
minute (mean solar)	second (mean solar)	+01 6.00*
minute (sidereal)	second (mean solar)	+01 5.983 617 4
month (mean calendar)	second (mean solar)	+06 2.628*
second (ephemeris)	second	+00 1.000 000 000
second (mean solar)	second (ephemeris)	Consult American Ephemeris and Nautical Almanac
second (sidereal)	second (mean solar)	−01 9.972 695 7
year (calendar)	second (mean solar)	+07 3.1536*
year (sidereal)	second (mean solar)	+07 3.155 815 0
year (tropical)	second (mean solar)	+07 3.155 692 6
year 1900, tropical, Jan., day 0 hour 12	second (ephemeris)	+07 3.155 692 597 47*
year 1900, tropical, Jan., day 0, hour 12	second	+07 3.155 692 597 47

Viscosity

centistoke	meter^2/second	−06 1.00*
stoke	meter^2/second	−04 1.00*

TABLE 15-2   Conversion Factors as Extracted Multiples of SI Units (*Continued*)

To convert from	To	Multiply by
foot²/second	meter²/second	−02 9.290 304*
centipoise	newton-second/meter²	−03 1.00*
lbm/foot-second	newton-second/meter²	+00 1.488 163 9
lbf-second/foot²	newton-second/meter²	+01 4.788 025 8
poise	newton-second/meter²	−01 1.00*
poundal-second/foot²	newton-second/meter²	+00 1.488 163 9
slug/foot-second	newton-second/meter²	+01 4.788 025 8
rhe	meter²/newton-second	+01 1.00*
	Volume	
acre-foot	meter³	+03 1.233 481 9
barrel (petroleum, 42 gallons)	meter³	−01 1.589 873
board foot	meter³	−03 2.359 737 216*
bushel (U.S.)	meter³	−02 3.523 907 016 688*
cord	meter³	+00 3.624 556 3
cup	meter³	−04 2.365 882 365*
dram (U.S. fluid)	meter³	−06 3.696 691 195 312 5*
fluid ounce (U.S.)	meter³	−05 2.957 352 956 25*
foot³	meter³	−02 2.831 684 659 2*
gallon (U.K. liquid)	meter³	−03 4.546 087
gallon (U.S. dry)	meter³	−03 4.404 883 770 86*
gallon (U.S. liquid)	meter³	−03 3.785 411 784*
gill (U.K.)	meter³	−04 1.420 652
gill (U.S.)	meter³	−04 1.182 941 2
hogshead (U.S.)	meter³	−01 2.384 809 423 92*
inch³	meter³	−05 1.638 706 4*
liter	meter³	−03 1.00*
ounce (U.S. fluid)	meter³	−05 2.957 352 956 25*
peck (U.S.)	meter³	−03 8.809 767 541 72*
pint (U.S. dry)	meter³	−04 5.506 104 713 575*
pint (U.S. liquid)	meter³	−04 4.731 764 73*
quart (U.S. dry)	meter³	−03 1.101 220 942 715*
quart (U.S. liquid)	meter³	−04 9.463 529 5
stere	meter³	+00 1.00*
tablespoon	meter³	−05 1.478 676 478 125*
teaspoon	meter³	−06 4.928 921 593 75*
ton (register)	meter³	+00 2.831 684 659 2*
yard³	meter³	−01 7.645 548 579 84*

SOURCE: Adapted from T. G. Hicks, *Metrication Manual*, McGraw-Hill, New York, 1972. Used with permission of the publisher.

NOTES

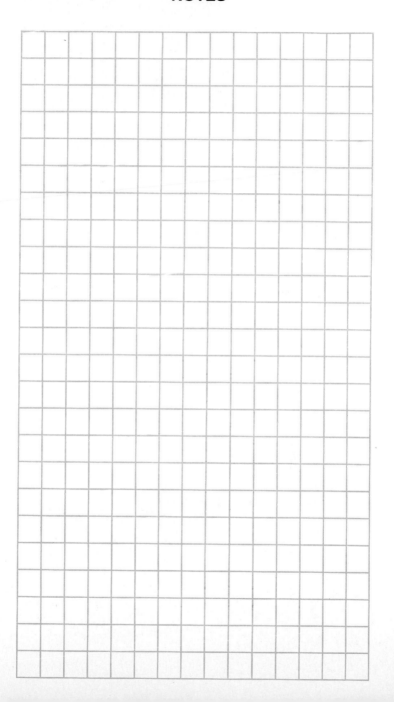

NOTES

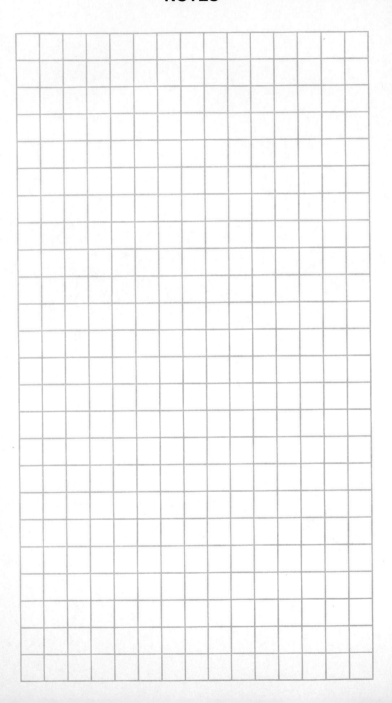

Index